Genetic Witness

ooooooooooooooooooooooooooooooooooooooooooooooooooooooooooooooooooooooooooooooooooooooo

Genetic Witness

Science, Law, and Controversy in the Making of DNA Profiling

JAY D. ARONSON

RUTGERS UNIVERSITY PRESS

NEW BRUNSWICK, NEW JERSEY, AND LONDON

Library of Congress Cataloging-in-Publication Data

Aronson, Jay D., 1974–
Genetic witness : science, law, and controversy in the making of DNA profiling / Jay D. Aronson.
    p.   ;   cm.
Includes bibliographical references and index.
ISBN 978-0-8135-4187-7 (hardcover : alk. paper)—ISBN 978-0-8135-4188-4 (pbk. : alk. paper)
    1. DNA fingerprinting—Great Britain—History.   2. DNA fingerprinting—United States—History.   3. Forensic genetics—Great Britain—History.   4. Forensic genetics—United States—History.   I. Title.   [DNLM: 1. DNA Fingerprinting—history—Great Britain.   2. DNA Fingerprinting—history—United States.   3. Forensic Medicine—history—Great Britain.   4. Forensic Medicine—history—United States.   5. DNA Fingerprinting—legislation & jurisprudence—Great Britain.   6. DNA Fingerprinting—legislation & jurisprudence—United States. W 611 AA1 A769g 2007]

RA1057.55.A76 2007
614.'10941—dc22

                                                                2007006029

A British Cataloging-in-Publication record for this book is available from the British Library.

Copyright © 2007 by Jay D. Aronson

Visit our Web site: http://rutgerspress.rutgers.edu

Manufactured in the United States of America

To Tamara and Ezra
with love

Contents

Acknowledgments

As I am faced with the daunting task of thanking the individuals and institutions that have made this book possible, I am reminded of just how incredibly lucky I have been during the time that I conceived and carried out this project. Friends, family, and colleagues have supported me in ways both little and big along the way.

At least in my humble opinion, a young scholar could not have asked for better mentors than John Beatty and Sheila Jasanoff. John is one of the nicest people I've ever met and the best teacher I've ever had. Throughout the past several years, he has not only asked me to be the best historian I can be, but also the best person I can be. For both, and especially the latter, I cannot thank him enough. Sheila is, quite simply, present on each and every page of this book, even when it is not entirely apparent. Through her research, teaching, innumerable long conversations, as well as the invigorating STS community that she has created at Harvard University's Kennedy School of Government, Sheila provided me with the intellectual resources necessary to write this book.

While in Minneapolis, I was thankful for the companionship of my fellow graduate students who supported me through two years of coursework. Perhaps the most important person at Minnesota, though, was Barbara Eastwold, who not only resolved seemingly endless bureaucratic crises for me but also managed to brighten my day every time I came to see her. I wrote the bulk of this book in Cambridge, Massachusetts, where I benefited a great deal from my interactions with members of the vibrant STS community at the Kennedy School. Kaushik Sunder Rajan, Jenny Reardon, and David Winickoff deserve special thanks for many interesting conversations and debates. Steve Hilgartner gave me my first course in conducting oral history interviews shortly before I began doing my research. More than seventy interviews later, his advice still comes in handy. Mike Lynch and Simon Cole have also provided me with a tremendous amount of support and have contributed in numerous ways to the creation of this work. Thanks must also go to Dick Lewontin, who provided me with office space when

I first arrived in Cambridge and served as my tutor in molecular biology and population genetics while I wrote this book. He was also the first person to read the completed draft of my dissertation. Those who know Dick won't be surprised to learn that he was more critical of my work than almost anyone else who read it. His comments dramatically improved my thinking, and he can take credit for making me treat science seriously. (Any errors, of course, remain entirely my fault.) Thanks also go to several people who have helped me at various stages of my career and with this book, especially Caroline Acker, Pnina Abir-Am, Jennifer Alexander, Michael Dietrich, Greg Gibson, David Hounshell, Jay Kadane, Sally Kohlstedt, Arthur Norberg, Mark Perlin, Joyce Seltzer, Alan Shapiro, Joel Tarr, Joe Trotter, and Susan Wolf.

It has been a pleasure to work with Rutgers University Press on this project. Thanks go to Audra Wolfe for helping me shape the story that is told within these pages, and Doreen Valentine for her tireless efforts to make my writing crisp and coherent. An author could not ask for better editors. Monica Phillips and Marilyn Campbell did an excellent job of turning my rough manuscript into a polished book.

Research on recent science, not to mention science in the legal system, poses significant challenges to the historian. To begin with, almost all of the participants in the story I tell are still alive today. Indeed, most are currently at the peak of their careers. Therefore, I must begin by thanking the sixty-plus individuals who took time out of their busy schedules to discuss their roles in the development of DNA typing. Special thanks go to George Sensabaugh from the University of California at Berkeley and William Thompson from the University of California at Irvine for devoting many hours over many days to help me understand the multiple interpretations of the history of DNA profiling.

Among the most difficult challenges to overcome are the sheer volume of relevant information and the difficulty in gaining access to the most interesting materials. Therefore, I must acknowledge the following institutions and individuals for their help with this process. The Cornell University Library Division of Rare Books and Manuscripts collection on the O.J. Simpson Trial and DNA Typing Archive proved to be the single most important source of documents during my research. This collection contains the entire set of *Frye* Hearing Transcripts from *NY v. Castro*, as well as various other transcripts, interviews, and other crucial materials. These items can be found in the "O.J. Simpson Murder Trial and DNA Typing Archive," #53/12/3037. I cannot begin to give adequate thanks to the people who had the foresight to put this collection together (especially Sheila Jasanoff, Steve Hilgartner, and Bruce Lewenstein).

Arthur Daemmrich of the Chemical Heritage Foundation provided access to publicity materials (brochures, pamphlets, fee schedules, advertisements, etc.) and protocols of Lifecodes and Cellmark. He obtained these materials while doing graduate research on the business side of forensic DNA analysis in the mid-1990s. Daemmrich recently transferred his collection to the O.J. Simpson

Murder Trial and DNA Typing Archive at Cornell, which he helped put together. Individuals wishing to access this collection should consult the archivist at Cornell's Rare and Manuscripts Collections.

The Hennepin County District Attorney's Office provided access to case files of *State of Minnesota v. Thomas Schwartz*, 1989 (SIP No. 89903565/C.A. No. 88–3195) and *State of Minnesota v. Larry Lee Jobe*, 1990 (SIP No. 88903565/C.A. No. 88–3301; Hennepin County District Court).

Richard C. Lewontin of Harvard University provided access to a multitude of materials relating to the debate over population genetics issues, as well as transcripts and court documents from *United States v. Yee, et al.*, from his private collection.

The National Academy of Sciences Archive, National Research Council Commission on Life Sciences: Board on Biology Archives (NAS Library, Washington, D.C.) provided access to nonclassified materials from the two National Research Council reports on forensic DNA analysis. Thanks to archivist Janice Goldblum for all of her help.

Kellie Nielan from the Florida Attorney General's Office, Daytona Beach, provided access to the "Answer Brief of Appellee" and "Amicus Curae" (by Andre Moennsens) in *Tommie Lee Andrews v. State of Florida*, 1988 (Case No. 87–2166; Fifth District, Florida Court of Appeal).

William C. Thompson of University of California–Irvine, provided me with access to materials related to the CACLD and the FBI as well as numerous court transcripts and other documents from his private collection.

Jan Witkowski from the Banbury Center at Cold Spring Harbor Laboratory provided access to his files relating to the Banbury Meeting on "DNA Technology and Forensic Science," the New York State Forensic DNA Analysis Panel, as well as a transcript of the *Frye* Hearing in *New York v. Wesley-Bailey*. These materials can be accessed through the Cold Spring Harbor Laboratory Archive, Box Containing Jan Witkowksi's Materials Related to DNA Fingerprinting and the Banbury Conference on "DNA Technology and Forensic Science," which is uncataloged at the present time.

The following records of U.S. Congressional Hearings provided a rich source of otherwise difficult-to-obtain documents, correspondence, and other materials: (1) "DNA Identification," Hearing before the Subcommittee on Constitution of the Committee on the Judiciary, United States Senate, First Session on "Genetic Testing as a means of Criminal Investigation," 15 March 1989, Serial #J-101–4 (Washington, D.C.: U.S. Government Printing Office, 1990); (2) "FBI Oversight and Authorization Request for Fiscal Year 1990 (DNA Identification)" Hearing, House of Representatives, Subcommittee on Civil and Constitutional Rights, Committee on Judiciary, 22 March 1989 (Washington, D.C.: U.S. Government Printing Office, 1990); and (3) "Forensic DNA Analysis," Joint Hearing before the Subcommittee on Civil and Constitutional Rights of the House Committee on the Judiciary and the Subcommittee on the Constitution of the Senate Committee on the Judiciary, 13 June 1991, House Serial 30/Senate Serial J-102–47 (Washington, D.C.: U.S. Government Printing Office, 1992).

Thanks also go to Graham Cooke in London and Peter Donnelly in Oxford for access to material relating to the use of DNA evidence in Great Britain.

It is an understatement to say that this book would not exist without the love and support that I have received from family, especially Karen and Richard Heilman, Amy Heilman, Muriel Baum, and my late grandmother, Norma Heilman. Everything that I have achieved, and everything that I will achieve, is born from their generosity, compassion, and kindness. The Dubowitz and Faiman families have welcomed me into their hearts and homes (around the world) in a way that exceeded my wildest expectations. Dubie and Jean shower warmth, affection, encouragement, and love upon me as if I were their own son, and Granny Bubbles (a.k.a. Marion Faiman) has inspired me with her energy, curiosity about my work, and zest for life.

Finally, this book is dedicated to the two most important people in my life. To Tamara for making me whole, allowing me to feel true love, being my best friend, challenging me to think deeply and differently about the world, making each and every moment of my existence special, and so, so, so much more. I love you more than life itself. And to Ezra, you are not only the cutest child ever to walk the face of the earth, but you are also my constant source of inspiration, joy, and wonder—you continue to teach me how profound love can be.

Genetic Witness

CHAPTER 1

Introduction

For nearly a century, the law enforcement community has been searching for a forensic silver bullet—a foolproof technique that can identify absolutely the perpetrator of violent acts from the physical traces left at the crime scene *and* provide a tool for tracking and surveillance of past and future criminals. Throughout this history, various techniques have emerged as potential candidates, including fingerprinting, anthropometrics (the systematic measurement of body parts), and the voiceprint. All of these techniques, however, were eventually shown to have technical and practical restrictions.

In 1984, the most recent of these contenders was invented in England by a geneticist named Alec Jeffreys: forensic DNA profiling. Heralded by judges and prosecutors as the "greatest advance in crime fighting technology since fingerprints," numerous press accounts proclaimed that the technique would "revolutionize law enforcement."[1] In many ways, it has. In the past two decades, DNA evidence has been used to solve countless violent crimes, putting thousands upon thousands of rapists and murderers behind bars and inducing guilty pleas from thousands more. The FBI's nationwide index of DNA profiles now contains more than 4 million criminal profiles and has led to more than 45,400 "hits" since it was put into place in 1994.[2]

Today it is widely considered to be the best method of forensic identification ever created. Even defense attorneys Barry Scheck and Peter Neufeld, who orchestrated the three most widely publicized challenges to the validity and reliability of forensic DNA analysis (*People v. Castro, United States v. Yee, et al.,* and *People v. Orenthal James Simpson*), now call DNA testing "a gold standard for truth telling."[3] According to former Attorney General John Ashcroft, it is a "truth machine."[4]

Law enforcement agencies, politicians, and prosecutors all rely on DNA evidence as one of the most important weapons in the fight against violent crime. Forty-three states require that all individuals convicted of felonies submit DNA samples to a central DNA database, and all fifty states participate in the FBI's

1

national DNA index. If politicians like former Governor George Pataki and New York City Mayor Michael Bloomberg have their way, New York will soon become the first state in the nation to collect DNA from all people convicted of any misdemeanor or felony, including juveniles.[5] This plan reflects a fundamental belief that the existence of criminal DNA indexes would both aid in investigations of current crimes and act as a deterrent to future crimes. It is also seen as progressive and crucial to the cause of safety and justice—so progressive, in fact, that even though no other states have adopted such legislation, the politicians involved are portraying New York as behind the curve in terms of crime-fighting innovations. As Mayor Bloomberg said, DNA evidence "is by far the most powerful and precise crime-solving tool since the advent of fingerprint identification a century ago. . . . When it comes to cutting-edge investigative technologies, New York should be leading the charge, not lagging behind. As long as we are lagging behind, make no mistake: People are being murdered and raped by criminals who should already be behind bars."[6]

At the same time, "Innocence Projects" have sprung up across the country, founded by defense attorneys and eager young law students, based on the notion that postconviction DNA testing of evidence can yield conclusive proof of the innocence of the wrongly convicted. Such projects have led to the release of nearly 200 individuals from prison, most of them poor and members of racial or ethnic minorities, based on the strength of DNA identification. In almost all cases, this genetic evidence has cast doubt on other forms of evidence, including mistaken eyewitness testimony, false confessions, the use of jailhouse snitches, and microscopic hair comparison. Postconviction exonerations have also revealed some of the biases and shortcomings of our criminal justice system: racism, prosecutorial success defined solely by convictions, and underfunded or incompetent indigent defense counsel. Activists have even used DNA evidence to launch a fundamental critique of the American criminal justice system. They argue that it "has opened a window into wrongful convictions so that we may study the causes and propose remedies that may minimize the chances that more innocent people are convicted."[7]

Today, as more and more genetic markers are being linked to specific racial and ethnic groups (or "biogeographical origin," as many scientists prefer to call it), law enforcement officials dream of one day being able to predict the physical appearance of a criminal based solely on genetic material left at the crime scene. Although such markers are still rarely used in forensic investigations, primarily because of their highly probabilistic nature and the difficulty of developing a useful visual profile of a person based on their complex heritage, a Florida biotechnology company called DNAPrint Genomics already offers two genetic tests that it claims bring law enforcement agents closer to this goal. The first, DNAWitness, provides investigators with "percentage of genetic make up amongst the four possible groups of Sub-Saharan African, Native American, East Asian, and European" ancestry groups, while the second, Retinome, claims to enable the inference of eye color from DNA.[8]

DNA identification's reach has extended far beyond the criminal justice system to fill a wide range of identification needs. The New York City Medical Examiner's Office, for instance, has used DNA typing to identify the remains of victims of the 11 September 2001 terrorist attacks, often matching tiny bits of charred flesh with saliva from the individual's toothbrush. DNA played a crucial role in identifying the badly damaged remains of victims of the July 1996 TWA Flight 800 crash and is being used to identify victims of human rights abuse and genocide in regions emerging from periods of terror and civil war. Thanks to DNA typing, we now know that the Vietnam veteran who was (until recently) interred in the Tomb of the Unknown Soldier was Michael J. Blassie.[9] As a result of DNA typing and the creation of a DNA registry, Department of Defense officials believe that unknown soldiers will soon become a distant memory.

So powerful is the allure of DNA profiling, with its roots in academic science and links to the code of life, that anecdotes abound that juries have become unwilling to convict defendants in its absence because they "know" from television programs and the media that it the best form of forensic science available. Forensic scientists, lawyers, and journalists have dubbed this phenomenon the "CSI effect." DNA profiling's widespread adoption is undermining the credibility of the technique that was once thought to be the gold standard of forensic science: fingerprinting.[10] DNA is causing scientists, lawyers, judges, and the general public to ask for the first time about the scientific underpinnings of fingerprinting, leading to what science studies scholar Michael Lynch calls an "inversion of credibility" between the two techniques. In 2002, a federal judge restricted the admissibility of fingerprint evidence in the case of *United States v. Llera Plaza* before changing his opinion ten weeks later.[11] Many commentators are even beginning to argue that DNA profiling ought to serve as the new model for the practice of forensic sciences like fingerprinting, bullet lead analysis, and tool mark analysis, among others.[12]

DNA profiling is indeed a good place to look for a guide to reforming forensic science, although not in a way that might seem intuitive to most people reading this book. As the history of DNA profiling demonstrates, identification technologies have limitations that only become apparent when they are applied in practice and are challenged by people who have a vested interest in pointing out their shortcomings. DNA profiling was a deeply problematic technology when it was unveiled. It was vulnerable to error and had not been subjected to peer review by scientists who did not participate in its development. Robust quality control and quality assurance measures were nonexistent, so there were few safeguards to prevent against error. Further, the interpretation of test results was based more on assumption than empirical evidence.

It took a dedicated group of defense lawyers using the mechanisms provided by the American legal system—expert witnesses, discovery, and cross-examination— to bring these deficiencies to light. As a result of defense challenges, scientists were forced to go back to their laboratories and professional societies to develop more robust methods and protocols, better quality control mechanisms, and more effective, inclusive peer review systems. They then brought these improvements back into court to be subjected to the scrutiny of the defense bar and their experts. Although this process was fraught with controversy, inefficiency, and personal antagonism, the end result was a technology in which match probabilities and molecular biological procedures were supported more by empirical evidence and peer review rather than assumption and secrecy.

DNA profiling as we know it today is a product of Anglo-American legal institutions interacting with science and technology over the course of a decade. Thus, if the development of DNA profiling is to serve as a model for other forensic scientists, then it must be acknowledged that there was *not* a linear, rational pathway from the academic scientific laboratory to the forensic laboratory to the courtroom. The development of DNA typing featured scientists weaving together technical claims with legal, social, and political ones, as well as lawyers, politicians, and judges making choices that would seem to require a great deal of scientific knowledge and expertise. The technique itself, standards of "good science," as well as the relevant expertise needed certify its credibility within the legal system, were significantly altered by this process. In other words, developing the technique for use in the criminal justice system was just as much about social engineering as it was about getting the science right.

Nor is it possible to point to a single cause of the resolution of the debates over DNA profiling, which came to be known as the "DNA Wars" to many participants and members of the media. Rather, the legal and scientific debates over DNA evidence were extinguished slowly through a combination of FBI initiatives; judicial decisions; federal legislation; a shrewd public relations move by two main participants in the debates over DNA evidence, former defense witness Eric Lander and the FBI's Bruce Budowle; technical changes in RFLP-based DNA profiling, as well as the development of an entirely new method for creating DNA profiles called Short Tandem Repeat analysis, or STR for short.

Finally, it is crucial to recognize the productive role that the adversarial legal process played in making DNA profiling a better technology. This view stands in marked contrast to the way that most commentators explained the existence of the controversy over DNA profiling, namely "culture clash."[13] According to this model, adversarial interactions are seen as an impediment to the successful uptake of knowledge and technology in the legal system. Not only are standards of evidence different in science and law, so are norms, values, and their ultimate purposes. For science, the search for ultimate truth and an ethos of progress dominate, while law can be characterized as beholden to process in the search for serviceable truth (i.e., good enough to make a decision), social control, and resolution of disputes. In the end, according to the culture clash model, scientific discussions get hijacked by the adversarial process. Lawyers use dirty tricks to challenge scientific techniques that provide evidence against their clients, scientists who are trained to seek consensus are forced to battle it out as expert witnesses paid by opposing sides, and judges and juries get confused or make bad decisions because they have been swayed by convoluted and scientifically suspect testimony.

It cannot be denied that all of these things took place in the development of DNA profiling. However, once one gets beyond the shrill courtroom pronouncements and questionable legal tactics, it becomes clear that the adversarial process served as a means to correct the problems that arose when DNA profiling technology was transferred from the laboratory to the courtroom. Thus, one of the main lessons that can be learned from this history is that efforts to curb adversarialism in the legal system might not be in the best interest of legal decision making or scientific advancement.

While the interactions of science and law certainly improved the validity and reliability of DNA profiling, the story told in this book is not entirely positive. Although the technique is better than it was when first introduced in 1987, many of the problems that emerged as the technique was challenged in the courtroom have not been resolved. Most important, overtly false claims of the infallibility of DNA evidence initially made by Lifecodes, Cellmark, and the prosecutors they worked with have survived unscathed in the courtroom and in society. There is a general unwillingness to admit that for all the improvements made to the technique over the past two decades, the potential for serious error in DNA evidence still exists. As a result, there is still no effective means for calculating error rates in DNA testing, no explicit standards for interpreting the complex results that emerge from biological stains in which there are multiple contributors, and, most important, no agreed-upon method for conducting proficiency testing in the dozens of DNA labs around the country.[14] Although most members of the forensic science community are willing to admit that laboratory and interpretative errors occasionally occur, they believe that the problem is too small to necessitate the implementation of costly monitoring systems.

The forensic community has been embarrassed by the discovery of deep-seated problems in DNA testing. While some have been the result of simple clerical errors, others have revealed overt fraud and inexcusable incompetence on a grand scale. The most well known of these cases is the Houston Police Department's Crime Laboratory, which was closed down in 2002 after it was discovered that employees regularly fabricated DNA and other forensic evidence in their labs and lied in court about the results of their work. An ongoing investigation of the Houston Police Department's (HPD) casework has revealed forty-three DNA cases "in which there are significant doubts about reliability of the work performed, the validity of the Crime Lab's analytical results, or the correctness of the analysts' reported conclusions."[15] While the Houston story is no doubt the most dramatic, other laboratories have not been immune from serious trouble, including the FBI, where an analyst had been faking the quality control portion of her work for more than two years. In another case, serious procedural and interpretation problems were discovered in DNA testing carried out by Virginia's main public crime laboratory. While these cases will be explored in detail in the conclusion, for now it is important to realize that DNA profiling is not the infallible, foolproof technology that many scientists and lawyers claim it to be.

What is even more alarming is that none of these mistakes were discovered by forensic scientists themselves or by the numerous layers of quality control and quality assurance that the public is told guarantees valid and reliable results. Instead, they were discovered by journalists, crusading defense lawyers, or advocates of civil liberties. As such, these problems may represent the tip of the iceberg of DNA errors and suggest that there may be a systemic malfunction in what is considered to be the new gold standard of forensic science. How has it come to pass that this standard retains such a potentially damaging vulnerability? As I will argue, the answer can be found in the technique's tumultuous history.

Science for Hire

Alec Jeffreys, a geneticist at the University of Leicester in Britain, invented the first usable version of DNA profiling in 1984. After its debut in the British legal system in 1985, the technique rapidly spread to the United States. From the start, there was intense competition in the American DNA profiling market, with two biotechnology companies—Lifecodes Corporation and Cellmark Diagnostics—introducing variations of the technology within months of one another. Both firms were business units of large, publicly traded, multinational chemical conglomerates, each with yearly sales of well over $20 billion, that were hoping to cash in on the biotechnology boom of the 1980s.[1] Both firms were also quick to file patents on their technologies and closely controlled who gained access to their genetic markers—a move that many academic scientists found contrary to the openness that they believed acted as the foundation of good scientific practice. The production of forensic evidence by private companies, however, was a new development in the world of law enforcement—up to this point, most forensic techniques were offered by public crime laboratories, especially the FBI.

Although both companies initially targeted the paternity market, law enforcement agencies quickly asked for help in solving difficult criminal cases. Even though DNA profiling was a new and largely untested technology, neither company had much difficulty convincing judges that DNA evidence was ready to be introduced into court. While defense attorneys occasionally undertook feeble efforts to question the validity and reliability of the technique, their limited understanding of basic genetics and molecular biology was no match for the impressive stable of experts assembled by the private companies and prosecutors. By the end of 1988, DNA profiling had been admitted into evidence without challenge in nearly two hundred cases.

THE FIRST TEST

When his plane touched down at London's Heathrow airport, fifteen-year-old Andrew Sarbah expected to sail through immigration. Although he spoke little

English, he was, after all, returning to the city of his birth. Andrew had been born in London to parents of Ghanaian descent, but after his parents separated he went back to Ghana with his father at the age of four. Eleven years later, he was returning to the United Kingdom to live with his mother and several siblings.[2] He carried an up-to-date Ghanaian passport showing that he had been born in London, as well as a British passport with a picture of him as a young baby. But at the immigration desk, his nightmare began when the authorities denied his application to enter the United Kingdom as a British citizen. According to press reports, there was a suspicion that Andrew's British passport had been tampered with and that the young man trying to enter the country was actually one of his cousins. However, Andrew was granted a "temporary admission" until his actual immigration status could be resolved.

The Sarbah family obtained legal counsel from Sheona York, a London immigration attorney. York quickly began collecting photographic and testimonial evidence to support Andrew's claim. The Sarbah family also obtained serological testing, which revealed that Andrew was Christine Sarbah's son with 98 percent certainty. The only other scientifically feasible explanation for the result of the test was that Andrew was a child of the same father and one of Christine's sisters. Although there was a good fit between the testimonial and serological evidence, the Home Office maintained that Andrew's passport had been tampered with. This did not bode well for the Sarbah family. Despite overwhelming evidence in his favor, Andrew's options were running out.

And then, just after York had filed one last appeal, a colleague showed her a newspaper article from the *Guardian* that described a new genetic technique for uncovering family relations.[3] It proclaimed that "scientists have discovered a method of identifying people by their genes—a genetic 'fingerprint' so precise it can even tell you who your father is." The test, developed by Jeffreys, was so powerful that it was able to "distinguish every individual, even the children of a first-cousin marriage."[4] With nothing to lose, York got in touch with Jeffreys to see if he was prepared to try out his technique on a real case.[5] Despite his expectations that "everything would go haywire and I'd have to say that it simply doesn't work and we've got no evidence," Jeffreys agreed to analyze biological samples from the Sarbah family.

How Does DNA Profiling Work?

In order to understand Jeffreys's new identification tool, it is first necessary to review some aspects of genetics and molecular biology. Jeffreys's method was based on the fundamental premise that, with the exception of identical twins, no two human beings have exactly the same genetic makeup. Although all humans share a nearly identical set of the three billion "letters" that make up our genome, each of us also carries with us a substantial number of mutations (i.e., changes to our genetic material) that make our DNA profile unique. Indeed, any two randomly selected individuals will have different letters at about one in one thousand sites. Some of these changes are harmful or even lethal to individuals (e.g., in sickle-cell anemia, Huntington's Disease, Tay-Sachs Disease, and cystic fibrosis, to name a few well-studied genetic diseases), while others either have no effect or confer some small advantage to the individuals who possess it.

Higher organisms are made up of a huge number of cells (the human body contains approximately one trillion), all of which are descended from a single fertilized egg. The genetic material, DNA, is found in the form of chromosomes of the innermost part of the cell, the nucleus. The fertilized human egg has twenty-three pairs of chromosomes, with one copy coming from the mother and the other coming from the father at the time of conception. During the course of both embryonic development and cell division throughout life, these chromosomes are completely replicated numerous times. One result of this process is that all cells from a single person, be they hair, skin, blood, semen, or muscle, contain the exact same DNA sequence.

Chemically speaking, DNA is a repetitive polymer of four different types of nucleotides: adenine (generally abbreviated as "A"), cytosine (C), guanine (G), and thymine (T). Under normal circumstances, DNA is composed of two strands that are linked together by bonds that form between the nucleotide of each strand according to a simple set of rules: cytosine pairs with guanine and adenine pairs with thymine. For example:

A T T C G G A A C T
T A A G C C T T G A

Each linked dyad (either A-T or C-G) is called a base pair, or bp for short, and this is the unit of measurement for DNA. The DNA section above is a 10 bp fragment. (Long DNA fragments are measured by the thousands, or kilobases. Thus a 7,470 bp long fragment would be written as 7.47 kb.)

Our DNA has many functions. The sections typically called "genes" contain the code for producing proteins and organizing them into cells, organs, and body parts. Other parts of the genome are necessary for the structural integrity of the chromosome. Still other parts have no clear function and are often called "junk DNA," and this is where most of the variation used for DNA profiling is found.[6] Within these noncoding regions of the genome, there are places where short sequences of DNA, sometimes called "minisatellites" or "core sequences," are repeated one after the other. Although scientists do not understand why these repeats occur, it has been shown that the number of repetitive sequences at certain locations in the genome is highly variable from person to person. As a result, they are called "variable number tandem repeats," or VNTRs. Each fragment of different length is called an "allele." This term is used widely in genetics to refer to the variants found at a particular physical location, or "locus," in the genome. At any given locus (i.e., location) where VNTRs occur, there can be dozens of alleles, each with a different number of repetitive units and therefore a different

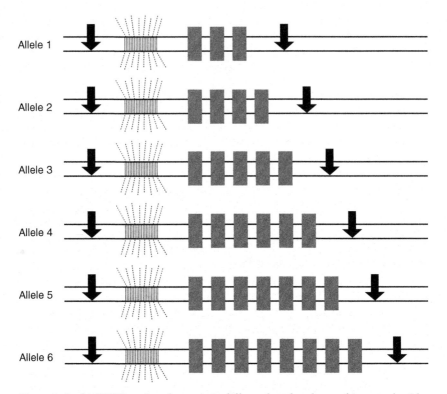

Figure 1. In this VNTR region, there are six different length polymorphisms, each with a different number of repeats. The arrows represent restriction sites where a particular restriction enzyme cuts DNA with great precision. Note that the more sequences repeated in tandem in a particular allele, the longer the fragment will be after it is cut by the restriction enzyme. In this diagram the probe binds to the DNA toward the left end of the fragment. In the early years of DNA typing, these probes were made using radioactive phosphorus (P-32), which meant that they gave off a small amount of radiation and could be visualized at a later stage of the profiling process. Illustration by Susan Heller Simon.

length. The presence or absence of particular alleles is what makes individuals identifiable through their DNA profiles (see fig. 1).

ALEC JEFFREYS AND "DNA FINGERPRINTING"

The discovery of DNA fingerprinting occurred unintentionally while Jeffreys and a group of his students were studying the evolution of mammalian globin genes. While comparing the globins of various sea-dwelling animals, they noticed a particular 33 bp sequence in varying repetitive patterns in almost all of the species tested. Upon further examination, these motifs seemed to be present in the genomes of most mammals. Furthermore, the patterns they generated appeared to be different in each animal. This was an exciting discovery. Jeffreys and his colleagues reasoned that these repetitive "minisatellites" might act as markers, pinpointing genes responsible for particular traits—the holy grail of human medical genetics in the 1980s. In other words, if a particular length polymorphism (a technical term that simply refers to different versions of the genetic code) shows up in most individuals with a certain trait or disease, then it is likely that that marker is very close in the genome to a gene that influences that particular physical outcome.

So, Jeffreys and his colleagues decided to set up an experiment to look for these length polymorphisms in humans. To do so, they took advantage of powerful molecular tools called "restriction enzymes." Restriction enzymes, which are obtained from naturally occurring bacteria, cut DNA at specific nucleotide sequences, generally either four or six bases long, called the "restriction site."[7] The specificity of a given restriction enzyme to a particular nucleotide sequence is quite strong. These restriction sites are found all throughout the genome, including active coding genes and the noncoding regions where most VNTRs are found. Restriction enzymes are useful because there are VNTRs in between many of the sites where they cut. Thus, by cutting up the DNA of different individuals, the resulting fragments will be of varying sizes depending on the number of repeats found in between each cleavage site. In order to visualize the fragments containing these repetitive elements, Jeffreys used special molecules called "probes." A probe is essentially a short section of single-stranded DNA that complements the sequence of the VNTR of interest. In other words, if this VNTR had the sequence ACTTG-CAAAA, then the probe for it would look like this: TGAACGTTTT.

The use of restriction enzymes and probes to find VNTRs was part of a larger multistep procedure known collectively as "Southern blotting."[8] Because it is the same process used in the kind of forensic DNA profiling described in this book, I will describe it in some detail here, using the schematic diagram in figure 2 as a guide. The Southern blotting process begins by obtaining a blood or tissue sample (Step 1) and extracting DNA from it (Step 2). The DNA is then subjected to a specific restriction enzyme, which cleaves it into fragments of various sizes—because of differing numbers of tandem repeats in between restriction sites (Step 3). These restriction fragments are then separated using a technique called "electrophoresis" (Step 4). Essentially, the cleaved DNA is loaded into a thick gel-like substance called agarose and then subjected to an electric field. Because these fragments will be of differing size and electric charge (due to polymorphism), they will travel through the gel at differing speeds. Thus, the smallest fragments will travel the furthest and the largest ones will travel the shortest distance. Up to this point, the DNA remains in its original double-stranded form. As such, it must be "denatured," or converted into single-stranded DNA, so that the probe can bind to its complementary sequence. In Step 5, the DNA from the gel is transferred to a piece of nylon membrane using a rather rudimentary wicking system. Once the transfer is complete, the single-stranded DNA becomes fixed into place on the membrane.

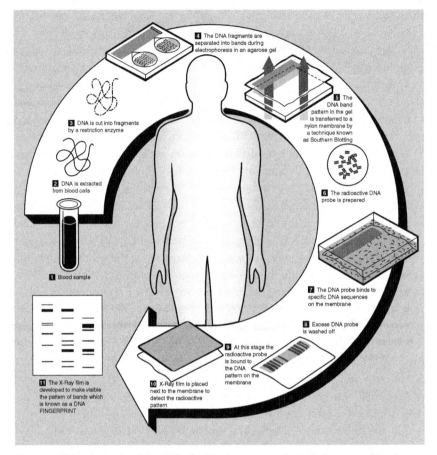

Figure 2. This schematic of the DNA profiling process, technically known as "Southern blotting," was created by Cellmark Diagnostics and widely distributed in their promotional material in the early 1990s. Reproduced with permission of Orchid Cellmark. Image supplied by Wellcome Library, London.

In Step 6, the probe is prepared by denaturing it so that it can bind to its target. Since DNA is most stable in its double-stranded form, the probe has a high affinity for its complementary sequence and readily binds to it. In Step 7, the nylon membrane is placed into a "hybridizing solution" containing various chemicals, salts, and numerous copies of the probe that was just readied for use. After letting it sit in the solution for a while, the nylon membrane is removed and the remaining hybridizing solution is washed off, leaving only the probe that has bound to a particular DNA sequence within a single restriction fragment (Steps 8–9). Since probes were made using radioactive phosphorus (a principle component of DNA) during the 1980s and into the 1990s, the repetitive DNA in question could be visualized by exposing X-ray film to the nylon membrane (Step 9). The end result of this process is an "autoradiogram" (often abbreviated "autorad") that contains many black bands often said to resemble a barcode (see fig. 3). The bands represent fragments that are different sizes because they contain different numbers of repetitive elements.

On a Monday morning in early September 1984, Jeffreys and a graduate student came into the lab to check the autorad; they were astounded by what they saw. The probes had picked up several minisatellites in the human genome, and surprisingly these minisatellites appeared to be uniquely distributed among individuals (i.e., each individual had a different pattern) and passed down through families in Mendelian fashion. Following the publication of this discovery in *Nature* in early 1985,[9] they worked out the frequencies in which various minisatellite fragments appeared in twenty "unrelated British Caucasians." In a second *Nature* paper, published later that year, they concluded that the probability that two individuals would have the same DNA fingerprint was less than 1 in 33 billion.[10]

In neither of these papers did the authors spell out that this was a technique in development that still had to be perfected before use in forensic casework. They did acknowledge that small DNA fragments were difficult to visualize and measure with accuracy and pointed out that new minisatellites arose at these hypervariable regions in about 0.4 percent of individuals. However, Jeffreys and his coauthors did not consider these issues grave enough to prevent the technique from being used in practical situations. This, it seems, was one of Jeffreys's main goals. Indeed, his careful choice of the name for the technique piggybacked on the longstanding credibility of traditional fingerprinting.[11] Only after several months would Jeffreys and others publicly discuss the limitations of the technique. But by this time, the development of the next generation of genetic identification technology was already well under way.

DNA TYPING ENTERS THE IMMIGRATION TRIBUNAL

On the day of the Sarbahs' final appeal to the immigration tribunal, Alec Jeffreys arrived with a copy of his original 1984 *Nature* paper and evidence that reputedly supported the claim that Andrew was the son of Christine (see fig. 3). Nervous about the possibility of having to evaluate this new, untested scientific information, and fully convinced by the other evidence provided by the Sarbahs, the chair of the Immigration tribunal demanded that the Home Office concede the case. Reluctantly, they did so, scoring the first major victory for DNA evidence in the world, even if the results of the DNA test were not the cause of the decision.

The British press jumped on the *Sarbah* case. Most major newspapers published stories about DNA fingerprinting, lauding it as new way to depoliticize controversial immigration decisions. Of these news sources, only *Nature* reported that the existence, not the probative value or validity of DNA evidence, was what persuaded the Home Office to drop the case.[12] This fact was quickly lost in the hype surrounding this new form of scientific evidence. The *Guardian*'s headline, for instance, proclaimed, "Son rejoins mother as genetic test ends immigration

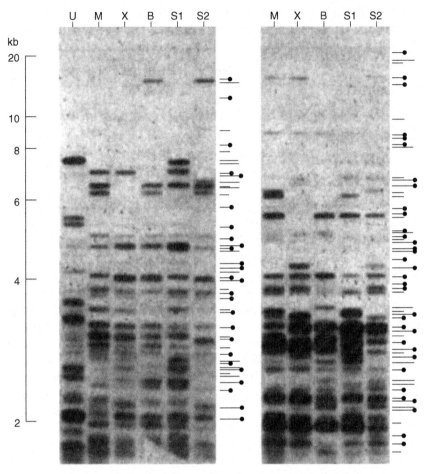

Figure 3. DNA fingerprints of the Sarbah family were created using two multilocus probes (named 33.15 and 33.6). *Left to right*: U, unrelated individual (in 33.15 run only); M, Andrew's mother; X, Andrew, the boy involved in the immigration dispute; B, his brother; and S1, S2, the two sisters. Fragments present in the mother's profile are designated by short horizontal bars at the side of the autoradiogram. Paternal fragments (those absent in the mother but present in at least one of the siblings) are designated by long horizontal lines. The bars with dots represent fragments transmitted to Andrew. His DNA profile contains only DNA fragments present in the profiles of his mother and his siblings, which proves with a very high degree of certainty that the familial relationship he claimed was true. According to Jeffreys, the probability that this result could have occurred by chance was less than 1 in 33 billion. Image courtesy of Nature Publishing Group.

dispute." Another story in the *Guardian* stated that while the Home Office authorities did not believe Sarbah, "Dr. Jeffreys' methods proved the boy right."[13]

As word of DNA fingerprinting made its way into the immigration-law and immigrant communities, Jeffreys and his colleagues at Leicester—which housed the only laboratory in the world carrying out the technique—became deluged with requests for help in resolving immigration disputes. "It was crazy, just impossible," remembers Jeffreys. "We just simply couldn't cope. It lasted precisely two years and it was two years I wouldn't like to live through again. We had enquiries by the thousand—box file after box file after box file of enquiries."[14] At the same time, Jeffreys's lab also began to do some paternity testing, and by early 1986 DNA evidence had been accepted by the magistrate's court as valid and reliable.[15] There is no definitive count of how many cases (each involving several individuals) Jeffreys worked on, but he estimates that it was about two hundred.

In response to the demand for DNA fingerprinting, especially in the realm of immigration, Jeffreys and the Lister Institute (which funded his research) agreed that Imperial Chemical Industries Plc (ICI)—the United Kingdom's largest chemical company and one of the biggest multinational corporations in the world—should be granted a license to develop the commercial potential of the technology. In early summer 1987, ICI opened a DNA testing laboratory that offered DNA testing to anyone willing to pay the £105 fee for each sample tested. In the Blood Test Regulations Amendment of 1989, DNA profiling was given official recognition as a legal method of determining paternity in immigration cases. However, this formal acceptance was something of an anticlimax: the Home Office had already begun to use the technique in forensic casework, and it had been accepted by numerous courts.

THE BLOODING

The first criminal investigation to make use of DNA fingerprinting also took place in the United Kingdom. It gained so much notoriety that an American crime writer Joseph Wambaugh, published a popular and entertaining account of the case entitled *The Blooding.* The story begins in November 1983, when a passerby discovered the body of fifteen-year-old Lydia Mann on a secluded path near the village of Narborough in Leicestershire. The girl's rape and murder created a swirl of fear in this quiet and peaceful part of England, and the police launched a major investigation. But in spite of the attention, the case went unsolved. Nearly three years later, in July 1986, a similar crime was carried out in the neighboring village of Enderby. This time, fifteen-year-old Dawn Ashworth was found in a similar location, stripped of her clothes and strangled.

As in the *Mann* case, the police launched a massive investigation. This time, however, they had a suspect—a seventeen-year-old boy who worked in the kitchen of a local mental hospital. Apparently, he had become very excited and agitated about the murder, which occurred near the hospital, and somebody alerted police about his odd fixation with the crime. Upon questioning, the young kitchen porter, who had a low IQ, confessed to Ashworth's murder. Seeking to ensure an easy conviction, the Leicestershire police asked Jeffreys to compare a sample of DNA from the kitchen porter with a DNA sample obtained from materials collected in the Mann investigation.

Jeffreys agreed to undertake the analysis in September 1986. He succeeded in extracting DNA from sperm collected from the three-year-old crime scene, but there was insufficient genetic material to carry out the multilocus DNA fingerprint test that he used in paternity and immigration cases. However, Jeffreys had been developing probes that could interrogate a single locus, an approach that required only minute quantities of DNA. Much to his surprise, this new technique produced results: the DNA profile of Mann's murderer did not match that of the kitchen porter. This was a setback for the police. Almost at a loss of how to proceed, they gave Jeffreys a sample of DNA from the second crime scene. The results were astonishing: the DNA from both murders was identical, but the profile did not match that of the kitchen porter.[16] Police had a serial killer on their hands, but could rule out their original suspect.

The police officials who received the news were shocked by the results of the DNA analysis, but they accepted it at face value because they felt they had no choice. The Home Office's rapidly developing Forensic Science Service (FSS) at Aldermaston confirmed Jeffrey's conclusions, and a few days before his trial, police dropped all charges against the kitchen porter. He was released from jail— the first criminal suspect exonerated by DNA evidence.

With no leads in the case and pressure mounting from the community to find the double murderer, Leicestershire police investigators took an unprecedented step: they requested voluntary blood and saliva samples from all men between the ages of seventeen and thirty-four living within the vicinity of the crime scenes. This, they hoped, would allow them to eliminate suspects from their investigation. Although the samples would initially be tested using traditional blood markers, all individuals who could not be eliminated in that way would be subject to DNA fingerprinting. From Wambaugh's account of this genetic manhunt, it is clear that police investigators were deeply skeptical of the reliability of the technique but felt that they had no other alternative. Three years of traditional police work had yielded no valuable information about the crimes.

Thus, the Leicestershire police force set about organizing the massive task of collecting samples from what eventually amounted to 4,582 young men, while the Home Office raced to finish their first DNA analysis laboratory in Huntingdon, Cambridge, in time to perform the actual DNA fingerprinting tests.

Once up and running, the Home Office set about DNA typing the more than five hundred individuals who could not be eliminated by standard blood typing. Much to the chagrin of investigators and forensic scientists, none of the samples matched the profile of Mann's and Ashworth's killer. The investigation seemed to have reached yet another dead end. But things took an astounding turn in mid-September, when police received a call from a woman who worked at a bakery in Leicester. She told them that one of her coworkers, Ian Kelly, had let it slip out over a lunchtime pint that he had fooled police during "the blooding" by giving a sample for a cake decorator employed at the bakery. The police arrested Kelly. Upon questioning, he quickly laid out the entire plot to the police. The cake decorator, Colin Pitchfork, convinced Kelly that he had already given blood for a friend who had been convicted of flashing in order to help him out. Reluctantly, Kelly agreed to cover for Pitchfork and gave blood in his name in late January 1987.

Based on this information, Pitchfork was arrested in his home, where he promptly confessed to both murders. Subsequent DNA testing revealed that his profile appeared to be exactly the same as murderer's profile. Pitchfork pleaded guilty in January 1988 and was sentenced to two life terms for the murders, plus additional time for rape and indecent exposure. The case was hailed as the first significant success for the use of DNA typing in forensic investigation. "Had it not been for genetic fingerprinting, you might still be at liberty," the judge told Pitchfork at sentencing. This quotation had a great deal of impact around the world and was reprinted widely.[17]

After the conviction of Pitchfork, DNA fingerprinting rapidly spread through the United Kingdom's forensic science community and was readily accepted in the context of paternity and immigration disputes in numerous courts and tribunals. After a period of disagreement over who possessed the rights to the technology (ICI owned the license, but the Home Office claimed that the technology was a public good), the Home Office and ICI agreed to divide the DNA testing market up so that ICI carried out all immigration cases and the Home Office carried out all forensic casework. Thus, in the United Kingdom, the Home Office's Forensic Science Service would be the dominant developer of forensic DNA analysis, ensuring that forensic evidence would remain the product of public laboratories for years to come.

CELLMARK CROSSES THE ATLANTIC

Hoping to expand the market for DNA testing worldwide, ICI was keen to set up testing laboratories in other countries, especially the United States, with its potential market for "hundreds of thousands" of DNA tests per year.[18] In October 1987, ICI Americas began offering DNA fingerprinting through a business unit called Cellmark Diagnostics USA. Cellmark's initial business plan was to concentrate on the paternity market, both to establish its reputation in the United States, and to build up a genetic database for the American population. They hoped to convince members of the family law community that their test could provide certain identification of fathers, and that armed with these results, they could more easily negotiate out-of-court settlements.

Although paternity was Cellmark's initial emphasis, shortly after opening, the company began accepting forensic casework as well.[19] According to one report, Cellmark received about a dozen forensic samples during its first month of operation. Until early 1988, however, Cellmark only offered multilocus testing, which was less than ideal for use in forensic casework because it required more DNA than was generally available in a forensic sample. Further, because multilocus probes yield so many bands so close together, the results of the test were difficult to interpret unless the question of identity was relatively clear-cut (such as in the case of paternity, when the child is a combination of parental patterns). Although the company would soon have a set of single locus probes available for use in forensic casework, this put them at a competitive disadvantage in the criminal justice market.

Once they did enter the American forensic market, ICI believed that they would have difficulty gaining admissibility for DNA fingerprinting because of strong admissibility standards and a highly adversarial courtroom. In an effort to gain firsthand experience on admissibility issues, ICI hired Daniel Garner as laboratory director of their American facility. Garner had gained significant experience in forensic science as chief of the U.S. Bureau of Alcohol, Tobacco, and Firearms' forensic laboratory for most of the 1980s.

Almost the moment that he arrived at the Cellmark's new facility, he began preparing for the introduction of DNA fingerprinting into the courtroom. In order to assist with this task, Cellmark hired two other Ph.D.-level scientists and developed a network of advisers around the country. In a recent interview, Garner stated that the first stage of the validation process was repeating many of the experiments done in the United Kingdom and developing additional quality control measures.[20] He also consulted widely with academic biologists and forensic scientists throughout the country to find out what additional validation work should be done and what issues may arise.

Equally important to ICI as its early scientific and technical staff was its management, business development, and marketing staff. If Cellmark was to make a significant amount of money on DNA fingerprinting, it had to become a market leader in the field as early as possible. The primary reason for this need was that most of the company's business would be based on precedent. Once a particular company's method was admitted into court in a particular jurisdiction (and this decision was upheld at the appellate level), prosecutors would want to use that technique since they would not have to hold an admissibility hearing (i.e., a formal courtroom inquiry into the validity and reliability of a scientific technique or body of knowledge), or worry that that the evidence would not be admitted at trial. As a result, gaining market share was of crucial importance. This fact is evident in the effort that went into making connections with the law enforcement and prosecution community. Through direct-mail campaigns, a newsletter called *Benchmark*, advertisements in major legal publications, presentations at professional meetings, and personal sales calls, Cellmark sought to convince potential customers that their technology was powerful and ready for use in the forensic market. In a mid-1988 marketing letter to attorneys, Anna Uchman, Cellmark's product manager for DNA fingerprinting, highlighted the technique's ability to provide "definitive" identification and a "conclusive" link between crime scene sample and suspect. Uchman also noted that DNA fingerprinting would speed police investigations and strengthen other forms of evidence in court. Her letter concluded: "DNA Fingerprinting^{SM} is fast becoming the test of choice for paternity establishment and forensic testing because of the certainty of the results."[21]

At about the same time, Cellmark came out with its first advertising campaign linked directly to the forensic market. Opening with the tagline "DNA Fingerprinting Links the Criminal to the Crime," and showing an image of handcuffs linked by the double helix, an advertisement for Cellmark stated:

> Police Departments. Prosecutors. Defense Attorneys. They're all turning to a new testing technology called DNA FINGERPRINTING that positively identifies suspects. With DNA FINGERPRINTING, you know who left the evidence—and who didn't. All by examining a suspect's one-of-a-kind genetic material.
>
> NOW CRIMINALS LEAVE MORE CONCLUSIVE EVIDENCE THAN EVER BEFORE. Because DNA FINGERPRINTING turns more of what they leave into evidence. A hair root. A drop of blood or semen. A bit of skin. If it contains DNA, it may be the difference between conviction and acquittal.
>
> CONCLUSIVE RESULTS IN ONLY ONE TEST! That's all it takes. Even if a sample is from a mixed stain. Or multiple sources. And because DNA is so stable, you can still get good results from samples that are months or even years old. . . .
>
> DNA FINGERPRINTING [is] the most powerful ID testing service available. The extraordinary breakthrough that has extensive validation. And acceptance by the scientific community.[22]

In all of these marketing tools, Cellmark went to great lengths to convince potential customers that DNA fingerprinting (which by 1988 involved dual analysis with single- and multilocus probes) produced conclusive results, and that the technique was well known and accepted by law enforcement, the legal community, as well as the general public. Cellmark also stressed that the results of DNA fingerprinting would provide a "one-of-kind" genetic identification on old or new crime scene samples. The company made no effort to warn potential customers of the inherent difficulties of performing DNA typing on forensic casework. They portrayed the technology as a black box in which even a minuscule amount of aged, and potentially contaminated, crime-scene evidence was fed in at one end, and highly accurate, individually identifying results came out the other side.

COMPETITION IN THE DNA TYPING MARKET: LIFECODES

While Cellmark had a persuasive marketing pitch and a proven technology, they did not have a lock on the American identity testing market. Although Alec Jeffreys and his collaborators at the Home Office and Cellmark generally received most of the credit for the invention and development of DNA typing technologies in the popular media, this claim is not historically accurate. Jeffreys did invent one specific type of DNA profiling, the DNA "fingerprint," but there were many concurrent efforts to develop marker systems for forensic identification in both Europe and the United States. The most successful of these endeavors was carried out by an American company called Lifecodes Corporation. Lifecodes began operating in 1982 under the name Actagen and was founded as a joint venture between three medical researchers from the Sloan-Kettering Cancer Institute in New York City and a large publicly held chemical company called National Distillers and Chemicals Corporation. The company had recently embarked on a significant expansion and reorganization campaign and hoped to exploit recent developments in the application of biotechnology to medical research and diagnostics.[23]

Actagen's initial research was not explicitly focused on identity testing, but rather sought to use recombinant DNA technology and genetic markers to aid in the diagnosis of cancer and infectious disease.[24] They also did a considerable amount of cytogenetics work. Actagen's lack of a specific focus was partly the result of generous funding from their parent company, and partly the result of uncertainty about which avenues of research would eventually succeed and which would fizzle. By 1986, however, it was clear to the company's management that identity testing was one of its most promising lines of research. At about this time, National Distillers underwent reorganization and was renamed the Quantum Chemical Corporation. Quantum quickly repurchased the half of Actagen it did not already own and changed the name of the company to Lifecodes to better reflect its primary research.

One of the central ideas floated by the founders of Lifecodes to their funders was the possibility of using restriction fragment length polymorphisms (RFLPs) in medical research, identity testing, and diagnostics. The basic idea was the same as that employed by Jeffreys and many other researchers in the world of genetics: find stretches of the genome that contain variable DNA sequences and figure out a fast, easy way to locate variants in individuals. These variants could then be used as markers to link particular genetic patterns with diseases, physical traits, behaviors, and even individual identity.

In an effort to pursue this genetic technology, the company hired a small group of molecular biologists, geneticists, and technicians. Lifecodes' first major hire, Michael Baird, was brought in to direct the company's laboratory. He received his Ph.D. from the University of Chicago, where he focused primarily on classical genetics analysis. From 1979 to 1982, he did postdoctoral work in the Department of Human Genetics and Medicine at Columbia University, where he learned the bulk of his molecular biology. Baird played a crucial role in the development of Lifecodes' DNA typing system and became the company's primary expert witness in early court cases involving DNA evidence. He was appointed vice president of Laboratory Operations in 1993. Shortly after Baird joined Lifecodes, the company hired Ivan Balazs as director of research and development. Balazs received his Ph.D. from the Albert Einstein College of Medicine and did research at Sloan-Kettering prior to being hired by Lifecodes. Baird and Balazs would go on to assemble a team of researchers, including population geneticist Kevin McElfresh, that would play a crucial role in the development of forensic DNA analysis methods throughout the 1980s and 1990s. Interestingly, however, none of the Ph.D. researchers responsible for developing Lifecodes' system had previous experience in forensic casework or law enforcement.

SLPs versus MLPs

As mentioned, there were many similarities between the approaches taken by Lifecodes scientists and Jeffreys, especially with respect to molecular biological techniques. The major difference was that Jeffreys's original DNA fingerprinting method involved using probes that bound to many loci in an individual's genomic DNA, while the Lifecodes method relied on highly specific probes that bound to only a single locus in the genome. As a result, instead of ending up with a complex pattern of bands that resembled a supermarket barcode, Lifecodes' method generally produced only two or three bands for each probe (see fig. 4). Although one probe was not enough to resolve individual identity (because a given locus had no more than a few dozen variations in the number of repeats),

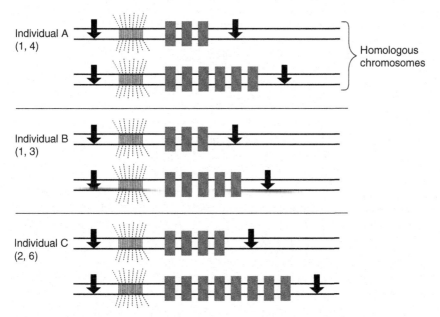

Figure 4a. Individuals receive one copy of each chromosome (except the Y-chromosome, which is passed down only through the male lineage) from their mother and one from their father. Thus, there are two copies of nearly every locus in the genome. Here, individual A has the alleles 1 and 4 at this locus; individual B has the alleles 1 and 3; while individual C has the alleles 2 and 6. Many people can share the same set of alleles (or genotype) at a given locus, but very few share the same set of alleles over many loci.

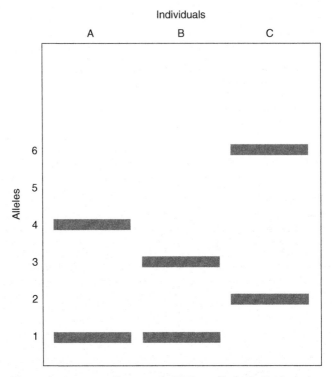

Figure 4b. An autoradiogram of a DNA profile for individuals A, B, and C. Because small fragments of DNA travel faster through the gel than large ones, allele 1 will travel the longest distance during electrophoresis, while allele 6 will travel the shortest distance. Illustration by Susan Heller Simon.

the use of several of these single-locus probes would build up a DNA profile that could theoretically lead to absolute individual identity.

This single-locus probe (SLP) method had several advantages over Jeffreys's multilocus probe (MLP) technology when applied to the forensic setting. Most important, the SLP technology was more sensitive than MLP. This meant that the SLP test could be carried out on a much smaller sample of biological material than the MLP test. Also, SLP results were much easier to measure, view, and explain to nonexperts since they consisted of at most a few bands. This factor was particularly valuable in cases where mixed biological samples had to be tested (a very common occurrence in violent crimes like rape and murder). Indeed, results from MLP testing of mixed samples were not generally interpretable. Thus, decisions about which technology was better were made both in the context of doing good science and developing a marketable product for the legal system. As Balazs said in a February 2002 interview, scientists at Lifecodes were well aware of Jeffreys's work but believed that they were developing a technology that was "more robust from a genetic standpoint . . . and that would give a powerful

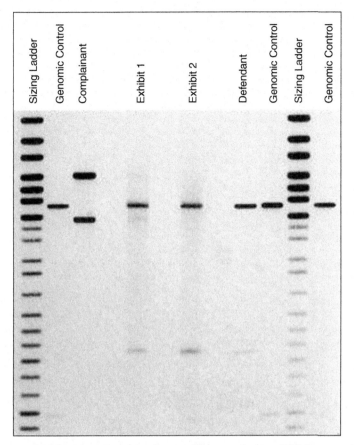

Figure 4c. A Tepnel Diagnostics autoradiogram illustrates the use of SLP in a sexual assault case. In the first lane is a size ladder used to determine the length of each fragment. This measurement is needed to determine how rare a particular fragment is in a given population. The second lane contains a known control sample that is used to ensure that the process is working properly. The third lane contains the DNA profile of the victim (or "complainant") for a particular locus. The fourth and fifth lanes contain DNA profiles taken from two pieces of evidence found at the crime scene. Exhibit 1 appears to contain biological materials from two people, while exhibit 2 appears to contain biological material from only one person. The sixth lane contains the defendant's DNA profile. The seventh and ninth lanes contain additional samples of control (probably tested at the same time as each of the evidence samples), and the eighth lane contains an additional size ladder in order to make it easier to measure the length of the fragments on the right side of the autoradiogram. Note that the defendant's profile appears to match the bands in exhibits 1 and 2, and the victim appears to match the bands found in exhibit 1. Thus, the DNA evidence suggests that he, or someone with the same profile at this locus, committed the crime, although errors such as contamination or sample mix-up could also lead to the same results. Image courtesy of Tepnel Diagnostics.

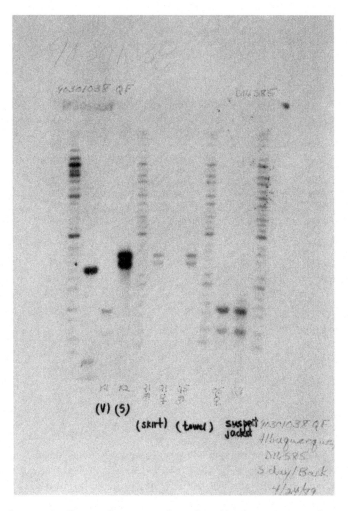

Figure 4d. An autoradiogram from casework conducted by the FBI in *State of New Mexico v. Jay Anderson*, the first case in which DNA evidence was litigated in the state. It is typical of the quality of autoradiogram at the time. Note how it differs from the promotional image in 4c. The gel is somewhat distorted, leading to a slight misalignment between the bands in the adjacent samples Q5-♂ and Q7. The first, fifth, ninth, and twelfth lanes contain size ladders. The second lane contains the known control sample. Lane K1 contains DNA of the victim. Lane K2 contains DNA of the defendant, Anderson. Lane Q1-♀ contains the profile of a semen stain found on the victim's skirt, while lane Q1-♀ contains the unsuccessful attempt to produce a profile from female cells recovered on her skirt. Lane Q5-♀ contains the profile for a semen stain found on a towel at the crime scene, while lane Q5-♂ contains the profile for female cells found on the same towel. Lane Q7 contains the DNA profile of a bloodstain found on Anderson's jacket. Anderson's bands appear to correspond to those of the male fraction of the semen stain found on the skirt and the towel, while the victim's bands appear to correspond to the female profile found on the towel and the blood found on Anderson's jacket. Note also that there is a faint third band in Q7. Image courtesy of William C. Thompson.

identification that was sufficient for the kind of applications that people in the legal profession would require."[25] In Baird's opinion:

> I really think that what caught everybody's eye with the Jeffreys's approach was the complexity of the pattern that you got. We had a profile that had only two bands per probe. He had a profile that had dozens of bands with one probe. And I think when people saw that, they said, "Aha, we've got a fingerprint, it's unique identification. End of story." But we had seen similar types of results looking at regions around the insulin gene, where if you reduce the stringency of your test hybridization or washing, you get a similar kind of pattern with dozens of bands. The problem was that to apply population genetic principles to that, we didn't know how to do it. Population genetic principles are looking at one gene at a time, in terms of doing calculations like the Hardy Weinberg equation, or combining results from independent loci, and all the other genetic tests that we were familiar with, both in paternity and forensics, used that approach, so we felt we needed to have something that would be able to go along with what was already done. And, to come up with this new system with a new math and a new way to do it, we didn't feel was the way to go.[26]

SLP technology simplified the visual inspection of DNA profiles a great deal and gave Lifecodes a competitive advantage in the forensic market until Cellmark came out with similar technology a few months after opening in the United States. SLP technology added a new layer of complexity to DNA typing, however, because test results could only be interpreted using population genetic principles and statistical techniques. This situation arose because a match between two DNA profiles could mean one of three things: that they came from the same person; that a mistake was made during some point in the DNA profiling process, such as contamination or sample mix-up, and the match was actually false; or, finally, that there was a coincidental, or "random," match between two different people who happen to share the same DNA profile.

As will be discussed in greater detail in the conclusion, because very little data exist on the rates of error in DNA profiling, it was, and remains, nearly impossible to calculate the probability that a match between two samples is false. Instead, in most cases, lawyers and scientists are primarily concerned about calculating a random match between two profiles. Although it is largely beyond the scope of this book, it is important to note that the probability of a random match only describes the chances that two randomly selected profiles from a particular population will match. It says nothing about the probability that the match is incorrect due to error, or about the probability of the innocence or guilt of the suspect. If the probability of a random match between a suspect and a crime scene sample is 1 in 100,000, this does not mean that the probability that the suspect is guilty is 100,000 to 1. Even if the suspect and the evidence DNA profiles match exactly, it is entirely possible that the presence of the suspect's DNA at the crime scene had nothing to do with the crime that was committed there.

For example, the fact that investigators find an exact match between a man and DNA evidence from a rape scene does not automatically imply that he committed the crime—he could, after all, be the victim's boyfriend or husband. The claim that the probability of a random match is equivalent to the probability of innocence is widely referred to as the "prosecutor's fallacy."

It should be noted that the corollary to this fallacy, the "defense attorney's fallacy," is that if probability of a random match between suspect and crime scene DNA is 1 in 100,000, and the crime took place in a city of one million people, they there must be ten people who share the same profile. Following this faulty logic, the defense attorney's fallacy leads to the conclusion that there is only a 10 percent chance that his or her client is guilty. This claim might be legitimate if it could be shown that all nine other men had equal access to the crime scene (i.e., that there is absolutely no other evidence linking the suspect to the crime), but this situation rarely occurs in practice.

Assuming that a false match is not at issue, scientists must choose between the two remaining possibilities. To do so, they need to know how common or rare the alleles in the DNA profile are within the appropriate reference population and whether they are associated with one another in a given population more often than would be expected by chance. Only when these two questions are answered can the scientist say anything about how likely it is that the two profiles come from the same person or are a random match. Although forensic laboratories and prosecutors initially presented these processes as being straightforward, they would become the dominant source of controversy after 1990.

The first step in determining the probability of a random match is deciding in which population one wants to make the calculation, since allele frequencies differ to a certain extent along racial and ethnic lines. It was well known both then and now that there is a risk of miscalculating the probability of a random match between an innocent suspect and the evidence DNA (i.e., the sample of biological material taken from the crime scene) if both the innocent suspect and the actual perpetrator shared a profile that was common in their own racial/ethnic population, but very rare in some other population that served as the allele frequency reference in the case (or vice versa). Forensic laboratories initially developed allele frequency databases for their probes based on the major racial and ethnic groups in the United States: Caucasian, black, and Hispanic. Until more ethnically and regionally specific studies were undertaken by population geneticists and forensic scientists, determining which database to use was fairly straightforward. If a rape took place in a predominantly white community in Iowa, the population in which one would want to calculate the probability of a random match between the sperm recovered from the victim and the suspect would undoubtedly be Caucasian. Similarly, if a murder took place in Harlem, then forensic laboratories would use the "black" database. If an assault took place in a predominantly Mexican American neighborhood in East Los Angeles, the allele frequencies from the Hispanic database would be used. In more complicated cases, such as a person of mixed race, or a crime that took place in a highly integrated region, forensic laboratories would calculate probabilities for multiple races and use the most conservative result.

Many population geneticists, however, argued that it was probable that there was a significant degree of substructure within populations classified according to American racial constructs. In other words, within that predominantly white community in Iowa, there may be pockets of people of Norwegian, Irish, and German descent who tend to marry people of their own ethnic heritage, rather than mating randomly with any Caucasian person in their home region. The frequencies of various alleles, or their patterns of inheritance, may be affected by such mating behaviors.

In order to calculate the probability of a random match within any of these populations using a simple rule of probability that depends on the assumption of independence, often called the "product rule," forensic scientists first had to establish that the alleles in question were inherited independently within that population. Although the product rule will be explained in greater detail below, for now it is important to know that it is used primarily to calculate the probability of a series of independent events occurring together. The product rule is nothing more than multiplying the probabilities of each of the events together.

However, one must take into account "conditional probabilities" in cases where the events are not independent. A conditional probability is one in which the probability of some event is dependent upon another event taking place. For instance, if one wanted to estimate the proportion of Europeans having blond hair, blue eyes, and fair skin simultaneously, one could theoretically conduct a survey of a subset of the European population to discover what percentage has each trait and then multiply all of these numbers together using the product rule. To illustrate the point, let us assume that each of the traits occurs in one in ten people according to the survey. Using the product rule ($1/10 \times 1/10 \times 1/10$), we would arrive at the answer that 1 in 1,000 Europeans have blond hair, blue eyes, and fair skin simultaneously. However, our experience tells us that there is something wrong with this answer: blond hair, blue eyes, and fair skin are not randomly occurring traits. If a European person has blond hair, the conditional probability of that person having blue eyes given that they have blond hair is higher than simply having blue eyes. This situation exists because the traits are positively associated in people of Northern European descent. Thus one in one thousand is almost certainly an underestimate of the true proportion of Europeans having all three traits simultaneously.

There are two kinds of independence that must be established before the product rule can be used without worrying about conditional probabilities: the inheritance of alleles within a particular locus and the inheritance of the loci themselves. Unfortunately, we cannot visually examine the extent to which genetic alleles tend to be inherited together (because they are invisible to the naked eye), so we have to use other techniques. The simplest way to determine the independence of alleles within a locus is to use the Hardy-Weinberg principle, a model in population genetics that makes certain assumptions about populations. It allows conclusions to be drawn about the expected frequency of genotypes (a technical term for genetic profile) when those assumptions are met. This principle states that under conditions of random mating, and given a set of very important additional assumptions—including nonoverlapping generations and a large population size—alleles at a particular locus should associate at random. Thus, under conditions of random mating, the relationship between allele frequencies and genotype frequencies is particularly simple. For example, in a locus with two alleles,[27] C and c, or "big C" and "little c," whose frequencies in a population are p and q (with $p + q = 1$), one expects the frequency of the homozygotic genotype (i.e., two copies of the same allele) CC to be p^2, the frequency of heterozygotic genotype (i.e., one copy of each allele) Cc to be $2pq$ (the factor of 2 is used since the female gamete could conceivably carry C and the male gamete could carry c, or it could be the other way around), and the frequency of cc to be q^2. If alleles at a particular locus are inherited at random, then the population is said to be in "Hardy-Weinberg equilibrium," often abbreviated "HWE." If there is bias toward particular combinations of alleles at a given locus, then the population is said to be in "Hardy-Weinberg disequilibrium." Geneticists have tested this principle in many cases and have found that it is remarkably accurate for genetic systems in which there are no advantages or disadvantages to having one allele versus another. It is crucial to realize that members of a population can mate at random for some traits, such as blood type, but not for others, such as height, skin color, or other visible markers.

As stated above, once the independence of alleles within a given locus has been determined, the same must be done for the loci that we are using to create a genetic profile. In other words, it is necessary to find out if there is any relationship between the inheritance of alleles at one locus and the inheritance of alleles at another locus. In genetics, the term "linkage" is used to describe the extent to which the alleles of different loci tend to associate during the process of gamete (sperm or egg) formation. Eggs and sperm are created through the process of meiosis, in which a cell with two chromosomes divides in half. During this process there is often an exchange of genetic material between the two halves of each chromosome. The most important thing to remember about linkage is that if allele A1 (at locus A) and allele B1 (at locus B) associate with one another (i.e., are found in the same gamete) at the frequency that one would expect by chance, they are said to be in linkage equilibrium. If they are nonrandomly associated, they are said to be in linkage disequilibrium. In order to use simple statistics to calculate the probability of a random match between two profiles, the alleles in the profile must be independent of one another (i.e., randomly associated). If alleles A1 and B1 are associated with one another nonrandomly, this makes it more likely that an individual with A1 will also have B1. Thus a DNA profile of A1, B1 would occur more frequently than one would expect by chance in a population in which they are in linkage disequilibrium. If there is a random association between alleles at different loci, then the population is said to be in "linkage equilibrium," often abbreviated "LE." If alleles at different loci are inherited nonrandomly, then the population is said to be in a state of "linkage disequilibrium." At least initially, the loci used in DNA typing were *assumed* to be in linkage equilibrium because they were located on different chromosomes, although this was contested by the defense in the *Castro* case, which will be described in the next chapter.

Assuming HWE and LE, the product rule can be used to calculate the probability of a random match between two profiles. As described above, the product rule states that in order to determine the probability of a set of events all taking place together, simply multiply their individual probabilities of occurrence. For example, if a match between two profiles is found at three multi-allelic loci, and the frequencies of the matching alleles within the relevant population are as follows:

Locus A, allele A1: 10%; allele A2: 15%
Locus B, allele B1: 5%; allele B2: 1%
Locus C: allele C1: 15%; allele C2: 3%;

then one would calculate the probability of a random match for this profile as 2 (0.10)(0.15) × 2 (0.05)(0.01) × 2 (0.15)(0.03) = (0.03)(0.001)(0.009) = 0.00000027, or roughly 1 in 3,700,000. It is important to remember that the use of the product rule depended on alleles being inherited in a random fashion. If all of this seems confusing, then the only important thing to remember is that in order to *easily* calculate the probability of a random match between two profiles it is crucial to know whether or not the components of those profiles are inherited independently. If the assumptions of HW apply, then the conclusion is that the events are independent.

Forensic laboratories initially utilized a simple test for HWE in which the expected homozygote frequencies under HWE were calculated and then compared to the observed frequencies. This test was based on "Wahlund's Principle," which posited that higher rates of homozygosity would be expected in populations containing distinct subgroups than in randomly mating populations. Thus, if the observed frequencies were similar to those expected under conditions of random mating, then it was considered safe to use the product rule in calculating the probability of a random match of that population. Initially, Lifecodes and Cellmark claimed both their racial/ethnic databases were in HWE and used the product rule without modification. In addition to this calculation, the forensic laboratories also based their assessment of HWE on two main assumptions: first, that the sources of the databases (generally blood bank donors) were a representative sample of a given race, and, second, that individuals within races mated at random.

Although these assumptions were initially unchallenged by the defense, geneticists hired by defense attorneys eventually began to question them. After a long legal fight, defense experts gained access to the raw data from the population databases and discovered that, contrary to the private laboratory's claims, there was an "excess" of homozygotes (i.e., individuals with two copies of the same allele) in their databases. In other words, there were more homozygotes than one would expect based on the Hardy-Weinberg principle. In the defense's view, this was evidence that the databases, and hence the population, were not in HWE. Forensic scientists and prosecution experts defended their databases by arguing that the "excess" of homozygotes was apparent and not real. Instead of being actual representations of human populations, they were the result of the technical limitations of Southern blotting. Specifically, very small fragments often ran off the end of a gel, and fragments of similar size tended to "coalesce" into a single indistinguishable fragment. As we shall see in chapter 6, defense experts would eventually claim that testing for population substructure using deviations from HWE was not scientifically tenable because there were too many situations under which substructure would not have a large enough effect to be picked up by this method. In opposition, they argued that the only way to uncover substructure was to undertake large-scale empirical investigations of allele frequencies in a wide array of subpopulations.

Early Development at Lifecodes

Unlike Jeffreys, whose invention of DNA fingerprinting occurred largely by chance in an academic situation, Lifecodes was building an identification technology essentially from the ground up for the purposes of commercialization. From 1983 to 1985, Lifecodes spent a considerable amount of time and resources developing SLP technology with the explicit aim of applying it to forensic DNA analysis. The first step in this process was choosing a set of genetic marker systems that were variable enough to produce unique DNA profiles when pooled together.

Once this decision was made, the research team under Baird and Balazs was also faced with the daunting challenge of validating the approach for use in actual forensic investigations and the courtroom. This process involved studying the effects of contamination and age on DNA; ensuring the reliability of results; and developing controls that could be used to show that the actual test was carried out successfully.[28] They also had to build databases and develop statistical techniques that would enable them to determine the frequency of occurrence of a particular variant at a given locus.

According to Balazs, this work was not carried out simply to satisfy the scientific community that single locus probe technology was valid and reliable. The requirements of the legal system also played a crucial role in the decision to do certain experiments and not others. Balazs told me that the multiple audiences for Lifecodes' validation work meant that they had to satisfy their own community, as well as a community of lawyers and judges that did not share the same norms and values. Much of the experimental work done by Lifecodes scientists between 1983 and 1985 was explicitly aimed at satisfying the basic requirements for introducing new scientific evidence into the court system.[29] This extensive validation process was bankrolled entirely by National Distillers/Quantum Chemicals, which clearly saw huge potentials for profit in DNA typing.[30]

Lifecodes also collaborated with Robert Shaler, who at the time was director of serology in New York City's Office of the Chief Medical Examiner, to test their method on actual forensic bloodstains that had been stored in a cold room for up to three years. Concurrent with this research, the company also entered into collaboration with the New York Blood Center in order to demonstrate that results from their DNA typing test were valid for the determination of paternity. Based on comparisons of DNA results with conventional HLA and blood protein analysis, they concluded in a November 1984 presentation to the American Society of Human Genetics that DNA typing could be used to replace, or increase the power of, the conventional tests.[31] The two articles in *Journal of Forensic Science*, along with this work in collaboration with the New York Blood Center, would serve as the foundation for Lifecodes' claim that their technique was valid and reliable when introduced into court in late 1987. In 1986 the company proclaimed that as a result of these experiments, "when investigators use this new test to determine whether two specimens are from the same or different individuals, they can be confident that their findings are reliable."[32] This work also made its way into Lifecodes' advertising campaign, with one early ad suggesting that the company's efforts from 1982 to 1987 ensured that results from the DNA-Print test could be relied upon by clients.[33]

Lifecodes Unveils the DNA-Print Identification Test

Once Lifecodes felt they had a product that was ready for market, company scientists and executives alike realized that they needed to make the forensic and law enforcement communities aware of their DNA test. Because Lifecodes had been primarily a research and development organization up to this point, the company had no significant experience or contacts within the world of forensic science. In an effort to obtain these missing ingredients, Lifecodes' management hired Shaler away from the New York Medical Examiner's Office to direct the company's forensic business development effort. His job was to provide the company with connections to the forensics community and to help the company's scientists "through the maze of the courtroom."[34]

One of Shaler's first tasks was to help the company develop marketing materials that would convince the forensic community that the DNA-Print would be a powerful new addition to law enforcement's arsenal of tools for determining individual identity. In their original promotional material, Lifecodes began by calling the DNA typing test an "exquisitely accurate tool" which allows for biological evidence to be "examined at its most fundamental level—the deoxyribonucleic acid (DNA molecule)."[35] They argued that this technology had three advantages over older forms of identification. First, it would be impossible for an assailant or murderer to alter their DNA profile in any way. If an individual attempted to alter his or her DNA profile, such an alteration would lead to no result, rather than the wrong result. Thus, there was no possibility of getting false positives with the test.[36] This notion would come to be widely repeated by almost every proponent of the use of DNA evidence in the legal system, and would become widely contested by those individuals who were more skeptical of DNA evidence. Second, Lifecodes claimed that DNA evidence could be recovered in many more cases than fingerprint evidence since it is found in almost all biological material and is significantly more stable than blood proteins. Finally, they pointed out that DNA-Prints were much easier to digitize for storage and transmittal than results from other types of identity tests.[37]

EARLY MARKETING EFFORTS AT LIFECODES

Like Cellmark, Lifecodes also undertook a significant public relations effort, which included sending representatives to professional meetings for lawyers and forensic scientists, as well as advertising in the journals and magazines that these groups were likely to read. Both companies sought to portray the ease of DNA testing and its superiority over older methods of identification. In a 1988 advertisement, Lifecodes proclaimed that its "DNA-PRINT™ Test makes proof-of-paternity child's play." The ad portrayed a young baby of unclear ethnicity holding a gavel with the caption "The verdict is in." The company sought not only to establish that the technique was valid, reliable, and ready for acceptance within the court system, but also that it was technologically simple. Lifecodes went further, claiming that the DNA-Print was "the only court-proven DNA-based test with an average inclusion of 99.9%," and pointed out that its staff included twenty Ph.D.'s. The advertisement concluded, "With such overwhelming evidence, isn't it time you made unprecedented accuracy your precedent?" No mention was made of the fact that the technique had yet to be tested in an adversarial court proceeding. While DNA evidence had been used in numerous civil proceedings, as well as one forensic case in which the defendant was acquitted, the technique had not yet been the subject of an admissibility hearing.[38]

◇◇◇

CHAPTER 3

DNA on Trial

The Andrews Case

The lack of an admissibility hearing changed for Lifecodes in the autumn of 1987, when twenty-four-year-old Tommie Lee Andrews was brought to trial for a rape in Orlando, Florida. He had been arrested prowling in a woman's yard in the wee hours of the morning in the southeastern section of the city. She lived in a neighborhood that had seen almost two dozen instances of rape, breaking and entering, or assault in the past year. The circumstances of these crimes were similar enough that police quickly concluded that they were dealing with a serial rapist. Andrews soon became the prime suspect. In almost all cases, a man would enter into a woman's home after midnight and attack her as she prepared for bed or slept. In all cases, he went to great lengths to cover the victim's eyes to prevent her from seeing him. Only the first of his many victims—twenty-seven-year-old Nancy Hodge—claimed that she was able to get a clear view of his face.

The perpetrator had a few particular habits, which included switching the lights on and off several times during the attack, as well as examining the woman's driver's license before leaving the scene.[1] The man also took great care to remove any evidence he left behind. One further characteristic that linked all of the rapes and assaults was that the perpetrator seemed to know the schedules of his victims very well. The first victim reported a prowler in her yard six weeks before she was raped. The rapist told another woman that he had observed her having sex with her boyfriend and wanted to do the same thing with her. As a result of this information, the police set up neighborhood patrols in the areas where the rapist had previously struck. These patrols paid off on the morning of 1 March 1987, when Andrews was arrested after a woman called the police to report that there was a prowler in her yard.

The following morning, Hodge came to the police station and identified Andrews as the rapist in a photographic lineup. Based on this identification,

Andrews was charged with sexual battery and armed burglary in the *Hodge* case, as well as a similar rape that had occurred only the week before he was arrested. Assistant state attorney Tim Berry was assigned to handle the case for the Orange County District Attorney's office. Berry immediately realized he would have difficulty convicting Andrews of the crime. While standard forensic blood group analysis showed that Andrews could have committed the crime, the results fit almost two-thirds of American males.[2] The limited fingerprint evidence in the case suggested that Andrews was at the scene, but it was not conclusive. Further, Andrews had an alibi—that he never left his home that night—with his live-in girlfriend and her sister as witnesses. According to most commentators on the case, this limited evidence would not have led to conviction. While it is important to temper this notion with the knowledge that such pronouncements were made knowing the role that DNA evidence would come to play in the case, one can be certain that a verdict of guilty was not guaranteed by the eyewitness account and traditional forensic evidence.

In a discussion with Berry about the case, his colleague, assistant state attorney Jeffrey Ashton, suggested to him that he should consider using a new kind of forensic identification technique—DNA typing—that he had become aware of a few months before. In his March 1989 testimony before the U.S. Senate Judiciary committee, Ashton said that he first heard about DNA typing in late 1986 through media reports of the Enderby murder investigation taking place in the United Kingdom. He said that he noted the particulars of the case with a great deal of interest and hoped the technique would be available in the United States.[3] Ashton went on to recount how his interest in the technique was rekindled in the summer of 1987 by a Lifecodes advertisement in a Florida lawyer's publication, entitled "He's Wearing His Daddy's Genes." Thinking that this was the company that did the forensic testing in the *Enderby* case, Ashton called Lifecodes to find out if the service was available in the United States. Although it turns out that he made the wrong association (Cellmark was the company that was associated with the case), Lifecodes informed him that they did indeed do forensic DNA analysis.[4] At the time, Ashton was not working on any cases in which such a test would be helpful. He did, however, have a hunch that "whatever case we were going to do this in needed to be something significant. It needed to be a case where identification was the only issue and a case where we could convince the court and our own office to spend the money to break into this new area."[5]

Florida v. Tommie Lee Andrews turned out to be just such a case. Berry and Ashton, feeling that they had no other choice, decided to give DNA testing a shot. Ashton contacted Michael Baird to discuss the possibility of having samples from several rape cases to Lifecodes, and Baird soon agreed that Lifecodes would do the tests. In August 1987, evidence was flown from Orlando to the company's Valhalla, New York, lab, where forensic scientist Alan Giusti performed DNA analysis. According to Lifecodes, the test showed that Andrews was beyond doubt the source of the semen in two of the six cases. Armed with this powerful result,

Berry was now left with the task of convincing the judge that this novel evidence should be considered admissible in a court of law.

The Admissibility of Novel Scientific Evidence

At the time of the *Andrews* case, the leading authority on the admissibility of scientific evidence in almost all jurisdictions was *Frye v. United States*.[6] In this 1923 case, the defendant appealed the trial court's decision to exclude results from a lie-detector test that was favorable to him. The Court of Appeals for the D.C. Circuit upheld the trial court's decision based on the following logic:

> Just when a scientific principle or discovery crosses the line between the experimental and demonstrable stages is difficult to define. Somewhere in this twilight zone the evidential force of the principle must be recognized, and while courts will go a long way in admitting expert testimony deduced from a well-recognized scientific principle or discovery, the thing from which the deduction is made must be sufficiently established to have gained general acceptance in the particular field in which it belongs.[7]

The requirement set forth in this statement—that the principles underlying an evidence-generating technique be generally accepted within a relevant scientific community—has come to be known as the "*Frye* rule," or "*Frye* standard." In practice, the *Frye* rule meant that a judge was charged with the responsibility of taking the pulse of a particular scientific community (or set of scientific communities) to determine whether or not a particular idea or technique was accepted by enough scientists to be considered valid and reliable in a court of law.

The merits and pitfalls of this particular method of determining the admissibility of scientific evidence have been the subject of vociferous debate within the legal community for several decades. The particular ways that these debates played out in the context of forensic DNA analysis will be explored in great detail throughout this book.[8] Proponents of the standard found it reassuring that the admissibility of novel scientific evidence would be determined based on what scientists believed, rather than on the opinion of a technically unknowledgeable trial judge. Critics of the standard charged that it was difficult to enforce in practice and worried that it would deprive courts of relevant evidence, especially when produced using "cutting-edge" scientific techniques. According to this view, not only is it difficult to determine the appropriate field that a particular form of evidence belongs to, but it is also nearly impossible to precisely define "general acceptance" (i.e., how much consensus is enough?). Also, many believed that the *Frye* standard was vague about exactly what aspects of the evidence must be generally accepted.[9]

Florida's Approach

While the *Andrews* case is rightfully seen as the first instance in which a person was convicted of a crime based on forensic DNA evidence, in many ways it stands alone in terms of the admissibility of this new form of scientific evidence. At the time of *Andrews*, the exact means by which Florida courts judged the admissibility of scientific evidence was subject to much debate with the state's legal community. The Fifth District Court of Appeals opening statement in Andrews's appeal hearing was "we begin by confessing some uncertainty as to the standard applicable in this state governing admissibility into evidence of a new scientific technique."[10]

Florida courts did not rely solely on the *Frye* standard when judging the admissibility of new forms of science-based evidence. Instead, many Florida courts, including the jurisdiction in which Andrews was tried, made use of the "relevancy approach." Professor C. McCormick set out this alternative to *Frye* in his 1954 text on evidence, in which he wrote the following: "'General scientific acceptance' is a proper condition upon the court's taking judicial notice of scientific facts, but not a criterion for the admissibility of scientific evidence. Any relevant conclusions which are supported by a qualified expert witness should be received unless there are other reasons for exclusion."[11] It should be noted that Florida courts were not free to admit any and all evidence. According to established precedent, they did have to evaluate the reliability of the evidence. But, it was ultimately left to the judge's discretion to determine reliability—general acceptance was only one factor of many that could be taken into account. He or she also had to appraise the potential that the evidence mislead or confuse the jury, and the actual connection between the proffered evidence and the factual question at issue in the trial.

According to the Florida Evidence Code, "if scientific, technical, or other specialized knowledge will assist the trier of fact in understanding the evidence or in determining a fact in issue, a witness qualified as an expert by knowledge, skill, experience, training, or education may testify about it in the form of an opinion; however, the opinion is admissible only if it can be applied to evidence at trial."[12] Thus, in *Andrews*, prosecutors were not forced to show that forensic DNA analysis was generally accepted in the scientific community. They merely had to find a witness who could satisfy the judge's concerns about reliability.

THE ANDREWS TRIAL: PART 1

The pretrial admissibility hearing in the *Andrews* case began on Monday, 19 October. The prosecution called Dr. David Housman to the stand to perform the function of the credible witness who considered the technique valid and reliable. Housman had been a professor of biology at the Massachusetts Institute of Technology since 1975 and had impressive academic credentials. He had received his B.S. and Ph.D. from Brandeis University, completed postdoctoral work at MIT in molecular genetics, and was on the faculty of University of Toronto for two years before returning to MIT. The author of approximately 120 papers at the time of the trial, he was also on the editorial board of a number of journals and reviewed grant proposals for the NIH.

Housman had recently been contacted by Berry and asked if he would be willing to serve as a consultant/expert in the case. After agreeing to do so, he was mailed a summary of the case and was visited by representatives of Lifecodes. Before the trial, he made an on-site visit to Lifecodes' laboratory, where he examined the company's protocols, on-site performance, and quality control measures. One issue that would emerge at a later stage of the trial was that Housman did not witness the actual tests being done on evidence associated with the *Andrews* case, since they had been completed before his arrival. The only thing he could do with the results was review the laboratory notebooks to determine whether or not it appeared that proper procedure had been followed.[13]

Although Housman had significant experience in academic and medical research, he had never been involved in a single forensic investigation. (At this point in the history of DNA typing, however, understanding of the contingencies of forensic practice was not considered crucial to being able to evaluate the validity and reliability of the techniques associated with DNA typing.) Housman's primary responsibility—at least from the perspective of the prosecutor and the judge—was to use his credibility as a well-established molecular geneticist to vouch for the validity, reliability, and routineness of DNA typing as a set of molecular biological techniques. As we shall soon see, both the nature and roles of experts on DNA evidence changed dramatically as it became a factor in more and more cases.

On the stand, Housman told the judge that he considered DNA typing to be a standard technique in molecular genetic research. He said that he used the technique routinely, between five to ten times a day in his lab, and that "it's done on a similar basis in laboratories around the world."[14] He went on to say that the technique is useful in diagnosing at least ten human diseases, as well as cancer, and is used in genetic research involving plants, animals, bacteria, and other invertebrates. He also told the judge that Lifecodes used the same methodology for doing forensic DNA analysis that had been used in the academic and medical research community for at least ten years, and that there had been numerous peer-reviewed publications that described the technique in this context. Based on this information alone, the judge determined that DNA evidence was admissible in the trial.

One major issue that was not discussed in any detail during this hearing was whether the application of DNA typing was any different in forensic investigations compared to the kind of controlled laboratory work that Housman did. The only comment he made on this topic was that although the process of extracting DNA from sperm was different from that used to extract it from blood, this would not lead to any major deviations in the ultimate result. This issue became prominent in the mid-1989 *Castro* trial.

On Tuesday, 20 October, Andrews's trial began in earnest with selection of the jury. The next day, the jury began hearing evidence from the prosecution. Once again, Housman testified first. He reiterated the testimony that he had provided a few days before and told the jury about the basic techniques and molecular biological tools underpinning the test. He took pains to point out that they were widely used within the scientific community and had not been challenged in the ten years since they were first published in peer-reviewed scientific journals.

Giusti and Baird testified on Thursday about the process used to analyze the evidence in this case, with Baird walking jury members through the test step by step. Then, Baird presented the jury with the actual results from the tests applied in the *Hodge* case and showed them that Andrews's DNA profile appeared to match the rapist's exactly. He then told the jury that the probability of a random match in this case was one in ten billion.

THE DEFENSE: NOWHERE TO TURN

Andrews's defense attorney Hal Uhrig felt almost powerless in the face of evidence that appeared to be so exquisitely certain. He says that there was "nowhere to turn" for an alternative source of information about DNA evidence. Although he called biology departments at what he believed to be some of the best universities in the country (including Harvard, Stanford, and New York University), he said that nobody seemed particularly interested in helping him out. When he was able to speak to geneticists and molecular biologists, they told him that they were unwilling to challenge Housman's testimony because he was such a well-respected, top-notch researcher. Uhrig recalled that Housman was like "an angel" in the scientific community. The prevailing opinion amongst scientists that he spoke to was that "if Housman has examined the evidence and he says that it's good, then it's almost certainly good."[15] Prosecution attorneys proclaimed that the defense's failure to call any experts to rebut their witnesses' testimony or to argue against the technique was proof that it was considered valid and reliable in the scientific community.[16]

Lacking any real expert advice, Uhrig decided to follow his intuition and challenge the aspect of the DNA evidence that was most puzzling to him: how could the probability of a random match to Andrews be one in ten billion when there were only five billion people on the earth at the time? When Berry asked Baird to explain that this figure was based on accepted statistical practice, Uhrig objected that such testimony was inadmissible. Although the challenge itself was not particularly sophisticated, it threw Berry off guard because he hadn't really thought about this issue at all. Unable to provide a legal justification for the admission of statistical evidence, the judge declared that the jury could not consider the probability statistic in reaching their verdict. Now the jury was left to decide the probative value of matching DNA profiles without the benefit of knowing how common or rare these profiles were. The result was a hung jury. The jurors did not feel they could trust Hodge's memory without other corroboration, and at least one strenuously objected to the use of this new technology to determine Andrews's guilt or innocence.

This lack of a verdict did not mean that Andrews would go free. In addition to the retrial, he was also scheduled to be tried two weeks later for the burglary, assault, and rape of a twenty-seven-year-old woman while her children slept in an adjacent room. In this trial, Ashton took the role of lead prosecutor. Just as in the first Andrews trial, Housman, Baird, and Giusti took the stand to testify about the validity, reliability, and scientific acceptance of DNA typing. Baird also discussed the validity and reliability of the statistical techniques being used. He claimed that the DNA test used by Lifecodes obeyed the rules of Mendelian genetics and met the requirements of Hardy-Weinberg equilibrium.[17] He then presented the results from the actual test in the case. He showed the jury the autorads from the case, inviting them to see for themselves that Andrews's profile matched the rapist's exactly. After being prompted by Ashton, he told jurors that the probability of a random match in this case was 1 in 839,914,540.[18] Not wanting to be caught off guard, Ashton had already prepared an argument for the admissibility of this probability statistic, complete with precedents for the admission of similar testimony. This strategy effectively closed off the defense's challenge on this issue.

Once again, Uhrig could find no geneticist or molecular biologist willing to testify that any aspect of DNA typing was not well established within the scientific community. Lacking the statistics maneuver that had worked well in the first trial, he attempted in vain to impugn the credibility of the three prosecution witnesses. He argued that Baird and Giusti could not be trusted because they had a financial stake in the decision to admit DNA evidence into trial. He went on to claim that Housman had also had such a bias because "he draws a paycheck by his virtue of doing five to ten of these [nonforensic] tests a week."[19]

Uhrig also asserted that the particular test in this case could not be trusted since somebody outside of the company had not actually witnessed it. He suggested that something might have gone wrong in the process of carrying out the test. For example, the reagents or enzymes could have been contaminated or spoiled, or the gel could have been improperly prepared. Prosecution witnesses and attorneys responded to these claims by making reference to Lifecodes' "very strict quality control program," as well as detailed working protocols, logs, and experimental controls used to ensure that the test results are correct.[20] According to Baird, Giusti, and Housman, any conceivable problem with any aspect of the test would be noticed at some point in the procedure. Prosecution lawyers summarized their testimony on this issue as follows:

If the gel is not properly prepared, the results could be affected, but would usually make the test not work, rather than getting wrong results. It is highly unlikely that a voltage fluctuation would affect the result. If the conductivity was wrong, again, you would get no results, rather than the wrong results. The same is true if a solution is made improperly or the pH level is wrong—no results as opposed to erroneous results. The use of the aforementioned control samples insures this process, as they would be affected too. The bottom line is that with this type of test you can have only one of two types of outcomes. You either see a pattern that is clear and recognizable, or [either] the DNA sample was not present or the test was performed incorrectly[,] and you get no result. Dr. Housman has seen this in tests performed by his students.[21]

This formulation of the technique was a clear attempt by prosecutors to make the technology itself seem error-free. Thus, the only type of error that could occur was human error.[22]

In his cross-examination and closing arguments, Uhrig presented no testimony to counter these assertions. While none of these statements had been scientifically proven (especially those dealing with the impossibility of error) or had appeared in a peer-reviewed article to date, they were taken as factual both by the trial court and the appeals court. At this stage in the history of DNA typing, there simply was no body of knowledge or expertise to point to when arguing against the claims made by proponents of DNA evidence.

Based on the DNA results presented during the trial, along with other evidence, the jury found Andrews guilty and sentenced him to twenty-two years in prison. On 2 February 1988, Andrews returned to court once more, this time for the retrial in the *Hodge* case. This trial was essentially a rehash of Andrews's second trial, with prosecution experts making claims that went unchallenged by Uhrig. This time, Uhrig's main challenge to the DNA test results was that Lifecodes had not examined *all* of Andrews's DNA sequence, but only a small fraction of it. Baird handled this challenge quite easily by explaining to the jury that only the highly variable regions of the genome were informative about individual identity.[23] On 5 February, the jury met for approximately ninety minutes and returned a verdict of guilty. Andrews was sentenced to an additional seventy-eight years in prison, bringing his total from the two trials to one hundred years.

On appeal, the Court of Appeals for the Fifth District of Florida found that the trial court had not erred in admitting DNA evidence under the standards of the relevancy/reliability test described in Florida Evidence Code §90.702. One of the central pieces of reasoning used by the court to reach this decision was that the DNA test is "essentially the same for all purposes" and has been used in molecular genetics, disease diagnosis, and medical research for almost ten years. Further, it had been the subject of numerous peer-reviewed articles. The court wrote that "this extensive nonjudicial use of the test is evidence tending to show the reliability of the technique."[24] The judges were also swayed by the prosecution witnesses' contention that any error in the test procedure would lead to no results rather than incorrect results. The court used this claim to dismiss out-of-hand the appellant's contention that the test was unreliable.[25] Further, the fact that the defense could find no expert witness to testify seemed to suggest to the court that the test was reliable and the results in this case were accurate and believable.[26] The court pointed to a civil case in New York in which DNA evidence had been admitted, and the English cases discussed at the beginning of this chapter, as evidence that the technique was already in the process of gaining judicial acceptance.[27]

In its conclusion, the court sought to distance DNA typing from the kind of supposedly scientific evidence that had been admitted into court in the past, only to be found to be unreliable or fallacious at some point in the future. "In contrast to evidence derived from hypnosis, truth serum and polygraph," the court opined, "evidence derived from DNA print identification appears based on proven scientific principles.... Given the evidence in this case that the test was administered in conformity with accepted scientific procedures so as to ensure to the greatest degree possible a reliable result, appellant has failed to show error on this point."[28] With this statement, Lifecodes had won a major victory in establishing its product as the premier forensic DNA identification test in the American legal system.

RAPID JUDICIAL ACCEPTANCE

While 1987 saw a trickle of cases in which DNA evidence was a factor, by early 1988 the floodgates had opened. Partly as a result of very positive media coverage of the *Andrews* case, in which it was portrayed as *the* decisive factor in his conviction, and partly as a result of Cellmark's and Lifecodes' aggressive marketing campaigns, the two companies were soon flooded with requests for the test. Indeed, by the end of 1988 Cellmark and Lifecodes experts had provided testimony about DNA evidence in more than thirty-five cases in fifteen states. They also provided evidence in a significant number of police investigations in which the defendant pled guilty upon learning that such evidence existed. Unfortunately, there are no solid statistics available on the number of cases in which this situation occurred, but at least one defense attorney estimated the number to be in the range of several hundred in 1988 and 1989.[29]

In any event, by early 1988 a series of cases were under way in which DNA evidence would be ruled admissible with relatively little challenge from the defense. In Florida alone, eight cases were brought to trial in which Lifecodes' DNA evidence was admitted based on the ruling in *Andrews*. In one of these cases, *Florida v. Martinez*, Lifecodes calculated the probability of a random match at 1 in 234 billion. Although defense counsel objected to this statistical testimony on the grounds that it was "nonsensical" to use a figure almost fifty times larger than the world's current total population, it did not present any expert witness testimony to suggest that there was anything wrong with the principles upon which this calculation was based.[30] As a result, both the trial court and the Fifth District Court of Appeals ruled that the probability evidence was admissible.

Maryland (where Cellmark dominated the DNA testing market) and Texas courts admitted DNA evidence in four cases, each with no serious challenge from the defense. In one Texas case involving a serial killer (*Texas v. Lockhart*), DNA evidence was admitted from a Florida case without any pretrial hearing.

DNA evidence from Cellmark or Lifecodes was also admitted without significant defense challenge in at least one case in the following states: Alabama, Colorado, Georgia, Idaho, Kansas, Michigan, Ohio, Oklahoma, South Carolina, and Virginia. As James Starrs, a professor of forensic science at George Washington University, was quoted as saying in *US News and World Report*: "It's like Chicken Little saying the sky is falling.... Mention DNA, and defense attorneys run for cover."[31]

This unquestioned acceptance of DNA evidence was not universal, however. While most defense attorneys throughout the country faced with the existence of DNA evidence felt powerless to challenge it seriously, there were several notable exceptions. In trials that took place in Albany, New York (Lifecodes), and Fort Wayne, Indiana (Cellmark), defense lawyers brought in experts to argue that DNA typing methodology had not been adequately peer-reviewed and that the statistical methods used were not scientifically valid or generally accepted. Prosecutors responded to these claims not only by bringing their own witnesses to the stand to contradict the defense testimony, but also by successfully convincing the court that the defense witnesses were outside of the relevant scientific communities that were responsible for judging the validity and reliability of the technique.

THE FIRST CHALLENGE

The combined admissibility hearing in the cases of *People of New York v. George Wesley* and *People of New York v. Cameron Bailey* (533 N.Y.S.2d 643, 1988) (I refer to the combined hearing as "*Wesley-Bailey*") represents the first time that the defense actually called witnesses to challenge prosecution claims about the validity and reliability of any DNA typing technique.[32] The *Wesley-Bailey* hearing, which began on 11 December 1987 and continued at various times through the summer of 1988, was convened to determine whether or not Lifecodes' "DNA-Print" system met the requirements of the *Frye* standard. Because the cases were taking place in the same jurisdiction (Albany County, New York) and dealt with the same fundamental issue (the admissibility of DNA evidence), it was decided to combine the *Frye* hearings to save time and money.

In each of these cases, the prosecution had asked the court for an order to draw blood from the defendants for the purposes of carrying out DNA testing. In *Bailey,* the defendant was charged with rape of a woman who allegedly became pregnant as a result of the attack. Prosecutors hoped to establish that Bailey was the father of the fetus, which had been aborted. In *Wesley*, the defendant was charged with murder, rape, attempted sodomy, and burglary of a seventy-nine-year-old woman. Both Wesley and the victim were clients of an organization that provided care for mentally disabled individuals. In this case, there was a great deal of forensic evidence linking Wesley to the murder (including blood-stained garments, hair samples, and fiber samples found in the defendant's and the victim's apartments). Further, the defendant implicated himself in the crime through contradictory statements he made to detectives.[33] Thus, DNA evidence was meant only to bolster an already strong case, not to serve as the sole source of evidence. In his opening statement, prosecutor Daniel Dwyer told the court that it wanted to introduce DNA evidence so that its admissibility would be established for future cases.[34]

By this point, prosecutors across the country were already in the process of building an impressive group of eminent scientists who were willing to testify in DNA admissibility hearings. To begin with, Lifecodes' and Cellmark's own scientists, many of whom were respected members of the scientific community, testified and provided assistance to prosecutors as they prepared for court. Further, extremely well-known and respected academic molecular biologists, geneticists, and medical researchers seemed more than content to testify on behalf of the technique, even if they knew little about what Lifecodes was actually doing in their laboratory. While many were independently contacted by prosecutors asking for assistance, others were recommended to prosecutors directly by the identification companies. Cellmark, for instance, put together a list of "Independent Expert Witnesses" and distributed it to prosecutors who were interested in using DNA evidence in the courtroom. At least initially, these "independent" experts served primarily to educate judges on basic aspects of molecular biology and population genetics, as well as to confirm the interpretations made by the private corporations. This list included Housman, Bonnie Blomberg (a researcher at University of Miami Medical School), George Sensabaugh (a professor of forensic science at the University of California, Berkeley, who would eventually become a major player in the disputes over DNA typing), Henry Lee (a forensic scientist in Connecticut who became famous as a defense witness in the O.J. Simpson murder trial), as well as three forensic scientists from various universities.[35]

This list was only the beginning of the options available to prosecutors, though. In *Wesley-Bailey*, for instance, soon-to-be Nobel Prize winner Richard Roberts agreed to testify on behalf the prosecution. (He won the prize in 1993 for his work on the isolation and development of restriction enzymes in the 1970s and 1980s.) Roberts told me that he was recommended to Dwyer by James D. Watson, co-inventor of the double-helix model of DNA structure, who had been his boss at Cold Spring Harbor Laboratory (CSHL).[36] At the time, Roberts ran a lab at CSHL that was responsible for isolating and maintaining a significant proportion of the restriction enzymes that were then available. The prosecution's other star was witness Kenneth Kidd, a well-known geneticist from Yale University. At the time, Kidd was a prominent leader in the nascent effort to map the human genome and ran a major human genetics lab that did DNA testing for diagnostic and medical research purposes. That Roberts and Kidd expressed little hesitation in testifying in court about the validity and reliability of DNA typing indicates that there was little debate within their scientific communities on this topic.

Defense lawyers, on the other hand, were faced with a very different situation. They found it nearly impossible to find academic scientists of the caliber of Roberts or Kidd to testify that any aspects of DNA typing were not generally accepted within the scientific community. One major reason for this situation may be that there was simply very little debate going on in the scientific community about the validity and reliability of the technique when applied to the medical research and diagnostic situations that most academic researchers were familiar with. In these nonforensic contexts, samples were often typed along with several others in a family group, providing built-in controls; the range of potential results was limited; and blood samples were usually abundant and could be retyped with little problem. Another reason is that a few very influential academic scientists—most notably Housman, Roberts, and Kidd—gave forensic DNA analysis their stamp of approval early on in the history of the technique. These endorsements may have served to inhibit other scientists from closely examining the forensic uses of the technique.

Most geneticists had seen laudatory accounts of the use of the technique in criminal cases in their local newspapers, and they had no reason to believe that there was anything wrong with the technique. One obstacle to the emergence of an early critique of the technique was that the majority of technical articles on forensic DNA analysis was published in specialized forensics journals such as *Journal of Forensic Science, Crime Laboratory Digest,* and *Forensic Science International.* This meant that the community who could easily engage in peer review was limited to people who read forensic science journals. Most of these individuals were crime laboratory personnel who had little knowledge of, or training in, genetics or molecular biology. Those that were published in journals widely read by most academic biologists (e.g., *Science, Nature,* and *American Journal of Human Genetics*) presented only a basic outline of the procedures used, initial results obtained in validation trials, and information about the frequency of given bands within the major racial and ethnic populations. Few of these early articles suggested that there were any problems associated with typing contaminated forensic DNA samples, measuring bands on autoradiographs, calculating probability statistics, the possibility of population substructuring, or any of the issues that would soon crop up about the validity and reliability of the technique. This meant that there was little opportunity for specific procedural and statistical claims of Lifecodes and Cellmark to be critiqued and skeptically examined through the process of peer review. There was simply no debate about any of these issues in any scientific journals throughout most of 1987 and 1988.

The absence of a critical dialogue about DNA typing in the published literature was exacerbated by the proprietary nature of most of the two private company's probes and databases, as well as the difficulty that most academic scientists faced in gaining access to forensic materials for validation work. Open access to the materials used to conduct DNA testing (especially the probes), as well as the databases used to determine the frequency of a specific allele, would become a major aspect of the controversy over DNA evidence in mid-1989.

This situation would only begin to change as a few defense lawyers spread throughout the country began to find academic scientists willing to listen to their pleas for help. The *Wesley-Bailey Frye* hearing represents the first time that a defense attorney brought in expert witnesses to challenge prosecution claims about the technique. For all of his effort, though, defense attorney Douglas Rutnik could not convince the judge in the case, Joseph Harris, that Lifecodes' DNA-Print procedure was not yet ready for use in the courtroom. The testimony lined up by the prosecutor was still far too overwhelming to convince the judge to seriously consider ruling against the admissibility of DNA evidence. The first prosecution witness to testify in the *Frye* hearing was Michael Baird, who spent a great deal of time educating Judge Harris about several fundamental aspects of biology. Just as in *Andrews*, his testimony began with a brief history of genetics, after which he explained the process by which Lifecodes carried out the DNA test, and concluded with an explanation of what it means when all of the bands in two DNA profiles match. At each step of the explanation, the prosecutor asked Baird whether or not the procedures and principles used by Lifecodes were generally accepted within the scientific community. Baird responded in the affirmative concerning all aspects of the test. Next, both the prosecutor and the judge questioned him about several specific aspects of the test. It is clear from his questions that the judge in the case possessed little knowledge about even the most basic aspects of genetics or molecular biology at the beginning of the admissibility hearing. For instance, at one point during Baird's review of the basic scientific details of the test, the judge interrupted to ask, "The term genes, what is the relationship with the term DNA?"[37] Later, he asked, "What is a chromosome, Doctor?"[38] These basic questions continued throughout the rest of the hearing, with prosecution experts serving the purpose of educating the judge.

One of the most interesting aspects of Baird's testimony was his concerted effort to set the terms of the discourse on the legal criteria for admissibility of DNA evidence. While there was an intense debate about this topic going on both in legal scholarship and in the courts at the time, Baird made the whole subject seem simple during his time on the stand. He argued that legal admissibility standards were synonymous with the standards of scientific validation. In doing so, he shifted the terms of the admissibility debate from the judge's interpretation of whether the technique was generally accepted within the scientific community to whether or not Lifecodes had followed generally accepted validation procedures in setting up their DNA typing unit.[39] Thus, the central issue of this *Frye* hearing was whether or not Lifecodes abided by its own rules and standards.[40] Baird described the criteria for admissibility as follows: "It can really be broken down into two parts, that being the method must be shown to be valid and also the laboratory that is doing the testing should also be shown to be valid."[41] He then listed four criteria that must be met for a method to be validated:

1. Publication in a peer-reviewed journal to ensure that "independent, outside scientists" have looked over an article before it is published
2. Replication [of results] by at least one other lab

3. You must show that the method used is valid for the particular type of sample tested

4. The possibility of "back-up testing" must exist[42]

Baird then went on to argue that Lifecodes had met all of these criteria. With respect to the first issue and second issues, Baird stated that articles that "address the kinds of DNA probes" that Lifecodes used had been published as early as 1980 by Wyman and White in the *Proceedings of the National Academy of Sciences*. He also mentioned articles by Alec Jeffreys and others on DNA fingerprinting, a Lifecodes publication in the *American Journal of Human Genetics*, Nakamura et al.'s 1987 *Science* article, as well as articles specifically addressing forensic issues by the British Home Office and Lifecodes in the *Journal of Forensic Science*. When the court asked Baird if these groups were "recognized within the scientific community," Baird responded that they were and that their results were consistent with Lifecodes.[43]

With respect to the third issue, he said that a number of different issues needed to be addressed to ensure the validity of the methods behind DNA typing. First, the "so-called laws of genetics" must be obeyed. He included in this list Mendelian inheritance and Hardy-Weinberg equilibrium (HWE). When asked by the prosecutor whether Lifecodes had tested the probes they used for HWE, Baird replied that they had indeed done so and that "basically what it shows is that the DNA patterns or prints that we get are very individualistic in that the pattern of DNA that you can obtain by using a series of DNA probes is as unique as fingerprints." (As described in chapter 2, Lifecodes tested for HWE in order to ensure that their database was free of population substructure. If the population that was represented in the database was not randomly mating, then one would expect to find divergences from HWE.) When asked by the prosecutor to "give us a statistical number," Baird replied that the test was well over 99.99 percent accurate.[44] With respect to the final issue, he said, "In terms of the DNA tests we do at Lifecodes, there are literally hundreds of laboratories around the world who could act as a back-up test. Any laboratory doing molecular biology or recombinant DNA would be well-versed in the technology that we use in terms of doing our DNA test."[45]

Baird also responded to questions about whether environmental factors could influence the outcome of a DNA test. He responded by saying that numerous peer-reviewed studies had shown that neither the substrate on which the evidence is found, nor heat, humidity, or UV-light (the most common sources of environmental insult in most forensic cases) have any significant effect on the DNA.[46] In this area of testimony, Baird went to great lengths to promote DNA typing as an error-free technology.

> From our experiments and from our observations, the substrate to which biological evidence has been affixed does not have an effect on the DNA molecule itself. So basically what we have found is that the DNA test will work, in that it will give you an answer. It will not give you the wrong answer. It may not

give you an answer at all if the DNA has been too degraded for some reason or there is not enough present. But you can't alter the pattern by any one of these known environmental effects.[47]

After completing his discussion of the validity of the technique itself, Baird went on to articulate his view that the laboratory carrying out the technique must also undergo some sort of validation for new scientific evidence to be acceptable. He testified that there were four major components to this process:

1. The scientific personnel doing the test and interpreting it
2. The existence of quality control mechanisms
3. There must be a blind trials testing program in place
4. The "experience of the laboratory itself"

According to Baird, Lifecodes had excellent quality control mechanisms "built into the test itself," which existed to "ensure the results have credibility."[48] This quality control regime included analyzing the DNA isolated from a forensic sample to ascertain its quality and quantity, making sure that the restrictions enzyme works properly through the use of control samples, performing maintenance on the equipment used to run the DNA test, as well as checking the quality of reagents. When the court asked Baird to clarify whether the quality control program was "controlled by machines or controlled by man and what is the possibility of some human error coming in," Baird replied:

> I have not ever seen a situation where you have gotten the wrong results because of a quality control error. There could be situations where you may get no result at all from an incorrect solution being provided or things like that, but in terms of getting the wrong result, in my mind, it is not possible by quality control error. Again, you have a control DNA that is built in to each one of your tests that allows you to look at the control sample and say that looks correct, I have seen that before, so therefore I know everything done in that test worked correctly.[49]

With respect to the third and forth point, Baird described the various "external blind trials" and other testing regimens that Lifecodes had participated in since 1985. He went over the laboratory's results from the California Association of Crime Lab Directors trial, pointing out that although they could not type all of the samples sent to them (because of poor quality), they correctly identified all of the samples for which they submitted a result. He also recapitulated the history of the company as it slowly moved into the forensic identity testing market, primarily in order to illustrate that the company was slow and deliberate in its development of the DNA-Print technology.[50]

On cross-examination, defense attorney Rutnik immediately sought to diminish Baird's credibility by pointing out the financial stake that both he and his company had in the court ruling DNA evidence admissible. After several minutes of Rutnik exploring this issue, the judge finally decided that he had had enough of this line of questioning and cut it off. By this point it seemed that the judge had already made up his mind that the technique was valid and reliable and did not think financial matters had any relationship to issues of scientific validity. In ending this discussion, he remarked to Baird, "Whether you make money or not, I take it the technology is scientifically reliable, is that correct?" Baird, not missing a beat, responded in the affirmative.[51] The judge also made a similar remark when overruling an objection from the prosecutor about the differences between Cellmark's and Lifecodes' probes. In that instance, he told Baird: "Well, most of the testimony here is that regardless of anything, that the thing is accurate altogether. I will let you answer the question."[52]

While Baird's testimony went essentially unchallenged in all previous cases involving DNA evidence, Douglas Rutnik, the defense attorney in this case, realized that it would be wise for Baird's testimony to be critiqued by scientists not associated with Lifecodes. Because there was not yet a pool of scientists willing to critique DNA evidence, he turned to New York State's well-established network of criminal defense attorneys for advice. Barry Scheck, a defense lawyer who had learned about DNA profiling while working on a postconviction case, recommended that Rutnik talk to Richard Borowsky and Neville Colman, two scientists whom he and his legal partner Peter Neufeld had consulted with on issues relating to human leukocyte antigen (HLA) testing and blood group analysis (both of which were forerunners of DNA typing).

Based on his discussions with these two men, Rutnik set out to challenge Lifecodes' DNA evidence on three major grounds. The first was the standard attempt to discredit the prosecution's witnesses as being biased and financially interested in the outcome of the trial.[53] As in previous cases, though, this tactic did not seem to work very well. The second major strategy used by the defense was to challenge the actual scientific and methodological underpinnings of Lifecodes' DNA test. Specifically, Borowsky argued that the population genetics studies carried out by Lifecodes were too limited to determine whether Lifecodes' racial/ethnic populations were in HWE (i.e., randomly mating, at least with respect to the inheritance of particular allele combinations with a specific locus), or the extent to which the alleles of the sites targeted by the company's probes were in linkage equilibrium (LE) (i.e., alleles at different loci associate at random). He argued that not enough people were in the database to make such judgments. In this case, Borowsky was never able to fully analyze Lifecodes' HWE calculations since the company would not release the raw data that would make it possible for him to do so. Borowsky also pointed out the elementary flaws Lifecodes' frequent assertion that the alleles of their probe targets were definitely in LE because they were not physically linked on the same chromosome.[54] Because Lifecodes had not adequately shown HWE and LE, Borowsky argued, they were not scientifically justified in using the product rule to calculate the probability of a random match between two DNA profiles.

The third major strategy employed by the defense was to convince the court that Lifecodes' methodology, probes, and protocols had not yet been subject to adequate peer review in the scientific community. If they could substantiate this claim, it would mean that Lifecodes' DNA profiling technology did not yet meet the criteria for admissibility according to *Frye*. In relation to this line of challenge, the defense argued that publication in a peer-reviewed journal did not necessarily constitute general acceptance of a theory or technique within the relevant scientific communities. In his testimony, Colman stated that he had a different definition of "peer review" than the prosecution witnesses. For him, it only meant that the article had been looked over by a qualified individual who ensured that nothing major had been forgotten and all of the basic elements of the article made sense. General acceptance, on the other hand, is a process that takes a long time, while peer review is a relatively confined process.[55] He argued that the technique was invented so recently that it had not yet had time to diffuse into the relevant scientific communities, and that the proprietary nature of the technique made it difficult for independent researchers wishing to review the company's finding to do so. He told the court that it would take somewhere between two and a half to four years to know whether Lifecodes' peer-reviewed paper on its methodology and probes would be generally accepted within the scientific community.[56] Borowsky pointed out that many of the papers cited by Baird as validating Lifecodes' probes never actually mentioned them explicitly. They merely discussed probes that were either similar in type or targeted a similar DNA sequence. Colman also criticized Lifecodes for entering into the record the manuscript of a paper that the company planned to submit to a peer-reviewed journal for publication. He told the judge that the document currently had no validity and that it was nothing more than a "collection of words on a page."[57]

Building on the theme that peer review does not constitute general acceptance during cross-examination, Rutnik forced Richard Roberts to concede that it was possible that the four probes that Lifecodes used had never actually been tested by an independent laboratory, even though they had gone through the process of peer review.[58] This tactic was a direct response to Baird's assertion that there were literally hundreds of laboratories that could *conceivably* verify the company's results and Robert's claim that DNA typing entailed no new scientific principles. Baird told the court that "any laboratory doing molecular biology or recombinant DNA would be well-versed in the technology that we use in terms of doing our DNA test."[59]

When Rutnik asked Roberts if he had actually tested Lifecodes' probes to confirm that they worked the way that the company reported in its published papers, Roberts responded: "I have not tested that personally. It's possible that the review[er] of the article tested it. But, it's something that could be tested. It's unlikely, though, it's unlikely that it's been tested."[60] The judge, who seemed to be confused about how peer review worked, then interjected: "Well, would there be some requirement in the scientific community to test the results of an article before the article is published in some reputable magazine like that?" Roberts responded to this question: "No. In general one looks at the quality of the experiments that are reported, the nature of the experiments that are reported, and the data that has been generated from it, and one accepts that on faith, unless one has good reason not to." Rutnik and Roberts then had the following conversation:

> RUTNIK: Well, isn't that a factor, Doctor, that one of the reasons for publishing your procedures in a peer review magazine is that it affords other scientists the opportunity to try the same techniques, or whatever process that has been described in that review article, peer article, in order to give it objectivity; is that correct?
>
> RICHARDS: That's correct, absolutely correct.
>
> RUTNIK: No, my question to you, Doctor, is if when you read these articles published by Lifecodes describing the four probes, to your knowledge has anyone ever tested these four probes, to see if they do what Lifecodes says they do?
>
> ROBERTS: Well, I don't know what the four probes are, so I can't really answer that question.
>
> RUTNIK: Isn't it a fact, Doctor, that these probes are patented and are in the sole possession of Lifecodes?
>
> ROBERTS: I don't know. I'm told that's the case, but I don't know from my own experience . . .
>
> RUTNIK: Doctor, let me ask you this: If, based upon your knowledge, no one has specifically tested the probes that Lifecodes used, would that alter your opinion as to the scientific reliability of the probes that Lifecodes uses?
>
> ROBERTS: I don't think so.
>
> RUTNIK: If no one else tested them?
>
> ROBERTS: It would not alter my opinion, because I think what they are doing, they are doing properly, that they are repeating work that has been done many, many times in other labs with different probes. There are many probes that have the properties of the ones that Lifecodes uses, it's not [as if] they are doing something unusual, they are doing something in fact which is exceedingly common.[61]

A few minutes later, Rutnik and Roberts had another interesting exchange that points to the defense's attempt to highlight some of the shortcomings of the peer-review process. Although Rutnik did not follow up on his discovery, Roberts acknowledged the very important role that trust and authority play in the "scientific method." Rutnik began this exchange by asking, "Is the independent checking of a laboratory important in your consideration as to the reliability of their testing procedures?"

> ROBERTS: The answer would be yes, but in a rather subtle way.
>
> RUTNIK: Would you explain?

ROBERTS: That is, when a paper is published in the scientific literature, labs don't normally go out and repeat that work just to show that it is correct. Rather what happens is that they build upon the work that has already been published, to plan new experiments. If you discovered when you're doing your next experiment that things are wrong, then you go back and check the previous results. So it's sort of a presumed-innocent-until-found-guilty situation. One assumes that the scientists publishing their work have been accurate and honest. Only when you find there are problems do you really go back and check.

RUTNIK: So there isn't any objective testing of what a person says he's done, it's only based on what he says occurred and other new scientific tests that are done?

ROBERTS: Yes, that's correct.

RUTNIK: That's the way it works?

ROBERTS: That's the scientific method, yes.[62]

This line of questioning continued a few minutes later when Rutnik asked Roberts how he could be certain that a random match could never occur if nobody had actually tested Lifecodes' probes. Roberts responded that despite not being an expert in population genetics, he felt that he could trust the company based on the fact that it employed "reputable scientists." He replied that they had indeed tested their probes:

That is, that they have done a test of their probes by taking a sample, and I think in the case of Lifecodes it's 3,000-and-some-odd people and so, they know the frequency with which the bands occur within that population of 3,600. And since I have reason to believe that people at Lifecodes are reputable scientists, based upon [sic] they've got a lot of publications and so on, they seem to do their experiments properly, then I assume they have taken reasonable selection in picking the 3,600 people, so that the results are scientifically reliable. But I'm not the person to look at the data and give you an expert opinion. You would need to ask a population geneticist, who's very well aware of allele frequencies among populations, to get an expert opinion as to whether they have chosen their samples properly.[63]

Rutnik, however, was not persuaded by Roberts's faith in the scientists at Lifecodes. He told Roberts that it disturbed him that the probes had not been objectively tested by an independent lab. Roberts, however, attempting to assuage Rutnik, replied that Lifecodes could be trusted and that there was no need for independent verification. Indeed, he told Rutnik that he would be satisfied simply

to see . . . a bar graph or some other tabulated data, similar to that found in the *American Journal of Human Genetics,* in which they report the results of the experiments that they have done with this population of 3,000-odd people, and the frequency with which various alleles appeared in the population. . . . If I saw that data, I would be quite content to believe it. I would not feel it necessary to have someone go out and duplicate it. I mean, these experiments are simple, you can teach high school children how to do these experiments, even lawyers, in a relatively small period of time.[64]

The prosecution responded to the defense challenge in a variety of ways. At the most direct level, the prosecutor brought Kenneth Kidd and Richard Roberts to the stand to contradict the two defense experts' testimony and to argue that Lifecodes' recently submitted article was scientifically sound. Based on this testimony, Judge Harris concluded that there was no need for the broader scientific community to approve of Lifecodes' work since these two scientists were world renowned in their respective fields and had reviewed the article in the context of this case.[65]

Baird also returned to the stand to contradict Colman and Borowsky's testimony. Not only did he state that in his opinion, and in the opinion of the scientific community, the databases used by Lifecodes were large enough to establish HWE and LE. Further, if these databases are analyzed properly, it is clear that the populations in question (specifically U.S. Blacks and U.S. Caucasians) were in Hardy-Weinberg equilibrium, and that the loci under study exhibited linkage equilibrium. However, he argued that since there are certain phenotypes that occur more or less frequently than expected by chance alone, a small correction in the probability of random match calculations should be made. According to Kidd, the necessary correction factor was exceedingly small, but that a factor of ten should be used to err on the side of caution. Thus, Kidd explained away Borowsky's population genetics concerns as minute problems that could be solved with a small correction factor. The judge bought Kidd's argument that Lifecodes' statistical evidence was reliable and therefore ruled it admissible.[66] He also accepted Kidd's refutation of Colman's claims about peer review. Kidd argued that the kinds of population genetics studies done by Lifecodes could not be replicated and had to be analyzed using different standards compared to more controlled (i.e., laboratory-based) scientific research. He further rejected Colman's assertion that the probes used in molecular biological research needed to be subjected to a rigorous prepublication review process. Instead, probes are peer reviewed as they are disseminated and used by the scientific community.[67]

The second major tactic used by the prosecution was to attack the credibility of the two defense witnesses. In an interview, Borowsky told me that in the six weeks between his testimony and the reconvening of the Frye hearing, "they made a thorough investigation of my background and probably interviewed just about everyone in the industry I ever spoke to."[68] He went on to say that a large portion of his cross-examination was an attempt to impugn his motives for being there.

The prosecutor employed the same strategy with Colman, who argued at length that Lifecodes' methodology could not be considered generally accepted within the scientific community because it had been published too recently. In addition to calling his expert witnesses to dispute Colman's assertion, the prosecutor also argued that Colman had no basis for making such a statement since he was not a member of the relevant scientific community that judged the validity and reliability of DNA typing. In response to this argument, Rutnik attempted to qualify Colman as such a witness by asking him whether there was a similarity between blood typing, which he had significant experience with, and DNA typing. The judge then directly asked Colman if he was an expert in forensic DNA typing or DNA paternity testing. Colman responded that he wasn't sure how to answer the question; while he knew a great deal about both topics and *felt* qualified to testify on them based upon his education, experience, and reading, they were not his "professional activity." He then attempted to establish his qualifications by stating that DNA fingerprinting was a product of his field of hematology.

Ultimately, the judge decided that Colman was an expert only in laboratory medicine and could therefore speak to matters of laboratory procedure, laboratory monitoring, and peer review, but he did not qualify him to testify on the general acceptance of DNA typing in the scientific community.[69] By limiting the scope of Colman's expertise in this way, Harris was able to completely ignore his testimony on DNA typing in rendering his decision on the admissibility of the technique. This is not to say, however, that Harris actually took Colman's testimony on peer review to heart in reaching his decision. Rather, he wrote Colman off entirely in a sarcastically worded paragraph that he concluded "on every point raised by him he was overwhelmingly refuted—both by the facts and by the opinions of experts with superior qualifications and experience." In his decision, Harris also pointed out that Colman had only authored twenty peer-reviewed papers, while Roberts and Kidd had published substantially more.[70]

With respect to Borowsky, Harris referred to him as a population geneticist specializing in fish and crustaceans, not humans, even though he had previously testified in several legal cases involving HLA and other types of blood group analysis.[71] At one point in his decision, Harris suggested that Borowsky was "obsessed" with a population genetics issue that Baird repeatedly denied was relevant.[72] He then went on to point out that Borowsky's claims had been refuted one by one by Baird and Kidd. The outcome of these maneuvers was that Judge Harris wrote off the defense experts' testimony by judging it to be irrelevant to the issues at hand.

At the conclusion of the *Wesley-Bailey Frye* hearing, Judge Harris ruled that DNA fingerprinting "is a scientific test that is reliable and has gained general acceptance in the scientific community and in the particular fields thereof in which it belongs." He went on to state that the prosecution had supplied sufficient evidence that Lifecodes' methodology was accurate, and that the defense's propositions were not supported by the evidence they presented.[73] As such, he ruled that DNA samples could be taken from the two defendants for testing by Lifecodes. Not shy, the judge then delivered the much-quoted opinion that the "overwhelming enormity" of the probability calculations associated with DNA typing, if accepted by courts, would "revolutionize the administration of criminal justice" by proving alibis false, reducing the importance of eyewitness testimony, and speeding up trials (since other forms of evidence would no longer need to be presented). "In short," he concluded, if DNA evidence is judged admissible, "it can constitute the single greatest advance in the 'search for truth,' and the goal of convicting the guilty and acquitting the innocent, since the advent of cross-examination."[74]

Examining Science in the Courtroom

A few general conclusions can be drawn from the introduction of DNA evidence into the legal system, and especially the *Andrews* and *Wesley-Bailey* admissibility hearings. The first is that, as numerous science studies scholars have noted, legal proceedings often serve as a forum for the deconstruction of scientific claims and methodologies by opponents of a particular form of scientific evidence.[75] To give but one example, defense attorney Rutnik deftly led the soon-to-be Nobel Prize–winning molecular biologist Richard Roberts to admit on the stand to one of the most fundamental claims made by ethnographers of science, namely that "scientific knowledge is established, assimilated, and transmitted by social trust and authority, rather than by the radical skeptical testing suggested by science's dominant public image."[76] Such an admission would never have occurred if Roberts was discussing issues surrounding DNA evidence with his colleagues, since the aspects of science that get scrutinized in such situations are limited by a common cultural identity and shared interests. Because Rutnik did not share a common scientific culture with Roberts, he asked questions that were much more skeptical than Roberts was accustomed to. This tactic forced Roberts to make damaging statements about many of the assumptions surrounding the scientific method that normally go unexamined.

At the same time, the admissibility hearings also served as a forum for the reconstruction of these entities by their advocates, as well as judicial fact finders. In this process, roles traditionally associated with either law or science became intertwined to the point that they could not be easily distinguished. For instance, in the *Wesley-Bailey* hearing, when Rutnik and expert witness Colman deconstructed the prosecution's claim that Lifecodes' methodology and probes had been subject to peer review by providing a more stringent definition of this process, the judge chose not to accept this argument. Instead, he rather arbitrarily decided what constituted scientific peer review (as well as whom the "peers" actually were) and determined that Lifecodes' technology had indeed been subjected to this process. Interestingly, he did not cite any particular scientist, scientific organization, or philosopher of science when formulating his definition of peer review. He was, no doubt, aided in this task by Lifecodes scientist Michael Baird's criteria for legal admissibility of novel scientific evidence.

Judge Harris's ruling demonstrates that despite the deconstructive nature of the adversarial legal system, there are pragmatic institutional mechanisms that limit the scrutiny with which scientific claims are analyzed. Many science studies and legal scholars have argued that the main reason for limiting the skeptical gaze of the adversary system is that social closure must ultimately be achieved in legal proceedings.[77] In this case, it was limiting the definition of expertise in such a way that the most damaging claims made by defense witnesses (and, indeed, the damaging testimony of prosecution witnesses) were rendered outside the scope of the relevant scientific questions set forth by Judge Harris.

Finally, in order for the prosecution to introduce novel scientific evidence into the courtroom, and the defense to have the ability to challenge it, expert communities must be constructed and mobilized within the criminal justice system. In the case of DNA typing, the prosecution had a head start in this process because they had access to the intellectual and professional resources of Cellmark and Lifecodes. This meant that there was initially little friction in the transfer of DNA typing from the laboratory bench to the courtroom. However, as the defense bar became increasingly successful in building a network of experts who were willing to testify against the prosecutions' experts and evidence, this situation changed dramatically: no longer would highly qualified prosecution witnesses testify without being challenged by equally well qualified defense witnesses; no longer would the stated and unstated assumptions of Lifecodes and Cellmark scientists go unexamined by the defense or the judge.[78] In other words, the defense would successfully show that the initial, rapid closure of questions surrounding the validity and reliability of DNA evidence was arbitrary and needed to be reexamined.

In the process, the interaction between prosecution and defense experts would lead to the emergence of various controversies that quickly spread from the confines of the courtroom to the pages of the world's most prestigious scientific journals. In the six years between the emergence of these debates and their final closure, judges, lawyers, and scientists would be forced to deal with the controversies on a case-by-case basis. As I will show in subsequent chapters, at the same time the notion of expertise (for both prosecution and defense) would continuously evolve as the shape and nature of the scientific controversies surrounding DNA evidence changed.

Challenging DNA

By the end of 1988, the results of DNA profiling had been admitted as evidence without reservation or doubt in more than eighty trials across the country and had been used to obtain confessions of guilt in countless more. Judges were inclined to repeat the claims of prosecution witnesses verbatim in their decisions while dismissing the protestations of defense witnesses as being irrelevant to the issues at hand. The prosecution's uncontested proclamation that DNA typing would produce the right answer or no answer at all showed up in most published judgments during this period.[1] The early success of the technique lies at least in part in the deep pockets of both Lifecodes' and Cellmark's multinational parent corporations. Both took in billions of dollars in revenue each year and were eager to capitalize on the biotech boom of the 1980s. Thus, they were willing to bankroll efforts to get DNA profiling into as many jurisdictions as possible, as fast as possible.

After failures in *Andrews* and *Wesley-Bailey*, the defense community realized that in order to successfully challenge DNA evidence, they could not rely on the strategies used in those cases. Defense attorneys desperately needed expert witnesses from the academic science community who could match the qualifications, reputation, and enthusiasm of prosecution witnesses. However, because the technique was not yet well known outside of the two private companies, there was not a pool of molecular biologists, population geneticists, or forensic scientists who were willing to testify on behalf of the defense. This trend led one legal commentator to describe DNA typing as the "unexamined 'witness' in criminal trials."[2]

This situation would finally change in early 1989, when a few defense attorneys around the country succeeded in locating academic molecular biologists and population geneticists who were willing and able to help them challenge DNA evidence in court. Yet locating these individuals was not enough—lawyers also had to convince judges that they possessed knowledge and experience relevant to the admissibility of a new forensic technique. This task was not easy in the face of intense prosecutorial opposition. This chapter will examine the defense community's efforts to create a network of experts who could challenge the testimony of the prosecution's experts. Particular attention will be paid to the emergence of a new expert identity in the case of *People of New York v. Castro* (545 N.Y.S.2d 985, 1989): that of the academic scientist who was knowledgeable enough about forensic science to be considered an expert by the judge even though he was not considered to be part of the forensic science community.

Defense challenges moved courtroom discussion from the certainties of DNA profiling in medical and diagnostic contexts to the uncertainties of forensic casework. This shift meant that new kinds of experts became relevant to the admissibility of the technique. Defense challenges to DNA evidence also catalyzed debates in the scientific community. Beginning in this chapter, and continuing throughout the remainder of this book, I will chart the growth of these various debates and follow them as they were transported from the narrow confines of the courtroom into the scientific community, various political arenas, the mass media, and, eventually, the general public. In the remaining chapters of this book it will become clear that because these debates were generated through interactions among science, law, politics, corporate culture, and popular culture, they were resistant to attempts by any single institution to end them. Despite the clamor taking place over DNA profiling, however, the legal system remained fairly well insulated from the disputes. For the most part, judges declined to declare that the technique itself was flawed and not ready for use in the legal system. Instead, they handed down remarkably conservative rulings that critiqued only single instantiations of forensic DNA analysis, rather than the technological system as a whole.

Scheck, Neufeld, and the Castro Case

Defense attorneys Barry Scheck and Peter Neufeld had recently taken over the defense of Joseph Castro from a court-appointed lawyer who felt overwhelmed by the DNA evidence in the case.[3] Castro was on trial for the February 1987 stabbing murders of twenty-year-old Vilma Ponce, who was six months pregnant at the time, and her two-year-old daughter.[1] Castro worked as a handyman in the Bronx neighborhood where the Ponce family lived and had been seen leaving their building by Ponce's common-law husband shortly before he found his wife and child dead in their apartment. Based on the identification, the New York City Police Department brought Castro in to question him about the murders. Although Castro denied involvement in the murders and there was no concrete evidence to link him to the crime scene, detectives noticed that there was a small bloodstain on his watch during the course of the interrogation. When questioned about the stain, Castro insisted that it was his own blood. Because there were no other serious leads in the case, prosecutor Risa Sugarman decided to send the watch to Lifecodes in July 1987 for DNA analysis. A few weeks later, Lifecodes reported back to her that, based on an analysis using three DNA probes, the blood found on Castro's watch matched Vilma Ponce's with a probability of random match of 1 in 189,200,000.

Over the course of their careers as defense lawyers, including long stints with the Legal Aid Society in the Bronx, where they met, Scheck and Neufeld had developed a healthy skepticism toward the validity and reliability of forensic evidence in criminal investigations. Further, both had come to believe that novel forensic techniques were all-too-often used in criminal trials before assurances of their validity and reliability were verified by scientists outside the forensic community.[5] They pointed to a long list of techniques, from paraffin testing (for gunpowder residue) to voice print identification to protein electrophoresis, that were immediately admitted into evidence and years later shown to be fundamentally flawed, either in design or practice. This realization occurred only after thousands of people had been convicted based on evidence produced with these technologies.[6] In their view, the legal system, and the defense bar in particular, did not do enough to ensure that a wide range of expert witnesses were heard from when the admissibility of a particular forensic technique was being debated.[7] Instead, witnesses were brought in by the prosecution essentially to legitimate the technology, rather than to probe its weaknesses.[8]

Scheck and Neufeld believed that this was the road down which DNA evidence was traveling and felt that they had to do something about it. They were especially concerned that prosecution witnesses in most early DNA cases failed to explain to judges that there was a crucial difference between the use of DNA typing for diagnostic or research purposes (which was generally accepted in the scientific community) and its use in forensic casework (which had not yet been subjected to rigorous scientific scrutiny).[9] Until *Castro*, prosecutors had successfully built their cases for the admissibility of DNA evidence on its wide use and general acceptance within other scientific and social arenas. As we saw in *Andrews* and *Wesley-Bailey*, they did so by bringing in eminent scientists like Richard Roberts and David Housman to testify that there was no difference between what Lifecodes' scientists did at their testing facility and the techniques used in academic and medical research laboratories around the country. It seemed to make no difference to judges that these scientists had no significant experience in forensic casework, since they were led to believe that there was no fundamental difference between forensic and nonforensic applications of the technique. As in the *Andrews* case, the appeals court judges concluded that the "extensive nonjudicial use of [DNA typing] is evidence tending to show the reliability of the technique" in the criminal justice system.[10]

By focusing primarily on the validity and reliability of the technique outside the legal system, Scheck and Neufeld realized, prosecutors ensured that the contingent aspects of forensic casework were largely dismissed, rather than discussed, by judges. As a result, when writing their opinions, judges did not wrestle with issues like contamination, degradation of DNA due to environmental factors, band shifting, or the difficulty of sizing faint bands on a gel. Instead, they simply repeated the unchallenged proclamation of the prosecution that you either get the right result or no result at all.

For Scheck and Neufeld, this claim was both demonstrably false and misleading to judges. In their view, there were several major differences between the research and forensic contexts that needed to be taken into account when judging the validity and reliability of DNA typing in the criminal justice system: First, in diagnostic testing, technicians have an ample supply of pristine blood that can be used to perform multiple experiments. In forensics, however, the unknown stain is usually so small that the scientist often uses up the entire sample in a single run. This situation leads to a lack of reproducibility when results are ambiguous or inconclusive. Further, these stains have often been degraded and contaminated by environmental conditions, which usually does not occur in diagnostic and medical research work. Second, in diagnostics and family studies, there generally is a built-in control, since parents and offspring are screened at the same time. No such control existed in forensic casework in which the donor of the crime scene sample is unknown. Third, the universe of possible DNA profiles is limited since diagnostics does not generally target highly polymorphic gene systems. Rather, they look at genes in which only a few variants (i.e., those that cause some sort of defect or disease) are of interest. Finally, in diagnostics there is usually no need to calculate the probability that a particular pattern will occur in a relevant population. Yet, forensic DNA profiling gains its power from establishing that a particular pattern is exceedingly rare in a given population.[11]

One major problem that Scheck and Neufeld had in advancing their views was that they were lawyers, not scientists. They needed to find a prestigious and credible scientist who could act as a conduit between them and the judge. This task was not as simple as it may sound. At the time of the *Castro* case, there were very few scientists in the academic community (i.e., outside of Cellmark and Lifecodes) who were familiar with the forensic aspects of DNA technology, and none had ever been qualified in such a capacity as an expert witness for the defense. Indeed, when asked about the defense community's ability to find expert witnesses, Scheck stated in a U.S. congressional hearing that usually when science and law interact, "the pattern is always you find your standard expert witnesses. There are no standard expert witnesses here."[12] As we saw in the *Wesley-Bailey* hearing, defense attorney Rutnik attempted to qualify Neville Colman as an expert in forensic DNA technology but did not succeed. Judge Harris decided that he was only an expert in "clinical medicine" and, as such, limited his testimony to matters of laboratory procedure, laboratory monitoring, and peer review.[13]

It is also important to remember that until *Castro*, "independent" expert witnesses (i.e., scientists not directly employed by or affiliated with Lifecodes or Cellmark) served primarily to educate judges about basic aspects of biology. They were valued for their ability to discuss the validity and general acceptability of DNA typing in nonforensic contexts, as well as to clarify any basic issues of molecular biological or population genetics theory. They very rarely discussed the mundane details of applying the technology in forensic investigations. This task was left to the scientists employed by private DNA typing companies. Not surprisingly, these individuals were reluctant to discuss the potential shortcomings of forensic DNA analysis and were not generally asked to do so by defense attorneys with little understanding of basic biology.

AGENT PROVOCATEUR

Scheck and Neufeld had just begun to look for scientists who could help them when Neufeld was invited to participate in a conference on forensic DNA analysis at the Banbury Center at Cold Spring Harbor Laboratory in New York. The aim of the November 1988 meeting was to provide a forum for multidisciplinary review of the scientific, social, and legal implications of forensic DNA analysis from many different perspectives. In addition to representatives of Lifecodes and Cellmark, the invited guests included population geneticists, molecular biologists, forensic scientists, lawyers, judges, and ethicists who were all willing to engage in a discussion about the use of DNA evidence in the courtroom.[14] The Banbury meeting was the first time that all of these stakeholders had the opportunity to talk to one another.

During a coffee break, Neufeld pulled aside a brash, young academic scientist named Eric Lander and asked him to have a quick look at the autoradiograms from the Castro case. Lander had just given a cautionary talk on the limitations of DNA profiling that deeply resonated with Neufeld. Lander was brought in by the organizers of the meeting as an *agent provocateur*—a brilliant scientist who not only knew a tremendous amount about the population genetic and molecular biology issues involved in DNA profiling, but also was not afraid to ask other participants tough questions.[15]

According to an article in the journal *American Lawyer*, Lander told Neufeld he would show him how science was done. Lander then proceeded to call over a few of the academic scientists at the meeting and asked them whether they believed the defendant's blood matched the crime scene sample. All three, as well as Lander, said that the results were inconclusive and should be redone.[16] Lander went so far as to call the autorads in the case "garbage."[17]

Based on this interaction, as well as Lander's generally critical stance, Neufeld and Scheck asked him if he would be willing to serve as a witness in the upcoming trial. Because of a very busy schedule, and some remaining doubt that his critique of Lifecodes' work could be incorrect, he declined. He did, however, agree to serve as an unpaid scientific tutor and consultant to Neufeld and Scheck. Initially, this meant that he would explain the basic aspects of genetics and molecular biology, look over Lifecodes' report in the case, and guide Scheck and Neufeld as they demanded materials from the company during the discovery process.[18] He was also willing to provide commentary on each day's expert testimony and suggest questions to be asked during the examination and cross-examination of witnesses.[19]

This arrangement would only last for a few weeks, however. Lander quickly became so distressed by what he saw that he agreed to write a fifty-plus-page report on the shortcomings of Lifecodes' results and testify as an expert witness on behalf of the defense. By the end of the trial, Lander had donated more than 350 hours to the case and spent several days on the witness stand.[20] As Scheck would later recall: "Eric Lander was right up our alley. Here was this fast-talking New York Jew from Brooklyn who loved to argue and knew a lot of law" because both of his parents and his wife were attorneys.[21]

Unlike the scientists who testified in previous cases, Lander did not draw an analogy between forensic DNA typing and diagnostics work in order to argue for the general acceptance of the technique in the scientific community. Instead, he said the closest thing to forensic DNA analysis going on in molecular biology at the time of *Castro* was the prospect of isolating DNA from long-dead organisms such as woolly mammoths, quaggas (an extinct form of zebra), Egyptian mummies, and prehistoric humans whose remains are preserved in peat bogs. About this work, he said,

> there's a lot of excitement in the scientific community about that. There's no general acceptance of doing that, but there's a lot of excitement about it because it holds prospects of learning things about smallpox infections of Egyptian mummies and asking if smallpox has changed in two millennia, just as DNA forensics is a very exciting prospect as well.[22]

Thus, Lander believed that although DNA profiling might one day be considered valid and reliable, that day was several months of hard work away.

In court, Scheck and Neufeld deftly wove together disparate bits of Eric Lander's somewhat eclectic qualifications and experiences to turn him into a kind of witness that only had legitimacy in the courtroom: an academic molecular biologist with specialized knowledge of the forensic context, but who did not belong to the forensic science community. In all major cases involving DNA evidence in the future, this knowledge would become a prerequisite for being chosen to testify, since Lander's testimony in *Castro* irrevocably shifted the nature of the debate surrounding DNA evidence. Thus, in a very real sense, Scheck and Neufeld fostered the emergence of a new form of relevant expertise within the American legal system, which clearly caught the prosecution off guard. As Lifecodes' Michael Baird recalled in a 1994 interview:

> Things were going pretty routinely in terms of presenting the background, presenting the data, presenting the information. Suddenly Eric Lander shows up for the defense and has a booklet that is numerous pages thick that has what he critiques as all kinds of problems with the case. The prosecutor in that case is like, "who is this guy? Where did he come from?" I knew nothing about him, I had never really heard of him before. I knew of some work that he had done as a computer specialist for Collaborative Research, but I didn't know him in terms

of human identity. The first time I met him was at the Banbury conference. . . . So anyway, the prosecutor calls up and says, "what do I do? This guy is coming in." You know Scheck and Neufeld spent half a day just on his credentials to show that this guy walks on water before the judge.[23]

Lander received his A.B. in mathematics from Princeton in 1978 and then went on to receive his D.Phil. in the same subject from Oxford University as a Rhodes Scholar.[24] Immediately upon returning to the United States, he was hired as an assistant professor of managerial economics at Harvard Business School, even though he hardly knew any economics. He was told that he could research any subject he chose, as long as he taught classes in mathematical and statistical theory.[25]

Soon after taking up the position, Lander became interested in biology through his younger brother, a neuroscientist, and set about teaching himself molecular biological techniques in the laboratories of colleagues at Harvard. At the end of this "moonlighting" period, he had gotten "sufficiently involved" in biology that he asked HBS to give him a leave of absence to spend a year as a visiting scientist in the lab of the well-known MIT biologist Robert Horvitz. Through time, he became "more and more involved in molecular biology, while my initial interest had been the application of mathematics, we spent four years, about, working at the bench without any mathematical application to become fully proficient" in laboratory techniques.[26] By 1985, Lander had become involved in a collaboration with David Botstein, the author of the first scientific article on RFLPs and pioneer in the field of DNA typing. This collaboration would prove to be very fruitful, leading to several papers on techniques to create a map of the human genome.

In 1986, Lander secured a fellowship at the MIT's Whitehead Institute for Biomedical Research in order to develop his ideas, and in 1987 he received a MacArthur Foundation "genius" grant to continue this work. In early 1989, Lander got competing bids for full-tenure positions in the departments of biology at Harvard and MIT. He ultimately accepted the MIT offer and went on to found the Whitehead Institute/MIT Center for Genomic Research around the same time, which was one of the world's first genome sequencing centers.[27]

Having demonstrated that Lander was a respected member of the scientific community, Neufeld then told the judge that he wanted to establish Lander's expertise in two specific areas: first, "human genetics in Southern blotting," and second, "math, statistics, and population genetics, population studies."[28] At this point, Neufeld made no special mention of the forensic context. Instead, he went to great lengths to show that Lander had not only hands-on experience producing Southern blots at the lab bench and scoring them using a computer digitizing program, but also considerable opportunity to use his own "scientific expertise" to interpret large numbers of autorads as the codirector of a genome sequencing laboratory at the Whitehead Institute.[29]

After demonstrating to the court that Lander was intimately familiar with the process of producing and interpreting a Southern blot, Neufeld shifted gears and began asking questions that sought to establish Lander's "recognition in the research in the field of genetics and population genetics." About three-quarters of the way into the qualification portion of Lander's testimony, Neufeld began the process of convincing the judge that Lander should be considered an expert in the molecular biological, genetic, and statistical aspects of the forensic application of DNA typing. He asked Lander to describe the roles of various fields in which he had demonstrated knowledge in forensic DNA typing. Lander responded:

> Oh, the field of forensic applications of DNA typing requires putting together molecular biology because you're going to take DNA and analyze it on the Southern blot. Mathematics, because you're going to draw quantitative judgments about whether things are the same, and then population genetics because if you should conclude they are the same, you have got to ask yourself the very difficult question, how surprised am I to see things being the same, given the distribution of possibilities in the population, and so to make such a judgment, one needs to know about the population structure of the United States, the variation between and among groups and subgroups.[30]

Then, prompted by Neufeld's question about whether he had ever been called upon to deliver a talk on the subject of the population genetic aspects of forensic DNA analysis, Lander described to the court that he gave a paper on the subject at the Banbury meeting, and that he had been invited by none other than Michael Baird to give a talk on the same matter at an upcoming meeting of the American Association of Blood Banks.[31] He also told the court that he had recently been commissioned by the U.S. Congress's Office of Technology Assessment (OTA) to write a report on the validity and reliability of forensic DNA analysis.

Neufeld, making sure that everyone understood what this appointment meant, proclaimed that Lander's selection was recognition by the OTA that Lander "has the unique qualification as an expert in both disciplines of statistical genetics as well as molecular biology, so it's an independent assessment that this man is qualified as an expert in those two areas."[32] Officially, the judge declared Lander an expert in genetics and population genetics, but it is clear from his reaction in subsequent proceedings that he also considered him to be familiar with molecular biology as well as the forensic context.[33]

A FUNDAMENTAL CRITIQUE OF LIFECODES

In their final report to the Bronx County District Attorney (22 August 1987), Lifecodes presented results that unambiguously matched the blood found on Castro's watch to Vilma Ponce's. The report explained that the company had hybridized a blot containing Ponce's DNA, her daughter's DNA, as well as DNA from the blood on Castro's watch, with probes for three RFLP loci (DXYS14, D2S44, and D17S79), plus a sex marker probe that detected a Y-chromosome locus. In each case, they reported an exact match between Ponce's blood and the blood on the watch. At the D2S44 locus, they reported that both were homozygous for a 10.25 kb band. At the D17S79 locus, both had bands of exactly 3.87 kb, 3.5 kb, and 4.83 kb (the probe produces between one and six bands at this locus), while at the DXY14 locus both had bands measuring 3.0 and 1.94. The report concluded that the probability that the blood on Castro's watch randomly matched Ponce was 1 in 189,000,000.[34]

Unlike previous defense attorneys who accepted reports like this as proof of their client's guilt, with Lander's advice, Scheck and Neufeld immediately demanded access to all of Lifecodes' raw data and materials relevant to the case in an effort to interrogate the company's claims. As a result of their inquiries, the defense team received autoradiographs, laboratory notebooks, computer printouts of test results and population frequency databases, as well as information on quality control and quality assurance programs. Analyzing this information with Lander's assistance, Scheck and Neufeld discovered several technical faults that called Lifecodes' ultimate conclusions into question.

To begin with, the defense team noticed that there were two additional bands in the DNA sample from Castro's watch at the DXYS14 locus that were not present in Ponce's DNA profile. Baird testified that these "extra" bands were "of a nonhuman origin that we have not been able to identify." When questioned about how he reached this conclusion, Baird stated that in his experience with this locus, the intensity of the bands tended to decrease in proportion to their length. Thus, he decided to ignore the two extra bands because they were not of the intensity that he would have expected to see in fragments of those lengths.[35] At Lander's suggestion, the defense brought two witnesses to the stand to argue that Baird's explanation was scientifically untenable.[36] Both Howard Cooke of the Medical Research Council in Edinburgh, Scotland, who had invented the probe used by Lifecodes to target the DXYS14 locus, and David Page, of MIT, testified that there was absolutely no correlation between the intensity of a band and its size. They also told the court that the probe in question could produce banding patterns with anywhere from one to six bands. Cooke further testified that in the absence of additional experiments, the two extra bands in the watch profile were evidence that bloodstain did not match Ponce.[37]

The defense team's next discovery was that, while Lifecodes' final report stated that both the DNA samples had a 10.25 band at the D2S44 locus, in reality, the band in the watch DNA profile was 10.16 kb, while the band in the Ponce's DNA as 10.35 kb. The 10.25 kb figure turned out to be an average of these two bands. A simple calculation led the defense team to uncover several serious problems with the methodology that Lifecodes used to declare a match and calculate its rarity within a relevant population. Specifically, while Lifecodes claimed to employ a rule that two bands must fall within ± three standard deviations (3 s.d.) of the average size of the two bands to be considered a forensic match,[38] 10.16 kb and 10.35 kb differ in size by 3.06 s.d. In scrutinizing Lifecodes' data, Lander discovered that another band (the lower band in the D17S79 locus) differed in size by 3.66 s.d., which was considerably in excess of the 3 s.d. standard. Thus, if Lifecodes had actually adhered to its own published standards, then they would have reported that the blood on Castro's watch did not actually match Ponce's.

As a result of this discovery, Baird was forced to admit that, despite the published protocol, which stated that Lifecodes used a computer digitizer to match bands based on ±3 s.d. standard, in reality they relied almost exclusively on visual observation to determine matches—that is, by comparing the bands of two profiles on an autorad.[39] The computer-based technique was used only after two bands had been deemed to match by sight in order to generate a numerical size measurement. This number, in turn, was used to determine the band's frequency within the relevant population. Interestingly, once engaged in determining the rarity of a particular match, Lifecodes abandoned its 3 s.d. matching rule and instead used a standard only two-thirds of one standard deviation.[40] For Lander, this was scientifically unacceptable. He told the judge:

> Whatever choice you make for your matching rule, when you go and tell a court what is the chance this would have arisen at random in the population, you had better be using the same matching rule. To do otherwise is to report a probability that is simply not true. If I go out and I catch matches with a ten-foot-wide butterfly net and I say I caught a match, and then I come to court and I say, and it was so rare that I caught this match, and I will prove it to you by showing that when I go out with a six-inch butterfly net, I never catch matches in the population, that would be absurd.[41]

While the prosecution tried to downplay this issue, for the defense it was a shocking revelation that they believed severely damaged the credibility of Lifecodes' evidence. When Scheck got Baird to admit on cross-examination that they did not follow their published protocol in forensic casework, he raised his hands as if he were a football referee signaling a touchdown.[42]

The next technical mishap uncovered by the defense team dealt with the use of controls in Lifecodes' sex chromosome test. In addition to the three loci discussed above, the company used a probe that targeted a specific region of the Y-chromosome called DYZ1. When Lifecodes ran this test, they concluded that all three forensic samples in the investigation (i.e., the blood on the watch, Ponce, and her daughter) were female because none showed a band at this locus. However, the defense disputed this conclusion because the control lane in the test also showed no band. While Baird initially claimed that the result made sense because the control DNA was derived from the female HeLa cell line (which was widely used in medical research), another Lifecodes employee, Alan Giusti, looked at the banding pattern in the control lane and stated that Baird was mistaken in his assertion. When questioned about the identity of the control sample, Giusti told the court that at the time of the Castro investigation Lifecodes was using the DNA of a male company employee named Arthur Eisenberg.[43] When he returned to the stand, Baird told the court that he had indeed been mistaken and that Eisenberg must have had a rare genetic condition known as a "short" Y chromosome, in which the DYZ1 region is absent. When defense witnesses testified that that any individual with such a condition would almost certainly be sterile, Baird went back to his laboratory to do additional experiments (DNA typing Lifecodes' employees), and found that the control DNA actually came from a female Lifecodes technician named Ellie Meade.

Several defense witnesses concluded that this mistake was illustrative of a failure of Lifecodes' quality control and record-keeping mechanisms. During the course of his testimony, Lander went so far as to call Lifecodes' sex test in *Castro* an "absolute failure." He continued: "you can't determine anything from this autorad about sex and it's a stunning failure because it's a locus that's two thousand times easier in principle on average to detect than single copy RFLPs."[44]

The defense also argued that Lifecodes had thwarted generally accepted scientific practice by continuing to use a probe in forensic casework that they knew was contaminated.[45] This issue emerged when Baird argued that the contaminated probe led to the production of a spurious 6 kb band on the autorad that contained the hybridization for D2S44 and D17S79 (these two loci were probed at the same time). Baird told the court that Lifecodes had gone to great lengths to ensure that the 6 kb band was indeed bacterial in nature and not human. They did so by running a variety of experiments, including reprobing the forensic DNA samples with synthetic probes (which did not contain the extra bits of bacterial DNA that presumably targeted the 6 kb band) and by probing these samples with a probe that targeted a specific bacterial DNA sequence. In the latter case, the 6 kb band was visualized, lending credence to the notion that it was indeed bacterial in nature.[46] Although Lifecodes went to great lengths to demonstrate that the band was indeed the product of contamination, the defense argued that these efforts were insufficient and unscientific. In their view, the company should have stopped using the probe as soon as they discovered that it was contaminated. Baird, however, told the court that such a process would have slowed the company down significantly, since it took a long time to make a new probe.[47]

Ultimately, the defense team argued that a comparison between the "elaborate procedures run retrospectively by Lifecodes in their attempt to exclude the six kb band on autorad 17" and "the absence of any effort to rule in the one or two different bands" observed in the DXYS14 locus" suggested that there was an "imbalance in the priorities" of Lifecodes.[48] Lander opined:

> There is a tremendous amount of effort put into ruling out the six k.b. band, which would distinguish the two samples. So, the effort is put into showing that there's not such a distinction and there is no effort, not even a trivial effort of putting the membrane back down on film for another forty-eight hours to rule in the possibility of a difference, which single difference could exclude the defendant. That strikes me as a funny sort of decision tree for a laboratory to follow. One makes decisions about what experiments to pursue. If one has a

hard and fast procedure, one needs to make relatively few decisions. Here, it appears to me, in my judgment there were decisions made along the way and there is an imbalance in the sort of decisions along the decision tree. If one makes that tree and pursues all of the lines saying "rule out possible difference" and does not chase down the path saying "rule in possible difference," one is putting one-self in the position of—well, one is running the risk of making mismatches.[49]

The final technical argument launched against Lifecodes by the defense was that their Hispanic database was not in Hardy-Weinberg equilibrium (HWE), the condition of randomly segregating alleles that is required for application of the product rule. After gaining access to the raw data gathered by Lifecodes in an article published in the *American Journal of Human Genetics* (in which the company claimed that all the loci they used in DNA typing were in HWE),[50] Lander found what he called "spectacular deviations" from HWE. While one would expect about 4 percent of a population to be homozygous for any particular allele if it was in HWE, Lander calculated that the Hispanic database used by Lifecodes had a homozygosity rate of 17 percent for the D2S44 locus, and 13 percent for the D17S79 locus. When population geneticists see an excess of homozygotes in a given population, they generally assume that that population is divided into subpopulations that do not mate with one another. This state results in reduced genetic variation within these subpopulations, leading to more individuals with two copies of the same allele at a given locus. So, for Lander, the inescapable conclusion from the high degree of homozygosity that he observed in Lifecodes' Hispanic database was that there was significant nonrandom mating among subpopulations within American Hispanics.[51]

In an effort to bolster his conclusion, Lander suggested that Scheck and Neufeld contact Phillip Green, one of the population geneticists who peer-reviewed and approved Lifecodes' *American Journal of Human Genetics* article without having ever seen the underlying data. When he saw the raw data, he agreed to testify on behalf of the defense that he would not have given his approval had he seen the database they used to make their calculations. On the stand, he testified that Lifecodes changed the way they calculated HWE without alerting readers, and that the "only reason I can think of is they may have done the test the correct way the first time and gotten the significant result, and just redefined their [terms] so they could get the result they wanted."[52] Although the issue of population substructure would develop into a major controversy in the months after *Castro*, for reasons he did not state, Judge Gerald Sheindlin did not consider population genetics in reaching his decision.

THE PROSECUTION RESPONSE

For the most part, the prosecution's rebuttal of the claims put forth by the defense was rather weak. The prosecution's primary argument was that even though the defense may have been technically correct in criticizing certain aspects of Lifecodes' protocol, the mistakes and oversights made by the company did not affect the final results in the analysis of the bloodstains. Further, they argued that it was impossible to achieve the kind of experimental perfection that Lander called for in his testimony because of the unique nature of forensic case-work. As Michael Baird stated in a 1994 interview, "the reality is that when you do a test on a forensic sample, it is what it is, and you have to interpret it. It isn't my fault if the sample is contaminated or mixed or shitty. It's not my fault. I'm just trying to interpret what's there. I think that's where things got lost in the transla-tion. We basically got everything that we could, but there's a disagreement about the interpretation, but that's the way that it was."[53] In other words, the prosecu-tion and Baird both hoped and expected that the validity of Lifecodes' DNA evi-dence would be taken on trust by all involved in the *Castro* case, even though there was no shared culture or agreed-upon standard of interpretation.

The prosecution also attempted to convince the judge that Lander was not familiar enough with the forensic context to make claims about what should and should not be done in the course of a criminal investigation. As soon as prosecu-tor Sugarman had the opportunity to cross-examine Lander, she asked him a series of questions that demonstrated his lack of significant experience in the forensic realm. Specifically, he admitted that he had seen only a few forensic autorads during his career, that he had very little experience working with dried blood samples, and that he had never attempted to extract DNA from the kinds of substrates commonly found at crime scenes. Further, she forced him to admit that he had never even visited Lifecodes and had had very little contact with employees of the company over the past year.[54]

Additionally, prosecution witnesses and lawyers argued that although the defense offered plausible solutions to problems inherent in forensic science, there was more than one way to solve a particular problem. This idea was espe-cially prevalent in Baird's testimony, who constantly proclaimed that controls used by Lifecodes were just as good as those prescribed by Lander. In a similar vein, prosecution witnesses also suggested that there was a subtle disjuncture between the ideals of science and the actual practice of science, especially in the forensic context. This difference, it seems, made Lifecodes' mistakes in some way excusable. For instance, when Pablo Rubinstein (who was the head of the immuno-genetics laboratory at the New York Blood Center, which had collaborated with Lifecodes on several projects in the past) was asked whether it was problematic that Lifecodes occasionally diverged from its stated protocols, he stated that he was ambivalent about the issue:

> You see, most labs work with many probes and when you do that then you, you are not so clear on how these probes work. You see them for only a few exper-iments and then it's, maybe, months until you use them again. Now, if you use the same probe over and over and over and you have a limited number of probes that you use, I don't know. I'm trying to be very honest. My initial

reaction is that for my own peace of mind, I would like to run [control] gels [to make sure that the probes are functioning properly]. But I don't know if I would do it because this is important. It's not just one more lane. There are a number of technical issues that come up when you do that kind of thing. So, that's as close as I can come to a definitive answer. I cannot say that you must run them under all conditions.[55]

He expressed a similar opinion when asked whether he believed the scientific community required Lifecodes to keep accurate and complete lab notebooks in order to determine whether or not their methodology was generally accepted. He replied that while the scientific community "expects" good record keeping, "nobody is going to go in their lab and examine their protocol books."[56]

The prosecution also went to great lengths to close the "black box" opened up by the defense team by erasing issues of contingency from the proceedings. This strategy worked very well for the defense and can be seen in a particularly revealing exchange between Lander and Sugarman during a cross-examination in which Sugarman tries to get Lander to agree that forensic DNA typing is essentially nothing more than Southern blot analysis, which is a generally accepted technique within the scientific community. Lander vigorously disagreed with this characterization and tried to prevent her from making such a claim. The exchange began when Sugarman asked Lander, "And it would be your opinion, sir, that the method of Southern blotting, taking into account from the way the judge asked you about it at the beginning, the total procedure of Southern blotting, would be that procedure that is used for DNA typing in a forensic case?" Lander replied in the negative, and the conversation continued as follows:

SUGARMAN: No?

L: The method of Southern blotting, standing on its own, is clearly not the procedure that would be used for DNA typing on its own.

S: We are talking about the procedure that you talk about from extraction to the preparation of the autorad?

L: They, too, would be components of a procedure.

S: But that component, to create the extracted DNA, take it all the way through the creation of the autorad, would be the procedure that would be used in that component.

L: No, it would not be. I am taking the—

S: Of that component, part of the procedure.

L: Part of the procedure, yes.

S: You are not going—

L: It would form part.

S: You don't envision the invention of a new way to extract DNA, take the restriction enzyme, do the probe, hybridization, and all of that?

L: That is not true. There is much research going on right now over the issue of hybridization solution. As I am sure you know, different hybridization solutions have different properties as to backgrounds. . . . There is the famous paper of Church and Gilbert on genomic sequencing, in which a new, and I think very powerful, hybridization solution, which may, indeed, be more appropriate for this, has been developed, and this is actively being investigated and used in the forensic community. It would be wrong to say, indeed I think incorrect to say, that Southern blotting, as defined by Ed Southern in 1975, will be the procedure used. Indeed, there is very important active research going on right now to ask what is the best specific set of protocols, specific set of solutions to use in Southern blotting in a forensic case, given the special needs of forensics.[57]

Toward the end of Lander's testimony, presumably when it became apparent to her that there were indeed serious problems with the evidence in the case, Sugarman began to dissociate Lifecodes' work in *Castro* with their work in other previous and pending cases. She repeatedly asked him if he was making a judgment about the general acceptability of Lifecodes' methodology on other cases or just in this particular case. Lander stated that he was only considering Lifecodes' work in *Castro*, since that was the only case that he knew about, and that was what he took to be the point of this *Frye* hearing.[58] At the end of this line of questioning, Sugarman asked, "Doctor . . . lending yourself to an opinion as to this case, your testimony is as this case stands, [the] procedures apply to this case and [in] your opinion do not meet those required in general acceptance in the scientific community. . . . And that is all, is that correct?" Lander answered in the affirmative.[59] Sugarman continued, "And you are, therefore, not rendering an opinion as to a case that Lifecodes may, in fact, do today, correct . . ., nor any other case that they did in June of 1987; is that correct?" Lander answered no, and having made her point she moved on to other issues.[60]

AN OUT-OF-COURT MEETING

One reason that the prosecution's rebuttal of the defense's arguments was so weak was that by the end of Lander's testimony, most of the prosecution witnesses had become convinced that his criticisms were in some sense valid. No witness was more affected by Lander's critique than Richard Roberts, the soon-to-be Nobel Prize winner who had testified to Judge Harris in *Wesley-Bailey* that he did not need to thoroughly examine Lifecodes' work because he trusted their scientists a great deal. Although Roberts had previously testified in several cases involving DNA evidence, he had never actually examined forensic DNA evidence in detail and was surprised by what the defense had uncovered. In the half dozen or so other cases he testified in he saw his role as primarily providing background information for the judge to use in deciding whether DNA evidence should be admissible in a court of law. In *Castro*, as in other cases, he relied on the prosecutor and Lifecodes to show him all of the relevant material he needed to use in deciding whether a particular result was valid and reliable. In a recent interview, he told me that it never occurred to him to ask if they were withholding any data from him. "I assumed they were showing me all they had," he said.[61]

Roberts's opinion changed dramatically, however, when he ran into Lander at an April 1989 meeting at Cold Spring Harbor, shortly after Lander had finished testifying in *Castro*. Lander told Roberts that he had been "duped" by the prosecution and suggested that he read the report that he had written for the case. Upon reading Lander's report, Roberts became indignant toward both the prosecution (for withholding information from him) and the American legal system (for condoning a system in which deception is an accepted practice) and decided that the scientific issues had to be settled as quickly as possible. He told me that he believed that the American legal system was not interested in truth. Rather, it was interested in who could be the most cunning adversary. As a result, lawyers in the *Castro* case practiced techniques of deception that prevented the kind of deliberations that go on in science. "I think that if there had been a demand for honesty right from the very start," he said, "then the problems would have come out rather easily."[62] Indeed, he went on to say that if he had had access to all of the data in *Castro*, he could have made "a perfectly sound judgment about what was acceptable, what was data that you could interpret, what was data that was really interpretable, and certainly should not have been used to try to falsify. I think, for the expert witnesses for the defense, they were in exactly in the same state. It was just that because they had this additional material, they were able to make arguments."[63]

Although he remained convinced that Lifecodes' ultimate conclusions were sound,[64] he decided that it would be a good idea for scientific experts from each side of the case to meet outside the confines of the courtroom to have a "scientific discussion" about forensic DNA analysis, as colleagues rather than adversaries, in a forum where there was "none of this lawyerly talk."[65] Eight of the ten witnesses in the case who were contacted agreed that this meeting outside of the courtroom was a good idea, but not all could fit it into their schedules (Baird was ultimately the only witness who did not participate). In the end, two witnesses from the defense, Lander and Lorraine Flaherty, would meet with two prosecution witnesses, Roberts and Carl Dobkin, on 11 May 1989 to discuss the case. The meeting was described in various press accounts as "highly unusual" and rebellious.[66] They subsequently issued an unprecedented joint statement concluding that Lifecodes' DNA evidence in the *Castro* case was not scientifically reliable. They further stated that if it were "submitted to a peer review journal in support of a conclusion, it would not be accepted."[67]

Ironically, though, the group disparaged the very legal process that brought them together in the first place and served as the basis for their statement: "All experts have agreed that the *Frye* test and the setting of the adversary system may not [be] the most appropriate method for reaching scientific consensus. The *Frye* hearing is not the appropriate time to begin the process of peer review of the data. Initiating the peer review at this time wastes a great deal of the court's and the experts' time. The setting also discourages many experts from agreeing to participate in the careful scientific review of the data."[68] Advocating for an alternative arena for resolving the issues that arose during the course of the *Castro* hearing, they called for the formation of a National Academy of Sciences committee to study the use of DNA typing in forensic casework.[69]

Roberts was especially critical of the legal system in his comments following the case. He told a reporter from *Science*: "The court system is adversarial and expert witnesses are encouraged to go further in their statements than they might otherwise be prepared to go. We all did so much better when we sat down without the lawyers and had a reasoned scientific discussion."[70] The implication was that the adversarial legal system was not the best place to evaluate new forms of scientific evidence. What Roberts seemed unwilling to admit, however, was that until he was confronted with information presented by a witness who only emerged because of the adversarial legal process (Lander), he was unaware of the problems associated with the technique and had almost complete faith in the work of a company that he knew very little about.

Upon being presented with the statement signed by their own witnesses, the prosecution initially decided that they would withdraw the DNA evidence in the case. Ultimately, they decided against this tactic and attempted to block admission of the joint statement on grounds of hearsay. The defense, however, counteracted this maneuver by calling the four experts to the stand to testify about their statement.[71] In this way, the joint statement became a part of the official record of the case. Although the prosecution attempted to mount a rebuttal case, their efforts were largely ineffective in salvaging Lifecodes' evidence. Realizing this, the prosecution conceded in their final brief that Lifecodes' tests were legally inadmissible and laid out a set of procedural safeguards that should be followed by all parties involved in the criminal justice system in order to ensure the validity and reliability of DNA evidence.[72]

THE RULING

As Judge Sheindlin discussed at the beginning of the pretrial admissibility hearing, he believed that previous courts ruling on the admissibility of DNA evidence had focused too much attention on the general acceptance of the concept of forensic DNA analysis within the scientific community. As a result, not enough attention had been paid to the specific DNA testing results in the actual cases under consideration. In his view, "passing muster under *Frye* alone is insufficient to place this type of evidence before a jury without a preliminary, critical, examination of the actual testing procedures performed in a particular case."[73] Thus, Sheindlin decided that the ultimate issue he would address was admissibility of the results of the specific DNA analysis performed by Lifecodes on the defendant's watch and the murder victims.[74] To clarify his position, Sheindlin broke the *Frye* standard down into the following "three-prong analysis" that took into account his concerns:

> Prong I: Is there a theory, which is generally accepted in the scientific community, which supports the conclusion that DNA forensic testing can produce reliable results?
>
> Prong II: Are there techniques or experiments that currently exist that are capable of producing reliable results in DNA identification and which are generally accepted in the scientific community?
>
> Prong III: Did the testing laboratory perform the accepted scientific techniques in analyzing the forensic samples in this particular case?[75]

While the first two prongs would deal with issues that were traditionally addressed by *Frye*, the third prong would deal with the specifics of the case. Although there was some debate about whether this issue went to the weight of the evidence rather than its admissibility, Sheindlin clearly believed that the performance of the technique in the particular case was crucial to its admissibility in that case.

On the issue of Prong I, there was little discussion. Sheindlin wrote that "the evidence in this case clearly establishes unanimity amongst all the scientists and lawyers as well that DNA identification is capable of producing reliable results."[76] Prong II, however, proved to be much more contentious over the course of the *Frye* hearing. While Sheindlin believed that it was possible to accept the general proposition that DNA typing can be done reliably, but still have doubts about the reliability of a particular test, the defense argued that it was nearly impossible to separate the issues in Prong II from those in Prong III.[77] In their view, any testimony addressing Prong III that suggested that Lifecodes' methodology was not generally accepted in the scientific community would necessarily have an impact on Prong II, since the company was such a major force in the DNA typing industry (indeed, for all intents and purposes, they were one-half of it). At various points during the proceedings, Neufeld, Scheck, and Sheindlin argued over whether a sharp distinction could be drawn between Prongs II and III. In the end, Sheindlin was willing to admit that the boundary between the two issues was blurry but proclaimed that he was competent enough to make it anyway.[78]

Sheindlin began his discussion of Prong II by noting, as previous judges had done, that most of the techniques and methodologies used in forensic DNA analysis were not new, since they had been used in diagnostics, clinical, and experimental settings for many years and had gained general acceptance in the scientific community. Acknowledging Lander's testimony (and the defense's main argument), however, he noted that "it is the transfer of this technology to DNA forensic identification that has generated much of the dispute" in this case.[79] He also noted that there was a disagreement among several witnesses over whether the technology was already admissible, or whether it would be ready for court by the end of the year, as both Lander and Flaherty had suggested.[80] At least in his view, none of the witnesses suggested that the technique would not be generally accepted at some point in the near future. Ultimately, Sheindlin decided that the testimony of the "highly respected and rather brilliant" prosecution scientists, as well as previous legal precedent, all supported the conclusion that forensic DNA analysis met the *Frye* standard.[81]

Presented with a statement signed by witnesses from both prosecution and defense that the DNA evidence in the case was not scientifically valid or reliable, as well as an admission from the prosecution that they agreed with this conclusion, Judge Sheindlin had very little room to maneuver in answering his Prong III question. To begin with, he had no significant expert testimony from which to craft a ruling that the evidence in the case was valid and reliable. Thus, with respect to the evidence that Lifecodes presented to prove that the blood on Castro's watch was Vilma Ponce's, Sheindlin concluded that it was "inadmissible as a matter of law, since the testing laboratory failed in several major respects to use the generally accepted scientific techniques and experiments for obtaining reliable results, within a reasonable degree of scientific certainty."[82] Where he did have some measure of latitude, however, was in deciding whether or not Lifecodes' evidence *excluding* Castro from being the source of the DNA on his watch was admissible, since exclusions were much less technically and statistically challenging than inclusions (i.e., they stop at the stage of visual analysis).[83] In the end, he ruled that the exclusionary evidence was admissible because it did not suffer from the problems inherent in making a match, declaring its size, and calculating the frequency of the band within the relevant population.

Upon learning of Sheindlin's ruling, the prosecution was satisfied that the damage to DNA typing had been limited to a single case, and that the judge had taken their procedural recommendations into account in his final judgment. In several statements that appeared in the press from members of the Bronx District Attorney's office, prosecutors proclaimed that the case was a "service to the criminal justice system" and that the ruling was a "victory of national importance" that affirmed the overall validity and admissibility of forensic DNA typing.[84]

At least in public, Lifecodes seemed relatively happy with the ruling as well. In a letter to the journal *Science*, written shortly after the *Castro* decision, Lifecodes' scientist Kevin McElfresh summed up the company's official position on the case. He wrote:

> From [Sheindlin's] decision, we think it is clear that he was able to see through a number of issues that the defense in the Castro case blew out of proportion. We agree that the inclusionary aspect of the data had some ambiguities that were a function of the sample as well as the probes and technology in use in 1987. When tried against 1989 standards, these data were not as compelling as they could have been. Unfortunately, the membrane on which the DNA was examined had been exhausted by repeated hybridization and could not be further analyzed with the use of probes and technology available in 1989, when

the case finally went to trial. However, that does not invalidate the results that were generated, especially when they are viewed in conjunction with all of the evidence in the case.[85]

In the *New York Times* story on Sheindlin's ruling, Lifecodes spokesperson Karen Wexler expressed a similar sentiment, adding that "if we did the same test today, we'd draw the same conclusion, but we'd try to explain the ambiguity."[86] Perhaps best summarizing Lifecodes' take on *Castro*, Baird explained to me in a recent interview:

> You know . . . it wasn't a problem with the test methodology or the DNA testing itself, it was really a situation where we were out-experted or out-attorneyed. And certainly those things are going to happen, I mean look at the OJ case. I think once the scientific community and the forensic community got to hear what happened and what the result looked like, they were more comfortable, and if you look at really the judge's decision, he didn't really come down and say that DNA testing was bad or wouldn't work. He just said in this case we can't allow the call that showed there was an inclusion.[87]

Thus, Lifecodes' official response was essentially that any problems they had in *Castro* were isolated, not systemic, occurrences. In the months following the trial, they spent a great deal of time assuring potential customers and the public that similar mistakes would not be made again. At the same time that Lifecodes downplayed the *Castro* decision, however, the company did make several significant procedural and methodological changes in their protocol to address the issues raised by Scheck, Neufeld, and Lander. Most notably, they instituted a computer-based matching system that did not rely on subjective visual criteria, altered some of the internal controls that they used, and changed the way that they calculated the frequency of a given allele in their population databases.[88]

While Lifecodes and the Bronx County DA were happy that damage to the reputation of forensic DNA analysis was about as limited as it could have been, Scheck and Neufeld were stupefied by Sheindlin's ruling. They simply could not believe that the judge was able to maintain an airtight boundary between Prongs II and III of his framework for determining admissibility. While they obviously appreciated that he had ruled the DNA evidence in the *Castro* case inadmissible, they were "bitter" that he declined to address the Prong II issue of the use of DNA typing in general. "It seems fundamentally wrong," Scheck said, "that [Sheindlin] avoids making findings about the methodological problems that Lifecodes had that would reveal to anyone reading the opinion that all their past cases are invalid."[89]

Responding to this comment, and echoing his stance throughout the trial, Sheindlin replied that it would have been inappropriate for him to address the use of the technique in other cases that Lifecodes had been involved in.[90] Thus, he created a boundary between the technological system of forensic DNA analysis and its use in specific situations. Much to the dismay of Scheck and Neufeld, this separation would come to be the dominant perspective from which future judges evaluated the validity and reliability of the technique. In other words, judges following Sheindlin's ruling started from the assumption that Prong III was the only issue in question. As a result, DNA evidence was ruled inadmissible in only a very small minority of cases at either the trial or appellate level after *Castro*.[91]

While the stakeholders in the DNA typing arena had clear reactions to Sheindlin's ruling, the press was not quite sure what to make of it. News accounts of *Castro* were in marked conflict with one another. *Science*'s story about the case was headlined "Caution Urged on DNA Fingerprinting" and argued that the technique had "failed its first serious challenge."[92] *Nature*'s headline, on the other hand, declared "Judge backs technique," and the accompanying story went on to state that "genetic fingerprinting thus emerges . . . with backing for its use" both to demonstrate exclusions and inclusions.[93] The *New York Times* took an intermediate position with the headline "Reliability of DNA Testing Challenged by Judge's Ruling." Highlighting the ambiguity inherent in Sheindlin's decision, this story noted that "both the prosecution and the defense claimed a victory in the complex ruling, and even the laboratory whose tests were criticized praised the decision." It further noted that there were "sharp differences of opinion over whether the Bronx ruling constituted a major precedent that would lead to the reopening of scores of other cases involving DNA fingerprinting."[94]

Whatever perspective articulated by the various media outlets, the bottom line is that the *Castro* case received a tremendous amount of publicity and put the issue of the validity and reliability of DNA typing in the national spotlight, where it remained until the end of the 1995 O.J. Simpson trial. The publicity surrounding the trial also began the process of transforming Scheck and Neufeld into two of the most well-known attorneys in the United States. In the weeks following *Castro*, the duo received dozens of requests from defense attorneys around the country for help in challenging DNA evidence.

The case also provided a great deal of public exposure for Eric Lander, who was also just beginning his climb to national notoriety, both within and outside the scientific community. Indeed, according to one article, Lander was invited to testify in fifty-seven cases in the six months following his testimony in *Castro*. He turned all of them down due to time constraints.[95] Lander's publication of an article in *Nature* in June 1989 summarizing the testimony he had given in *Castro* just a few months before had a profound impact on the legal and scientific landscape of forensic DNA analysis. In addition to alerting the academic scientific community to the problems associated with forensic uses of the technique, his participation in the case made it more acceptable within the scientific community to testify in courts on behalf of the defense.

On 15 September 1989, Joseph Castro appeared before Judge Sheindlin and pled guilty to the murder of Vilma Ponce and her two-year-old daughter. He also admitted that the blood on his watch was not his, but that of Ponce. This confession emboldened Lifecodes' Kevin McElfresh to belittle the efforts of the defense in *Castro*. For McElfresh, the correlation between Lifecodes' result and Castro's confession indicated that "perhaps defense attorneys are beginning to be confronted with having to accept the reality of scientific data that is valid, reliable, and powerful."[96] Essentially, McElfresh was arguing that sound methodology and adherence to the accepted practices of science was not as important as the ultimate test result. In his view, in the production of DNA evidence the final result seemed to matter more than the process used to obtain it.

STATE V. SCHWARTZ

At about the same time that Lifecodes' DNA typing system was being scrutinized in *Castro*, defense attorneys and scientists in other parts of the country were gearing up to challenge Cellmark's technology. The most important of these cases revolved around a shocking crime that put citizens and workers in the Twin Cities of Minnesota on edge. On 27 May 1988, nineteen-year-old Carrie Coonrod was brutally murdered in a Minneapolis parking ramp at 9:00 A.M. while on her way to a job interview. Eyewitnesses reported seeing a man wearing a plaid shirt leaving the scene in a car soon after at a "high rate of speed" and noted the car's license plate number. Based on this information, police went to the listed address of the vehicle and arrested the owner, Thomas Schwartz, in connection with Coonrod's death. Upon searching his home, investigators found a blood-soaked pair of jeans. A few days later, they also found a blood-soaked plaid shirt that was missing three buttons near the scene of the crime. Three buttons identical to the remaining shirt buttons were recovered during a search of Schwartz's car. Preliminary serological work by the Bureau of Criminal Affairs revealed that the blood on Schwartz's jeans and the shirt was the same blood type as Coonrod's.[97] Although there was already strong evidence pointing toward Schwartz's guilt, on 24 June 1988, Steve Redding, the prosecutor in the case, and the Minneapolis Police Department decided to bring evidence from the case to Cellmark Diagnostics for forensic DNA analysis.

On 27 September 1988, Cellmark sent its final conclusions to the Minneapolis Police Department. Based on four single locus probes (g3, MS1, MS31, and MS43), the company stated that the DNA banding pattern from the plaid shirt was entirely consistent with Coonrod's DNA profile. They claimed that the frequency of this particular pattern in the general Caucasian population was approximately one in thirty-three billion. Although Cellmark stated that the banding pattern from the jeans was consistent with Coonrod's profile, for reasons not provided they could not reach a definitive conclusion about this match.

Like Neufeld and Scheck, Assistant Public Defender Patrick Sullivan was determined not to be intimidated by the complexity of molecular biology or the certainty expressed as minuscule random match probabilities. Despite having no background in science, he wanted to do his best to critically examine the technique from the defense perspective. His desire to challenge DNA evidence was not motivated so much by personal history, but by the realization that both Cell-mark and Redding hoped to make *Schwartz* "the test case for all of Minnesota." Further, he believed that they assumed that the case would proceed like most of the others involving DNA evidence around the country. He said, "I could see—it was pretty clear that my job was just to roll over and play dead, and this would happen, and I wouldn't do anything about it, and there was nothing I could do about it—it was just a foregone conclusion that I would lose this [admissibility] hearing. And so I additionally thought that I don't want to be the guy who lets this in for the entire state of Minnesota, for every case that comes down the pike later."[98]

Sullivan began his research by searching the mass media for articles about forensic DNA typing. The search turned up several articles on the *Wesley-Bailey* case, as well as a few news articles on preparations for the *Castro* trial.[99] Sullivan also located several articles from the *Fort Wayne Journal-Gazette* about the Indiana murder trial of Frank E. Hopkins, in which a long *Frye* hearing took place, and Cellmark received its first defense challenge.[100] In this case, Hopkins's defense attorney, Charles F. Leonard, focused on Cellmark's quality control regime and the potential for human error in the test. Although unsuccessful in his challenge, he did get Cellmark representatives to admit that they switched from the company's stated laboratory protocol to an alternative protocol designed to yield more DNA from biological evidence in the middle of the analysis of crime scene samples. He also forced a Cellmark forensic technician to admit that she had not followed proper procedure when labeling evidence. This mistake caused her to misidentify a sample, ultimately leading to an accidental mismatch in her analysis.[101]

While in the process of reading press accounts, Sullivan was contacted by a local newspaper reporter who was interested in the technique and was planning to write a story on the *Schwartz* trial. The reporter suggested to Sullivan that he contact William C. Thompson and molecular biologist Simon Ford, who were colleagues at the University of California-Irvine's multidisciplinary School of Social Ecology. Thompson, who had received a Ph.D. in psychology from Stanford and a J.D. from Boalt Hall (University of California, Berkeley), became interested in forensic DNA analysis in mid-1987 while doing a National Science Foundation–supported study that examined the use of statistical evidence in criminal courts. His collaboration with Ford, who was engaged in forensic science research at the time, began when he asked for help in understanding basic aspects of biology and genetics. Thompson and Ford began meeting regularly to discuss what they believed to be the most pressing issues surrounding the technique. After working through Alec Jeffreys's original set of papers, studying Cellmark's later work on single locus probes, and talking to numerous local forensic scientists and lawyers, they came to believe that Cellmark was making claims that its scientists could not support.

The reporter also gave Sullivan an article by Thompson and Ford called "DNA Typing: Promising Forensic Technique Needs Additional Validation" that they had recently published in *Trial* magazine.[102] In this article, as well as in a more substantial article published a few months later in the *Virginia Law Review*, Thompson and Ford laid out their fundamental argument against the admissibility of DNA evidence. They wrote that

- Claims about the certainty of DNA evidence were "unfortunately, exaggerations" since all matches had probability statistics associated with them[103]
- The technique had not yet been standardized, so there was no way to ensure that the work done by a specific laboratory complied with a generally accepted methodology[104]
- Autoradiographs were much more difficult to interpret, and bands were much more difficult to measure, than Cellmark and Lifecodes were willing to admit. They coined the term "slop" to describe the variation in appearance of autoradiograms caused by the contingencies of forensic casework[105]
- There was a significant possibility of laboratory errors, such as contamination or making an erroneous "call," and human error, such as sample mix-up, at several points during the DNA typing procedure[106]
- There were serious problems with the way that private laboratories calculate and present statistical probabilities, especially because they used unverified assumptions about populations and the independence of various alleles within them[107]
- Adequate validation and reliability studies had not been done, primarily because the private laboratories were not only policing themselves but also doing proficiency testing under ideal laboratory conditions, rather than under the more messy forensic conditions[108]

Thompson and Ford concluded their *Trial* article by stating that the stakes in the decision whether or not to admit DNA evidence into court were high. "On the one hand," they wrote, "there is a danger that excessive caution will prevent valuable evidence from being admitted in a timely manner. On the other hand, there is a danger that evidence accepted quickly and uncritically will later prove less reliable than promised. . . . It is an extraordinarily powerful and promising innovation, but the complexity of the techniques may hide some dangerous pitfalls and, in routine forensic use, it may fail to live up to the high expectations of its proponents. Until additional validation studies are done, the legal profession would be well advised to approach the new technique with caution."[109] Upon reading the article, Schwartz contacted Thompson, who not only gave him additional advice on how he should go about challenging Cellmark's DNA evidence in the case, but also provided him with materials and names of potential witnesses. Thompson also agreed to come to Minneapolis to testify in the *Frye* hearing on behalf of Schwartz.

The pretrial admissibility hearing in *Schwartz* took place in stages from December 1988 to February 1989. Over the course of twelve days in court, twelve witnesses testified, producing nearly 1,300 pages of transcript. Redding employed a two-part strategy in his efforts to convince the judge that Cellmark's DNA evidence was admissible. First, he argued that there was considerable ambiguity over whether the admissibility of novel scientific evidence in Minnesota courts was governed by the *Frye* standard or the relevancy approach, and that the relevancy approach was better because it could be "easily applied" and was consistent with the Minnesota and Federal Rules of Evidence.[110] The major advantage, he argued, was that it provided for the court's discretion in the final determination of the admissibility of scientific evidence and "allows the trial court to thoroughly examine the logical and legal basis for admitting [such] evidence. Reliability is examined through the expert opinion testimony . . . [and] the rule also allows the court to examine the opinion of the scientific community by considering its novelty and the scientific literature."[111] Although he argued that the technique was admissible under either standard, the relevancy approach was far better.

Based on this view, he then proceeded to call seven witnesses to the stand to argue that the technique was not novel, since it had been used in medical research and diagnostics for several years, and that it was highly reliable in these situations.[112] Several of these witnesses—including Bonnie Blomberg, from the University of Miami Medical School; David Goldman, who directed the genetic studies section at the National Institutes of Health; and P. Michael Conneally, who used DNA typing to study the inheritance of disease—argued that Cellmark's probes were available to researchers and that they were generally accepted as reliable within the scientific community.[113] Other prosecution witnesses, including Walter Rowe, a professor of forensic science at George Washington University, testified that Cellmark's quality control procedures were excellent.

Sullivan's strategy was to focus on the issues laid out by Thompson and Ford in their articles. In addition to these issues, Sullivan also argued that there were serious flaws in the way that Cellmark compiled its population frequency databases. He also had Seymour Geisser, who was professor and director of the School of Statistics at the University of Minnesota, and Rollin Richmond, who was a professor of biology at Indiana University at the time, testify that the only way to determine whether or not this database was in Hardy-Weinberg equilibrium was to examine the raw population data. Sullivan then argued that Cellmark actively sought to prevent the defense experts from gaining access to this data in an effort to cover up these potential flaws. He claimed that the company did so "to protect itself from embarrassment and to gain a competitive advantage."[114] In his view, Cellmark's protective tactics prevented defense experts from checking the databases for HWE and the independence of alleles. The prosecution answered this charge by pointing out that Minnesota Rules of Criminal Procedure dictate that Cellmark was only obligated by law to disclose information made in connection with the particular case at hand and not all of its available data. Therefore, the company and prosecution were in compliance with Minnesota regulations, and the company did not withhold any information from the defendant.[115]

Second, Sullivan argued that Cellmark did not follow the procedure generally accepted in the scientific community for doing RFLP analysis. Specifically, he claimed that Cellmark's DNA evidence was inadmissible because it met neither the requirements of the FBI's recently published validation protocol nor the FBI's recently formed Technical Working Group proposed guidelines for quality assurance on DNA typing.[116] Sullivan made this argument despite the fact that the FBI's protocol was nothing more than suggestions at the time and had not yet been codified, widely disseminated, or generally accepted within any scientific community.

With significant help from Thompson, Sullivan also explained to the judge that Cellmark had made several errors in a blind proficiency test carried out in late 1987 and early 1988 by the California Association of Crime Laboratory Directors (CACLD). In what remains one of the few examples of such testing in the history of DNA profiling, the CACLD created a test to gauge the ability of Lifecodes and Cellmark to accurately identify the source of biological materials. Their goal was to "provide sound advice to [local law enforcement agencies in California] regarding the value and limitations of a service available from private vendors before they have acquired the skills to provide it in their own laboratories."[117] Although Cellmark's mistakes were primarily clerical in nature—they mixed up one sample in the test and presented the results in a confusing format that made it seem as if there were several false inclusions—the CACLD proficiency test highlighted the threat of error in DNA profiling.

THE QUESTION OF ADMISSIBILITY

On 17 February 1989, at the conclusion of the *Frye* hearing, Judge Davis ruled that, despite not being "ready to hail this type of DNA testing as the best scientific discovery ever made," the prosecution had adequately demonstrated that forensic DNA analysis was generally accepted within the scientific community.[118] Further, he ruled that Cellmark's application of RFLP technology in forensic casework, including its probes, databases, probability calculations, quality control procedures, and protocols, were consistent with generally accepted scientific practice and capable of producing accurate, reliable results.

In any other state, this decision would have put at least a temporary end to the issue of the admissibility of DNA evidence. Minnesota, however, had a unique legal procedure that allowed a judge to "certify" controversial questions of law that were both "important and doubtful," as well as necessary to the resolution of the case at hand.[119] This procedure meant that the Minnesota Supreme Court would have the opportunity to rule on the admissibility of DNA evidence before

Schwartz, a case of first impression, proceeded. The ultimate outcome was that it created an immediate appellate decision that prevented the issue of admissibility from having to be relitigated in the future. Thus, instead of proceeding to the trial in *Schwartz*, on 17 February 1989 Judge Davis asked the Minnesota Supreme Court to issue a definitive ruling on the admissibility of DNA evidence.

The certified question from Judge Harris arrived at the Minnesota Supreme Court in the fall of 1989. The first obligation of the court was to decide what standard they would use to determine the admissibility of DNA evidence. In its opinion, the court proclaimed that it was "unconvinced by the state of the need for or the wisdom of overruling these prior decisions [in which Minnesota courts used the *Frye* standard to judge novel scientific evidence], we reaffirm that the admissibility of novel scientific evidence is determined by the application of the *Frye* standard."[120]

In the time between the *Frye* hearing and the Supreme Court hearing, Sullivan decided to radically alter the strategy he would use to argue against the admissibility of DNA evidence. At the trial level, he argued that forensic DNA analysis had not yet been generally accepted as valid and reliable within the scientific community because there were so many competing protocols, methodologies, and probes. In other words, a single set of standards had not yet been created, peer reviewed, and disseminated in such a way that a judge could use the standards to determine whether or not the DNA evidence in a particular case met them. For various reasons, he decided at the last minute to abandon that argument when presenting the defense case before the Supreme Court. Specifically, he decided to adopt a Prong III–like approach from *Castro* and shift his focus from DNA profiling in general to Cellmark's specific performance of the technique in the *Schwartz* investigation.[121] Thus, when the chief justice asked him the certified question "May evidence of 'DNA fingerprinting' test results be admissible in a criminal proceeding?" he replied "yes," rather than the negative answer that everybody had been expecting, including Redding.[122] As he told me in a recent interview, when he said yes, all of the Supreme Court justices "kind of leaned forward. Then, for the next twenty or thirty minutes, I had them."[123]

It is clear from the Supreme Court's ultimate decision that Sullivan's assessment is accurate. The Supreme Court followed his argument, answering that although forensic DNA analysis was generally accepted as valid and reliable in the scientific community, "specific DNA tests are only as reliable and accurate as the testing procedures used by a particular laboratory."[124] The court opined that it was troubled not only by Cellmark's error in the CACLD blind trial, but also by the possibility of an "ambiguous match" that results when two samples are determined to match even though there has been significant band shifting (see fig. 5).[125] The court also found that Cellmark's testing procedures were "deficient in several respects," especially because the company had not met the requirements set forth by the FBI or the CACLD for laboratories intending to perform forensic DNA analysis. The court noted that Robin Cotton, Cellmark's director of research and

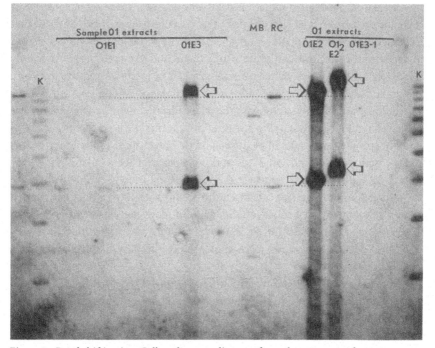

Figure 5. Band shifting in a Cellmark autoradiogram from the 1989 case of *Commonwealth of Massachusetts v. Robert Curnin* (Worchester, Mass.). The autoradiogram shows several profiles derived from the same evidentiary semen stain found on the victim's nightgown (Sample01). Each of the lanes in the areas labeled "Sample01 extracts" contain DNA extracted and purified under slightly different conditions. Lane MB contains the profile of the victim, and Lane RC contains Curnin's profile. The lanes marked K contain the size ladder. A comparison across all lanes reveals an upward band shift in the extracts labeled 01_2E2 and 01_2E2 when compared to $01E1$, $01E3$, and Curnin's profile. The dotted line and arrows indicate the degree of mobility shift between the same samples processed in different ways. Cellmark ultimately declared a match between the bands in Curnin's profile and the semen found on the nightgown without producing at trial experimental confirmation of the differential mobility of the DNA extracted in different ways. Image courtesy of William C. Thompson.

development, admitted that the company "has not comported with all of these standards."[126] Never mentioned in the judgment, however, was the fact that such standards simply did not exist at the time that Cellmark began performing DNA typing and were only beginning to be disseminated at the time that the tests were done in the Schwartz case.

The justices were also convinced by Sullivan's argument that Cellmark had not disclosed all of the information and materials necessary for the defense to make a scientifically complete investigation of the company's methodology, databases, and ultimate results. Borrowing language from Sullivan's brief, the court opined that Cellmark had not fully complied with the defense's request for the raw data from their population databases because "trade secrets may be at stake for the commercial laboratories." The court also agreed with Sullivan that Cellmark had not adequately published their methodology in peer-reviewed journals for the scientific community to evaluate, and that their probes were only selectively available to certain researchers.[127] Thus, in answering the question, the court issued a ruling remarkably similar to Judge Sheindlin's in *Castro*. They concluded that, "while we agree with the trial court that forensic DNA typing has gained general acceptance in the scientific community, we hold that the admissibility of specific test results in a particular case hinges on the laboratory's compliance with appropriate standards and controls, and the availability of their testing data and results. We answer the certified question accordingly. Because the testing laboratory in this case did not comport with these guidelines, the test results lack foundational adequacy and, without more, are thus inadmissible."[128] The Supreme Court's decision in *Schwartz* was a major victory for Sullivan. At least until deemed otherwise, DNA evidence produced by Cellmark was inadmissible in Minnesota.

On 10 January 1990, Redding announced to Judge Davis that the state would proceed with its prosecution of Schwartz without DNA evidence, claiming that waiting for the evidence and then conducting a new *Frye* hearing would not be fair to Carrie Coonrod's family. "She was their only daughter," said Redding in the *Minneapolis Star-Tribune*, "and they would like to get this phase of their nightmare over."[129] On 6 February 1990, two days into jury selection in his actual trial, Schwartz pled guilty to first-degree murder. Judge Davis sentenced him to life in prison.

The Perils of Private Forensic Science

As news of the *Schwartz* decision spread from Minnesota to the rest of the country, Bill Thompson and Simon Ford emerged as two of the most important members of a nascent network of individuals who provided information and resources for other defense attorneys hoping to challenge DNA evidence.[130] They served numerous functions on behalf of the defense community. First and foremost, as shown in *Schwartz*, they made defenders aware that it was possible to challenge DNA evidence and provided robust strategies for doing so.[131] They also provided the strategy for challenging the evidence, as well as the material and expert support necessary to do so. Along with Laurence Mueller, their biologist colleague from the University of California-Irvine, they became known alternatively as the "Irvine Mafia" or the "Combine from Irvine."

Their work upset the private companies a great deal, especially in the context of a series of cases taking place around the country that involved access to the raw data underlying Cellmark's statistical calculations of the probability of a random match. In early 1989, Thompson and Ford were contacted by a defense lawyer from Washington State, Peter Connick, who hoped to challenge DNA evidence in the upcoming serial rape trial of his client, Richard Cauthron (*State of Washington v. Cauthron*, 846 P.2d 502, 1989/1991).[132] Thompson, Ford, Mueller, and Seymour Geisser agreed to testify at the trial and asked Connick to demand access to all of Cellmark's raw data and autorads related to the Cauthron investigation, as well as information about the CACLD blind trial, proficiency testing, validation studies, and the population database they used to determine frequency of alleles at each loci.[133]

Cellmark was alarmed by the defense's request for so much raw data and case material and brought in a private lawyer to work with the prosecutor to fight the defense's discovery motion. In addition to attempting to obtain a protective order on certain aspects of their protocol and raw data (so that they could not be shared with other defense attorneys around the country), Cellmark's lawyer also argued that the defense should pay for all expenses incurred by the company during the discovery process, even though the defense was given a limited budget by the judge. According to Cellmark, the defense's demand for the population database was "unduly burdensome . . . and could not realistically be fulfilled in less than a two-month period."[134] Critics of private DNA typing laboratories, however, retorted in this case, as well as several others, that such tactics were aimed at shielding the company's protocol and population databases from the scrutiny of the scientific community.[135]

The extent to which the private companies shielded themselves from scientific scrutiny was not just an issue for defense attorneys and their experts. Efforts undertaken by the forensic science communities of both California (the CACLD) and New York yielded similar concerns. Indeed, the final report of the New York State Forensic DNA Analysis Panel, a committee created by the governor to provide recommendations on regulating DNA evidence in the state, lamented that the proprietary nature of the business meant that "sweeping claims of accuracy, stating that the probability of error is one in a million, or in some cases one in a billion," were "suspect" because they could not be double-checked by independent scientists.[136] Not only were the population databases used to reach such conclusions not publicly available, but the methods for calculating such probabilities were only beginning to be published. In response to this situation, the report stated the following:

> Private laboratories are reluctant to share information about their procedures, and they have generally adopted a proprietary stance and treated their protocols as trade secrets. . . . Yet the laboratories' scientists claim, as they must under *Frye* and most of its progeny, that their techniques are generally accepted as reliable in the scientific community. It is difficult to reconcile the practice of cloaking a methodology in secrecy with the claim that the methodology is widely accepted. Until private laboratories allow their procedures to be reviewed by the general scientific community, it will remain impossible to evaluate their merits.[137]

Ultimately, the judge in *Cauthron* ruled DNA evidence admissible and simultaneously ordered Cellmark to hand over its population database, but did not specify the form in which it should be presented. Defense witnesses were upset when they received a paper printout of the database, presented in the form of tables, rather than a computer copy that could be used to run statistical tests for Hardy-Weinberg equilibrium and linkage equilibrium.[138] As a result, the defense could not challenge DNA evidence in this case, which linked semen found on the clothing of five rape victims to the defendant. Cauthron pled guilty to several counts of rape and was sentenced to a long prison term, but then appealed the case.[139] In 1993, the Supreme Court of Washington reversed the 1989 decision and sent the case back to the lower court, arguing that while the results of DNA evidence are themselves legally admissible, the trial court judge had not adequately examined the admissibility of the statistical evidence of a match in the case. At retrial, Cauthron was convicted once again on all seven counts, but he subsequently appealed this sentence on the grounds that he was unfairly sentenced.[140]

The defense witnesses in *Cauthron* finally got their chance to examine the raw data underlying Cellmark's Hispanic and Caucasian databases in two subsequent cases, *People of California v. Axell* (1 Cal.Rptr.2d 411, 1989/1991), which took place just a few miles from the University of California-Irvine campus, and *State of Delaware v. Pennell* (584 A.2d 513, 1989).[141] Although these cases will not be explored in detail, it should be noted that the persistence of the defense led to the discovery that Cellmark's Hispanic and Caucasian population frequency database did not appear to be in Hardy-Weinberg equilibrium. Upon analysis, Mueller and Geisser found that there were significantly more homozygotes in the database than expected by chance alone, which suggests that there is some sort of substructure in the population.

Although Cellmark argued that this phenomenon was more apparent than real (they claimed that it was the result of many of the smallest alleles running off of the gel after the long period of time needed to fully separate the various bands), the company ultimately discarded its original 600-person Caucasian database in favor of a much smaller 250-person database.[142] When Cellmark introduced this new database in *Pennell*, they argued that it produced much better results than the larger one and that the percentage of homozygotes was in line with what one would expect if the population was in HWE. When cross-examined by the defense about the basis for this statement, though, Cellmark's population geneticist, Lisa Forman, could not produce any notes or calculations because she did the analysis on a computer that lacked a "functioning printer."[143] The judge in *Pennell* ultimately ruled that while Cellmark's procedure for matching DNA samples were generally accepted as reliable, "the statistical probabilities . . . have not been demonstrated to be reliable based upon the evidence adduced to this point to the extent that such large numbers [in this case the probability of a random match was 1 in 180 billion] should be expressed to the jury with their potential for an extremely prejudicial effect." In *Axell*, similar testimony was heard from both sides. The judge in the case also ruled that Cellmark's procedures were reliable and stated that any problems with the statistical probability calculations in DNA typing were seen to affect the weight of the evidence rather than its admissibility.[144] In both cases, although problems with population genetics and the statistics used to calculate the probability of a random match were noted, the convictions were subsequently upheld on appeal.[145]

JUDICIAL CONSERVATISM

The rulings in *Castro*, *Schwartz*, and the other cases described were remarkably conservative given the issues that the defense teams raised in both cases. All focused on the inadmissibility of specific test results rather than the shortcomings of the technological system as a whole. As such, they provided no guide for judges in other jurisdictions who may have felt compelled to rule DNA profiling generally inadmissible into evidence. Only specific results could be prevented from entry into the courtroom.[146]

The defense community did succeed, however, in convincing judges that it was crucial to ask what could possibly go wrong with DNA profiling in the forensic context, rather than simply looking for evidence that the technique was generally accepted in the scientific community for research and diagnostic purposes. This shift meant that experts increasingly had to answer questions not just about certainties, but also about the uncertainties associated with the technique. In the process, the very notion of what counted as expertise was significantly altered to include people who were more likely to testify on behalf of the defense.

The defense scored another major victory in the increasingly negative light in which the popular press portrayed DNA profiling after *Castro* and *Schwartz*. The technique went from being considered foolproof to potentially fallible almost overnight. To give three examples of this trend, the *Washington Post*'s article about the *Castro* case was headlined "A Smudge on DNA Fingerprinting? N.Y. Case Raises Questions About Quality Standards, Due Process," while the *St. Louis Post-Dispatch* reported "DNA 'Fingerprinting' Questioned: Geneticist Says Test May Be Less Reliable Than First Believed."[147] On the "CBS This Morning" television program, anchor Faith Daniels introduced a segment on forensic DNA analysis as follows: "More and more court cases are being decided on the basis of genetic evidence, what's called *DNA* fingerprinting. It's often treated in court cases as indisputable evidence, but is it really reliable?"[148] Proponents of the use of DNA evidence could no longer declare that the technique was infallible without inciting at least a small amount of skepticism on the part of judges, not to mention the general public.

As a result of *Castro* and *Schwartz*, the private DNA laboratories, as well as the FBI and numerous other professional organizations, lawyers, politicians, and scientists stepped up their efforts to ensure that these fiascos were not repeated and that the credibility of DNA evidence was not further eroded. Because forensic

DNA profiling was a hybrid technology involving forensic scientists, academic biologists, statisticians, lawyers, judges, corporate executives, and others, there was no clear answer to the question of who should write protocols, set standards, create guidelines, or develop rules for interpreting test results. There was even less agreement on how these instructions should be enforced or even if they should be officially enforced at all. While all of these communities made useful recommendations and suggestions, no definitive decisions could be made because they did not share a uniform definition of who had the authority to set rules on DNA profiling.

CHAPTER 5

Public Science

By mid-1989, Lifecodes' and Cellmark's claims about the validity and reliability of their DNA evidence had been seriously challenged, although not totally undermined, by a nascent network of defense lawyers and academic scientists. In admissibility hearings in *Castro, Schwartz,* and other cases, defense experts argued with some success that the private companies' DNA typing regimes were fundamentally flawed, both in design and practice. To begin with, they claimed that the private companies had rushed their DNA evidence to court and in the process evaded adequate peer review by all of the relevant scientific communities. They also argued that Cellmark and Lifecodes had shielded their protocols and probes from rigorous scrutiny by claiming that they were proprietary. Although judges most likely could have ordered full disclosure of all information pertaining to the two companies' DNA profiling systems, in the first few cases in which the defense challenged the validity and reliability of DNA evidence, they did not. Instead, the private laboratories made these materials available only to a small group of trusted scientists who formed mutually beneficial research collaborations with the companies. As a result, the defense argued, there were still serious flaws in the techniques and methods used by Lifecodes and Cellmark.

Most important, though, defense lawyers and their experts (especially Eric Lander) lamented that the work of Lifecodes and Cellmark was fundamentally unscientific because there were no common technical, procedural, and interpretative standards that could be used to evaluate specific DNA typing results in court. The defense community did not seem particularly bothered that such standards were generally not universal or explicitly stated in the academic scientific community or that academic scientists routinely violated well-defined technical protocols and procedures when they felt justified. They seemed to harbor an implicit belief that academic scientists could be trusted while corporate scientists (or scientists working in service of a particular client, and not the truth) could not. Well-defined standards agreed upon by the scientific community as a whole were necessary to tell whether a private lab had done its analysis properly and with the appropriate level of scientific rigor, since one could not assume that they would do so on their own. Although the private companies pointed out that they had undertaken significant efforts to validate and standardize their systems before coming to court for the first time, the defense community argued that proper validation and standardization could only be carried out by disinterested scientists outside of the laboratories actually doing the DNA testing.

A significant portion of the law enforcement and forensic science communities also shared these concerns. In addition to the need to shield their evidence from claims that there was no generally accepted standardized methodology for creating and interpreting DNA profiles, law enforcement agents and prosecutors hoped that they would eventually be able to send their crime scene samples to a public crime laboratory like they did with other forms of evidence. This situation would not only reduce the cost of DNA testing to police organizations but also enable the law enforcement community to take control of the technique from private industry. Further, as more and more police departments and prosecutors began to use DNA evidence on a regular basis, there was increasing hope that a nationwide DNA databank, similar to that created for fingerprints, could be developed in a few years' time. Thus, by 1989, two issues were becoming increasingly important in the debates over forensic DNA evidence: standards and who had the authority and expertise to set them.

Because the Federal Bureau of Investigation (FBI) was almost synonymous with forensic science in the United States, many people within the forensics community looked to them to develop procedural standards and standardized materials that would be used in forensic DNA laboratories around the country. Although other federal organizations, such as the National Institute of Standards and Technology (NIST) and the Food and Drug Administration (FDA) were mentioned as possible regulators of DNA profiling by people outside of the forensics community, the FBI quickly emerged as the agency that would carry out this task. The FBI sought to develop and implement a standardized DNA typing regime that could withstand the rigors of the adversarial American legal system, compete with Lifecodes' and Cellmark's system for market share, and serve as the basis for a nationwide DNA database.

The FBI's central task was to create a network of crime laboratories, scientists, and technicians who all followed the same set of rules and used the same materials, reagents, and techniques to carry out their analyses. Although several organizations weighed in on the issue of standards, because the FBI had traditionally served as a source of standards in the American forensic science community, it was able to use the private companies' missteps and the promise of a nationwide DNA databank to ensure that its DNA typing regime became the standard by the early 1990s. In doing so, however, the FBI would become vulnerable to the same sorts of defense challenges that plagued the private laboratories, especially with respect to the issue of what constitutes adequate scientific peer review.

The standards that the FBI produced were shaped by, and ultimately institutionalized by, the bureau's own value systems and views of how to regulate forensic science. Specifically, the FBI believed that only the forensic community could evaluate and regulate forensic science because the circumstances under which forensic laboratories operate are unique to the criminal justice context. Thus, the FBI took Scheck and Neufeld's argument about the nature of forensic casework vis-à-vis medical research and diagnostics (i.e., that it was fundamentally different) and used it to buffer themselves not only from the intervention of Lifecodes and Cellmark, but also from many of the defense experts that Scheck and Neufeld worked so hard to cultivate. FBI officials argued that nobody outside of the forensic community could adequately understand the conditions that made forensic casework a special case: the nature of forensic samples, the working relationships that forensic analysts have with criminal investigators, and the necessity that evidence stand up in court. For the same reasons, the FBI also believed that only other members of the forensic community could determine the best way to conduct proficiency testing, individual credentialing, and laboratory accreditation to ensure the reliability of evidence being produced by forensic laboratories.

Thus, although the FBI claimed numerous times that their standardization efforts were purely technical, the bureau explicitly sought to simultaneously construct technological systems and social systems that would work together to ensure the validity and reliability of the forensic DNA evidence. In doing so, the FBI somewhat reluctantly made itself the obligatory passage point for both technical and social exchanges.[1] While this system proved to be extremely robust, many members of the defense community complained that the bureau had woven its shortcomings—especially insularity and an unwillingness to listen to people outside the forensic science community—into the fabric of the technology. Recent DNA laboratory errors and scandals such as those in the Houston Police Crime Lab and the FBI laboratories demonstrate that these problems are not merely academic quibbles. They have had real consequences and have caused errors in cases where peoples' lives were at stake.

Although the FBI's role in the development and regulation of DNA profiling may seem inevitable in retrospect, it is important to realize that very different routes could have been taken with enough political will. The most important alternative to FBI control over the technique can be seen in New York's efforts to ensure that only reliable DNA evidence was introduced into courtrooms around the state. For a variety of reasons, the introduction of DNA evidence into New York courts occurred at a much slower pace than in other states, partly because defense lawyers in the state were quicker to challenge forensic evidence than those in many other states. Indeed, New York courts had been the site of bruising battles over the reliability of numerous forensic techniques over the past few decades. As a result, both prosecutors and defense attorneys agreed (in what the *Manhattan Lawyer* described as a "rare instance of harmony") that "a slew of scientific and legal questions must be answered before DNA testing becomes as accepted as the fingerprint."[2]

In an unusual stance for the forensic science community, the New York State Crime Laboratory Advisory Committee (NYSCLAC), an organization of crime laboratory directors from around the state, took the lead in putting together a diverse group of people to set regulations for DNA profiling carried out by both public and private labs in the state. In a 10 November 1987 letter to Laurence T. Kurlander, New York State's commissioner of criminal justice services, the chairperson of NYSCLAC advocated strong state regulatory control over DNA profiling:

> There is so much potential for benefit to the criminal justice system that great care and careful planning are clearly required to insure that premature or improper application of the technology do not destroy its credibility in court.... Because of the importance and the technical and economic difficulties of proper application of DNA technology, I feel that the State should closely oversee this critical area. Overzealous police, prosecutors or labs should be discouraged from application to the wrong case or application of methods that cannot be shown to meet well defined standards of acceptance in the scientific community.[3]

The chairperson, Howard Harris, went on to report that he was planning to invite a wide variety of active participants to the committee's next meeting in order to discuss what should be done about DNA profiling. He noted that the committee hoped to "have representatives of the eager and cautious persuasion to give both sides of the story."[4]

This concern for inclusiveness continued as the state took a larger role in regulating DNA profiling, starting with the diversity of the committee set up by the governor to determine what path the state should take. Unlike the FBI, which sought to explicitly marginalize most nonforensic scientists, the New York State Panel on DNA Fingerprinting included members from a wide range of disciplines: three crime laboratory managers, two Ph.D.-level molecular biologists, two prosecution lawyers, two defense lawyers, two law professors (one with a defense perspective and one with the perspective of the state criminal justice bureaucracy), two law enforcement officials, one judge, and a representative from a state police association. Only representatives of population genetics and statistics were missing.

The final *Report of the New York State Forensic DNA Analysis Panel* was published on 6 September 1989 did a remarkable job of integrating the concerns of all of the various participants into a single coherent document, especially in light of what one participant told me was a high degree of combativeness among the broad range of stakeholders involved in the deliberations.[5] One of the most striking features of this report was the critical stance it took on DNA typing. Although it was quick to point out that the technique had "untold potential for criminal justice," this optimism was tempered by the reality that while DNA technology "capture[s] the imagination," it is, without a doubt, "new science in the making."[6] As a result, the panel overwhelmingly advocated a cautious approach, which was, as previously mentioned, already the path taken in New York, and outlined potential problems in the legal, scientific, and political realms.

In its recommendations—which included the establishment of a state board to set minimum procedural standards, minimum quality control and quality assurance standards, minimum qualifications for laboratory workers, and a system of accreditation for all laboratories proffering DNA evidence in New York courts—the panel explicitly called for active participation from all stakeholders in the criminal justice system, including defense attorneys and legal academics. In general, the panel recognized the need to open up the peer review and regulation of DNA profiling to people outside of the forensic science community. In their view, having scientists who were familiar with forensic science but not a part of it was a valuable way to ensure the credibility and reliability of DNA evidence. It also seemed to be a way to ensure the credibility of the board.

Remarkably, many of the most far-reaching recommendations of the panel were eventually put into place in New York State after the passage of Executive Law Article 49-B, entitled "Commission on Forensic Science and Establishment of DNA Identification Index" in July 1994.[7] That said, for reasons that will become very clear later, the FBI was incensed that it was going to be placed under the authority of a state regulatory agency, especially one that was populated so heavily by nonforensic scientists. After intensely lobbying New York governor Mario Cuomo, which stalled passage of the law for some time, a compromise was reached whereby the state's regulations did not apply to "any laboratory operated by any agency of the federal government, or to any forensic DNA test performed by any such federal laboratory."[8]

Decision Making at the FBI

Now that we have seen what kind of approach the FBI rejected—that is, one that involved state oversight of the production of DNA evidence, led by a diverse group of scientists, lawyers, judges, law enforcement agents, and academics—we will see what the FBI considered to be a more acceptable alternative. It is crucial to realize that this process was not one of bringing together existing stakeholders and getting them to agree on a unified set of standards to use in their laboratories. Rather, they had to define the very actors who would fit into the network that they were in the process of creating. The bureau decided largely by fiat that forensic scientists based in public crime laboratories bore the primary responsibility for setting standards for, and carrying out, forensic DNA analysis. Ironically, these actors generally had little or no experience with molecular biology or genetics and had to be trained from scratch by the FBI to deal with forensic DNA analysis.

Conversely, the most integral actors in the early history of DNA typing— namely, the scientists employed by Cellmark and Lifecodes, defense lawyers, and the nascent group of defense experts who challenged DNA evidence—were actively excluded from the FBI's network. The FBI made no attempt to engage them or to build bridges between the forensic science community and other obvious stakeholders. They simply excluded everyone they did not want in their network, leaving only organizations and individuals who shared their particular commitments about forensic science and who would do things their way. Ultimately, public DNA labs came into being as a part of the standardizing process.

In December 1988, more than a year and a half after Cellmark and Lifecodes entered the DNA identification market, the FBI began offering forensic DNA analysis to law enforcement agencies around the country. The FBI's interest in the possibility of using genetic marker analysis in forensic casework, however, began in late 1984, at about the same time that Alec Jeffreys invented DNA fingerprinting. Their first foray into this field did not show much promise and was abandoned.[9] In late 1985, however, scientists at the FBI Academy's Forensic Science Research and Training Center (FSRTC), based in Quantico, Virginia, became aware of Alec Jeffreys's recently published work on the discovery and use of probes that targeted hypervariable regions of the genome in forensic casework. At this point, James J. Kearney, who was assistant section chief of the FSRTC, assigned FBI scientist Bruce Budowle to travel around the United States and England to try to "determine what was actually going on with DNA testing."[10] Budowle had recently joined the FBI after receiving his Ph.D. in genetics from Virginia Tech and soon dominated the development of DNA technologies at the bureau.

After more than a year spent visiting Jeffreys's laboratory, various English crime laboratories, Cellmark's facilities in the United States and the United Kingdom, as well as Lifecodes, Cetus (a biotechnology company that was developing PCR for commercial applications), and numerous university laboratories in both countries, Budowle reported back to Kearney about the various protocols, methodologies, and probes being used.[11] According to Kearney, "the more we heard about it, the more excited we got."[12]

Based on Budowle's report, in mid-1987 FSRTC decided to put together a research team that would conduct experiments to improve the efficiency and reliability of existing DNA typing methodologies used by other institutions and companies and to simplify them for transfer to state and local crime laboratories.[13] One of the DNA research group's first accomplishments was to compile a comprehensive list of fourteen "validation protocol steps that should be ascended at the research level before DNA typing techniques will be certified for use on case evidence in our laboratory." More than half of this road map, which was based in part on the FBI's process for validating traditional blood protein analysis, dealt with the preimplementation development of the technique itself. These steps included choosing and perfecting a particular method on both ideal and forensic samples, evaluating its reproducibility both within and between labs to ensure that similar results could be achieved no matter where the technique was being used, establishing allele distribution databases to calculate the probability of a random match between two samples, and carrying out experiments to determine the effects of sample age and environmental degradation to ensure that the system was compatible with forensic casework. The last several steps dealt directly with the issue of establishing validity of the FBI's entire infrastructure for doing DNA testing on actual forensic samples. These steps included the following:

- "Examine DNA profiles in non-probative evidentiary stain materials as examples of stains that are likely to have been exposed to a wide variety of adventitious substances and climatic extremes. Because all possible contaminants and environmental conditions that might affect DNA typing profiles cannot be addressed experimentally, special attention will be directed to a comparison of DNA profiles derived from victim's liquid blood versus victim's blood deposited on typical crime scene substrate."
- Set up mechanisms for the on-site validation of methodology.
- "Publish results of experimental studies in peer reviewed journals and present data at scientific meetings. (These mechanisms provide a forum for public criticism of the methodologies.)"
- "Train case-working unit personnel in all aspects of the performance of DNA typing methods and interpretation of typing results."
- "Establish formal training sessions for, and engage in collaborative testing procedures with, scientists in state and local forensic science laboratories. This program will not only serve to broaden the base of laboratories capable of performing the analyses, but also will demonstrate that the methods are technologically stable."[14]

It is clear from these steps that the team intended to take an extremely conservative approach to validating its technology for use in the FBI laboratory, as well as public crime laboratories around the country. In doing so, they hoped not only to "establish the scientific validity of DNA typing methods as they are applied in the examination of evidentiary materials," but also to "directly address" the requirements of the *Frye* standard.[15]

Even the most ardent critics of the rapid implementation of DNA typing in forensic casework complimented the FBI on its stated desire to take a systematic and methodical approach to validation. Eric Lander, for example, stated in March 1989 testimony before the U.S. Congress that he was impressed with the approach taken by the FBI in setting up their DNA typing scheme and that he hoped the bureau would continue to maintain those high standards as they brought cases to trial. In his view, "the FBI has been extremely open and solicited comments and criticism from the scientific community. This sort of openness is how science moves forward, and I'm very happy with it."[16] Similarly, Barry Scheck commented at the Banbury meeting that various forensic scientists had told him that the FBI was doing all of the things that it needed to do to ensure the validity and reliability of their technologies and methodology.[17] Scheck and Neufeld's praise of the FBI did not last long, however. As they learned more about FBI's work, it became clear to them that by the time the FBI was actually deciding what technologies to use and what standards to implement, they were open only to the extent that they would listen to anyone who would tell them what they wanted to hear. Further, as the FBI sped up the process of implementing DNA profiling in order to compete with Lifecodes and Cellmark, they did not fully follow through with many of their steps that they initially laid out. For instance, they did not publish the results of their experimental studies in peer-reviewed journals until more than a year after they began accepting casework from law enforcement agencies around the country.

By early 1988, the members of the FBI's DNA research group felt that they had made enough progress in their preliminary investigations, and they published a summary of their work in the January 1988 edition of *Crime Laboratory Digest*. In this article, the group "introduced crime laboratory personnel to the principles of DNA analysis that enable the detection of [genetic polymorphisms]."[18] They began by outlining the basic properties of DNA, probes, restriction enzymes, and the Southern blot process. The article then described single-locus probes and pointed out that these probes "have been applied successfully to the examination of body fluid by Giusti *et al.* (1986) and Kanter *et al.* (1986)." Despite the fact that both of these references were to Lifecodes' initial publications about their work, no explicit reference to the company was ever made, and Cellmark was only mentioned in the context of Jeffreys's probes.[19] In this article, the DNA research group also listed the fourteen steps to validation described earlier.

Because *Crime Laboratory Digest* is an FBI-published research journal dedicated to communicating the latest findings in forensic science to state and local crime laboratories around the country, its primary readership is forensic scientists and technicians in public crime laboratories. The journal had very little circulation in the academic community and was certainly not read by many of the scientists involved in the debates over the validity and reliability of DNA typing and associated statistical calculations. The issue of peer review and circulation of ideas would come to play a crucial role in the defense challenge to the FBI's DNA evidence and several cases, most notably *Yee*.

In their initial article, Budowle and his colleagues made absolutely no mention of the fact that both Lifecodes and Cellmark had already begun offering forensic DNA analysis to law enforcement agencies and prosecutors around the country. Yet, as FBI official John Hicks commented several times in the late 1980s, "developments in the private sector caused us to accelerate our efforts."[20] In a March 2003 interview, Hicks stated that the bureau was approached on several occasions by representatives of Cellmark and Lifecodes, who offered to place trained technicians in the FBI laboratory to begin immediate casework.[21] The FBI, however, continuously attempted to distance themselves from directly advocating one particular company's method over another. "Always in government practice," he stated, "you don't want to get into a position where you're promoting a particular product or service."[22] That said, Hicks and several other FBI scientists reiterated to me several times that although the privatization of the DNA typing realm was a concern, it was certainly not considered important enough to cause the bureau to begin doing DNA testing before they were absolutely ready to do so.[23]

TECHNICAL CHOICES AND MARKET CONSEQUENCES

In their *Crime Laboratory Digest* article, Budowle and colleagues gave no indication that they were in the midst of making an ostensibly "technical" decision that would irrevocably alter the landscape of the DNA typing market forever. Specifically, the FBI team decided to adopt another restriction enzyme (HaeIII) than either of the ones used by Cellmark (Hinf I) or Lifecodes (Pst I). On the surface, this decision seems rather mundane, or even trivial, since HaeIII was just as widely used in molecular biology as the enzymes used by the two private companies. Its importance only becomes clear when one realizes that the distinguishing characteristic of any DNA typing system is the size of the fragments detected by the various probes, and that fragment size is completely dependent upon the choice of restriction enzyme. Thus, by choosing HaeIII over the enzymes used by Lifecodes or Cellmark, the FBI made their system incompatible with either company's existing DNA typing regimes. While both companies were free to adopt the new enzyme system and could still sell their products to public crime laboratories, it meant that all of their validation work would have to be redone and that their systems would no longer be automatically accepted as a matter of precedent in numerous jurisdictions.

Explaining the choice in *Crime Laboratory Digest*, Budowle and his colleagues provided extensive scientific and technical justification. For instance, they noted that because HaeIII cut DNA more frequently than Hinf I or Pst I, it would create much smaller fragments. The advantage of smaller fragments is that they would tend to spread out more than larger fragments during electrophoresis, making it easier to determine their exact size.[24] The only nontechnical justification they provided for choosing Hae III was that it was relatively inexpensive.[25]

According to Dan Garner, Cellmark USA's first laboratory director, and several other sources I spoke to, Cellmark, at least, was deeply disappointed by the FBI's decision. As the FBI began to enter the DNA typing arena, Garner said that Cellmark attempted to persuade them to adopt the Hinf I system, since it was already being widely used in the United Kingdom and Europe. This choice would not only have been a major business coup for Cellmark, it also would have allowed immediate international compatibility. However, at least in Garner's opinion, the FBI's decision was not based entirely on logic or scientific merit. He said, "I think [Budowle] was just trying to slow us down to let the FBI catch up, and then let it take a leadership role."[26]

Whether the decision to go with Hae III was an explicit attempt to gain control of the DNA typing realm or not, this was the ultimate outcome of the decision. It is important to note, however, that the FBI did not wish to put either company out of business. Indeed, the FBI team chose to use probes, reagents, and other products produced by Lifecodes and Cellmark, as well as several other companies (it should be noted that the FBI used three probes produced by Life-codes, but only on made by Cellmark). Lifecodes and Cellmark, however, were no longer in a position of fighting to become the industry standard. From 1989 on, they increasingly found themselves following the FBI's lead.

<h2 style="text-align:center">TRAINING AND DISSEMINATION</h2>

By April 1988, the FBI's DNA typing regime was complete enough to begin train-ing the first members of what would become the FBI's casework-oriented DNA Analysis Unit. During this period, most FBI trainees, who came from other parts of the laboratory, lacked the basic scientific knowledge necessary to understand genetic profiling, so they first had to be provided with a forty-four-hour short course on molecular biology, taught by members of the University of Virginia fac-ulty on a contract basis (they did not participate in the creation or testing of the FBI's DNA profiling technology). Trainees then attended a series of lectures on DNA typing presented by the DNA research team and engaged in significant practice with the actual techniques.[27] Soon after receiving initial training, these analysts became central to the process of making sure that the various technolo-gies and techniques chosen by the FBI were valid and reliable for use in forensic casework. This process was very similar to that carried out by both private com-panies and focused on making sure that the results of the test would not be affected by external environmental factors at the crime scene, such as sunlight, temperature, substrate from which biological sample was recovered, and time elapsed since the crime took place. Experiments were also done to determine the impact of various biological and chemical insults that a crime scene sample may be exposed to, such as bacteria, detergent, gasoline, or organic solvent. The final process of validation was to perform forensic DNA analysis on the remnants of 122 actual cases that had previously been submitted to the FBI for serological examination. This research was published in the October 1988 edition of *Crime Laboratory Digest*. In this article, Dwight Adams concluded that "the results strongly support the FBI Laboratory protocol for DNA analysis by RFLP tech-nique as valid, reliable, and reproducible."[28] No special mention was made of the fact that the FBI used technicians straight out of training to validate their system.

Beginning in April 1988, the FBI also began offering a four-month training course at the Forensic Science Research and Training Center dedicated to teach-ing scientists and technicians from state and local labs the basics of molecular biology and genetics, as well as to give them hands-on experience with DNA typ-ing techniques. The main purpose of this endeavor was not only to transfer the FBI's technology to state and local crime labs, but also to "provide the technical resources to address the validity and reliability issues associated with DNA test-ing as quickly as possible."[29] According to several state and local crime laboratory directors I spoke to, whatever they thought of the FBI's choice of technology, they were impressed by the bureau's ability to rapidly disseminate their DNA typing system across the country. As Jan Bashinski stated in a February 2002 interview:

> I'm astounded at how well-controlled the RFLP technology ultimately was and how comparable the data was across—because I did electrophoresis for years, and it's not that straightforward. So, they did really a yeoman's job of getting that whole technology into a really controllable package. In the process, however, they made a lot of decisions—they had to make decisions—but, in many people's view, in somewhat arbitrary ways, and they did not necessarily pay enough attention to the valid and creative input of other people. And so, that's just two sides of the same story. But, they should be given a lot of credit for taking the bull by the horns, and creating a framework within which some standardization could occur.[30]

Similar to the role that FBI analysts played in the validation of the bureau's protocol, the Visiting Researcher program provided the FBI with the labor it needed to perform experiments to determine which procedures and protocols should be implemented nationwide, as well as the development of population frequency databases. Hicks said that, at the time, the FBI "didn't have the resources . . . to do that quickly, so this was a way to address the resource issue, and like I said, they were not only trained on the technique, but gained significant experience and refined their personal technique with the use of the procedure."[31] Hicks stated in a March 2003 interview that he did not find it troubling that the very same people with little or no experience with forensic DNA analysis were also the individuals performing the experiments that were used to optimize the bureau's testing regime.[32]

Also in April, the FBI decided to publish a special supplemental issue of *Crime Laboratory Digest* dedicated to "DNA Implementation," which described the basic aspects of forensic DNA analysis, gave an idea of the cost and challenges associated with implementing the technology at the local level, and listed various articles and books that dealt with DNA analysis. The first page of the issue pointed out that it was being distributed to approximately three hundred crime laboratories across the country. No mention was made of this issue being sent to members of the academic scientific community, or even forensic scientists outside of the major public crime labs.[33] This meant that the same group of people who created the FBI's DNA profiling regime, and were its primary users, would also bear a large part of the responsibility for peer reviewing and critiquing it.

This is not to say, however, that the FBI did not seek any outside counsel in developing their DNA profiling system. FBI officials knew that they had to have the support, although not necessarily the input, of people outside of the crime laboratory community in order for their technique to gain acceptance in the courtroom. In this supplement, the FBI also announced plans for an FSRTC-sponsored seminar on DNA technology to be held from 31 May to 2 June 1988. Ultimately, they invited more than one hundred forensic scientists, molecular biologists, geneticists, law enforcement agents, and lawyers to attend the meeting. Because there was as yet no significant critique of DNA profiling in either the legal or the scientific community (the Banbury meeting did not occur until November 1988), there were no participants who were inherently inclined to criticize the FBI and its work, with the possible exception of representatives of the private laboratories. Among the participants were David Housman from MIT, who regularly testified on behalf of the prosecution in DNA cases; Raymond L. White, a geneticist from the University of Utah School of Medicine who discovered many of the probes ultimately used by the FBI; Russell Higuchi, from Cetus; Ian Evett, from the British Home Office; Brian Parkin, from the Met Police Lab in London; Alexander Markham, from Imperial Chemical Industries (ICI); Ivan Balazs, from Lifecodes; Ed Blake, from Forensic Science Associates; Robin Cotton, from Cellmark; Dale Dykes, from the Memorial Blood Center in Minneapolis; George "Woody" Clarke, from the San Diego District Attorney's office, who became one of the most knowledgeable prosecutors in the country on DNA-related issues); and several FBI scientists. Topics at this meeting were heavily weighted toward the technical end of the spectrum. Clarke was the only speaker to address issues of law and ethics in his brief talk on some of the civil liberties issues raised by the creation of a networked national DNA databank, in which DNA profiles would be stored from convicted offenders and victims in unsolved cases together with basic information such as which law enforcement agency uploaded the profile.[34] This setup would allow law enforcement agencies from around the country to compare DNA evidence left at a crime scene with known violent offenders and also check to see if perpetrators of the crimes they were investigating could be linked to any other crimes around the country, potentially speeding the identification and capture of a serial offender.

According to Roger Castonguay, who was the assistant director in charge of the FBI Laboratory, two important topics were to be discussed at this meeting: the establishment of standards for DNA typing within the forensic science community and the creation of a nationwide DNA databank.

> Clearly it would be advantageous to law enforcement on a national scale to coordinate the development of such systems with the establishment of appropriate controls and standards to permit the effective exchange of DNA identification profiles. To accomplish this, the community must agree upon standards which provide a common language and thereby facilitate the exchange of critical investigative information and, at the same time, build flexibility into this system to accommodate changes as this technology continues to evolve. The success of this effort will depend heavily on a strong professional commitment to the needs of law enforcement along with a spirit of cooperation and mutual support within the forensic community.[35]

Again, no mention was made of either Lifecodes or Cellmark, or the fact that they were already using different markers and restriction enzymes in their DNA typing regimes even though representatives of both companies were present at the meeting. This omission suggests that the FBI did not invite them to obtain valuable input, but because it would look bad if they did not.

The Technical Working Group on DNA Analysis Methods

As apparent from Castonguay's comments, the issue of standardization loomed large in the minds of FBI scientists. What they needed was a quick and efficient way to achieve this goal. Several FBI forensic scientists—most importantly, John Hicks—realized that the notion of a databank "met several critical needs":

> One is, clearly if you are going to build a databank, you had to have standard-
> ization of your methods and techniques so the data was compatible. Second, in
> any type of databank, you want to have a high emphasis on quality control to
> be sure you have reliable data in the system. . . . Thirdly, it would clearly be a
> potentially very powerful investigative tool.[36]

Perhaps the most important development to come out of the May–June 1988 FSRTC-sponsored meeting was the formation of the Technical Working Group on DNA Analysis Methods (TWGDAM), an informal gathering of forensic scientists who met for the first time in November 1988. According to the group's mandate, the primary purpose of TWGDAM was to "reach a consensus as to the DNA methodology to be used in the nation's crime laboratories in order to provide a medium to exchange DNA testing data." In an interesting display of a general trend against strict regulation, and for self-policing, in the forensic sciences, TWGDAM was not given "any regulatory authority nor is it a certifying or accrediting body. The results of the TWGDAM activities or publications produced by the working group should be considered as suggestions or guidelines to assist individual crime laboratories in the establishment of their DNA programs."[37] Although I will discuss this issue further in the next chapter, for now it should be noted that TWGDAM was explicitly conceived of as a forum for members of the forensic community to develop standards that would lead to a nationwide DNA database. The idea was that voluntary adoption of the guidelines, rather than explicit adoption of regulation, was the only way to proceed. As FBI Director William Sessions wrote in mid-1989, "The FBI has no plans to initiate or otherwise become involved in licensing or certification processes for DNA testing laboratories. It is the position of the FBI Laboratory that a strong proficiency testing program as an element in a comprehensive quality assurance program is the most efficient and effective way to assure reliable and consistent test results."[38]

Sessions's stance was somewhat disingenuous, though. What he did not mention in his letter was that the FBI held a tremendous amount of power in the realm of forensic science. In fact, it was the primary source of training, support, standard setting, and technology transfer for crime laboratories and police agencies across the country.[39] Additionally, it provided these services at no direct cost to law enforcement agencies; the taxpaying public picked up the tab. Unlike Cellmark or Lifecodes, the FBI did not have to sell its methodologies or justify its decisions based on some prevailing market logic. Not only did the FBI have a substantial audience that was ready and willing to adopt its products and services, it was also taking the lead on the development of a nationwide DNA database that demanded a single standard DNA typing methodology.

Of the thirty-one participants in the first TWGDAM, eighteen came from public crime labs or law enforcement agencies, including two from Canada; eleven came from the FBI, including three of the four members of the original DNA typing research group; and only two were members of the academic community. The latter category included George Sensabaugh, from the University of California-Berkeley, who was collaborating with law enforcement officials in California, and Ray White, who had developed many of the probes chosen by the FBI for use in their DNA typing regime and was one of their primary scientific advisors.[40] It is important to note that no representatives from either Lifecodes or Cellmark, the two organizations with the most experience in doing forensic DNA analysis, were invited to attend. Although individuals from these companies were consulted by TWGDAM participants, they were actively excluded from having an official voice in the forum created to establish such guidelines.

While the FBI certainly never advocated an explicit regulatory function for itself or TWGDAM, once the bureau made the decision to establish and run a nationwide database, it was clear that any jurisdiction wishing either to load profiles to the database or check the database for matches to an unknown profile recovered from a crime scene would have to conform to the bureau's chosen standard. It should come as no surprise, then, that Sessions reported to U.S. Congressman Don Edwards that TWGDAM was "supporting the nationwide adoption of the FBI's DNA test protocol as a national standard," and that many state and local crime laboratories had announced their plans to use the FBI protocol for forensic DNA testing.[41]

By subsidizing TWGDAM, controlling about a third of the membership, not bringing in representatives from Lifecodes or Cellmark, as well as setting the agenda, the FBI put itself in the position of unofficially regulating forensic DNA analysis. Although everybody I spoke to from the FBI vigorously denied that their goal was regulation, in the sense of setting rules that *must* be followed by everybody, the FBI did just this through the development and promulgation of guidelines and suggestions for protocol validation and the requirement that they be followed if access to the national DNA databank was to be granted. This strategy was necessary for widespread acceptance of TWGDAM's views because forensic scientists bristle at the notion that their work can be regulated from outside in the same way as the work of a diagnostics laboratory or a drug manufacturer. Calling rules "guidelines" or "suggestions" allowed local labs to retain the sense of independence and flexibility needed to accept being regulated in any way, even by fellow forensic scientists.

EARLY CONGRESSIONAL DEBATE

TWGDAM was not the kind of regulatory group advocated by scientists like Lander, the members of the New York State Panel on Forensic DNA Analysis, or defense attorneys like Scheck, Neufeld, or Thompson. They believed that self-regulation in this context was impossible and preferred oversight by people outside of the forensic community. While there was not agreement on exactly who should perform this duty—Lander argued that it was the responsibility of the academic scientific community, while most defense lawyers argued that it should be undertaken by any one of several existing federal agencies that regulated similar technologies—there was general agreement that the FBI could not effectively regulate forensic DNA analysis. As a result, a significant debate emerged over who possessed the necessary authority and expertise to either guide or regulate, depending on one's view of the governance of forensic science, the production of forensic DNA evidence. These disagreements were made public in numerous forums, ranging from the courtroom to the pages of *Nature*, but they were most publicly carried out on Capitol Hill in the meeting rooms of various congressional committees.

Congressional interest in forensic DNA analysis became apparent in March 1989, when the Subcommittee on the Constitution of the Senate Judiciary Committee held its first hearing on "Genetic Testing as a Means of Criminal Investigation" and the House Committee on the Judiciary's Subcommittee on Civil and Constitutional Rights devoted a significant portion of its hearings on the FBI's oversight and funding request for 1990 to the bureau's role in developing forensic DNA typing in the United States.[42] At about the same time, the Senate Committee on Labor and Human Resources commissioned the now defunct Office of Technology Assessment (OTA) to set forth "a range of options for action by the U.S. Congress on five policy issues" surrounding the technique: standards, funding, the establishment of computer databanks of DNA test results, standardization of DNA analysis techniques, and privacy concerns.[43] The OTA had been created in 1972 to provide Congress with nonpartisan information about, and analysis of, complex scientific and technical issues of the day without telling legislators which way they should vote. Although the OTA was widely considered to be a solid research organization, many of their reports angered big business in industry. The OTA was shuttered by the 104th Congress in the wake of the "Gingrich Revolution."[44]

The OTA staff, headed by Robin Y. Nishimi, was under no pressure from Congress to secure a particular point of view, so they were able to assemble a diverse group of individuals to assist them in writing the report, each of whom brought unique interests and perspectives to the table. Participants included Michael

Baird, from Lifecodes; Eric Lander; Lisa Foreman, from Cellmark; Janlori Gold-man, from the American Civil Liberties Union; Henry C. Lee, the forensic scientist from Connecticut who became a household name during the *Simpson* trial; Michael Rutnik, the defense attorney in *Wesley-Bailey*; and various lawyers, forensic scientists, and law professors. Tom Caskey, who would ultimately work closely with the FBI on the development of the DNA profiling system currently used today, headed the advisory panel.[45] Unlike TWGDAM, no representatives of the federal government or the FBI were on the panel. Although the OTA report had little impact on policy decisions that would be made in subsequent years—at least partially because the FBI ultimately had no incentive to seriously engage with the report when setting standards—more than any other single document or report, it captured the controversy that was brewing over who possessed the authority and expertise to regulate DNA profiling.

After outlining the various debates that had emerged over the validity and reliability of the actual technique of DNA typing, as well as the population genetics issues involved in declaring a match, the OTA report moved on to what it claimed were more contentious issues. According to the OTA, setting standards for DNA testing was the "most controversial and unsettled" issue. The OTA identified two types of standards: technical (i.e., "proper reagents and gel controls; electrophoresis conditions; rules to match DNA banding patterns; the extent that computer-assisted matching should be permitted; and population data to compute the likelihood of matches") and operational (i.e., record keeping and proficiency testing). Perhaps because the full effect of the FBI's decision not to adopt the restriction enzymes used by either company had not yet become apparent, the OTA believed that operational standards were more controversial because "historically, attempts to regulate laboratory practices in any sector have met with resistance." As such, the OTA stated that "technical standards that allow flexibility for laboratory-to-laboratory variations need to be evaluated. Clearly defined rules and procedures—objective and scientifically based—should be established, set, and, most importantly, followed."[46] Thus, without directly saying so, the OTA report suggested that the kind of regulation created by the FBI (guidelines and suggestions that did not necessarily have to be followed) was highly problematic and needed to be fixed.

With respect to quality assurance, the OTA addressed regulatory—either by professional societies or federal and state governments—and nonregulatory means of policing forensics labs. While the report did acknowledge that it was within the purview of the federal government to step in and regulate DNA testing laboratories, it concluded the section on regulation in a less than positive light: "Some feel, however, that legislation like the Clinical Laboratory Improvement Amendments of 1988 . . . is more of a short-term solution—that, in fact, court conflict, as is presently occurring, sharpens the examination and evaluation of forensic DNA typing and will ultimately ensure quality by defining its boundaries. Moreover, questions are raised whether high quality necessarily follows from mandatory regulation."[47] OTA then proceeded to outline some nonregulatory actions that the federal government could take, including authorizing research projects in the forensic sciences, particularly those that are cross-disciplinary and "apply newly emerging basic research tools to real-world casework"; encouraging "consensus conferences" to develop and recommend protocols as well as to address outstanding controversies in forensic science; and creating oversight committees like the NIH Recombinant DNA Committee to monitor forensic DNA testing.[48] Once again, the OTA report suggested that the process used by the FBI to develop standards (i.e., the closed world of TWGDAM) was simply not adequate to create an effective long-term solution to the problem of ensuring valid and reliable DNA evidence.

The OTA report also pointed out that there were already various nonregulatory structures in place to deal with issues in forensic DNA testing, including the National Institute of Standards and Technology, a neutral Federal agency that has the potential to instill confidence in the techniques involved in DNA testing. Another nonregulatory federal initiative mentioned by the OTA and already in operation at the time was the FBI's TWGDAM. While some praised TWGDAM as being a "nucleus around which national expertise will develop," others portrayed it as not sufficiently open to noninterested parties.[49] In concluding this section, OTA acknowledged that "nothing is routine during the course of a forensic investigation," and therefore any quality assurance program must be flexible and not place undue burden on the already overwhelmed and underfunded public laboratories. Finally, they argue that any solution must ultimately rest on the implementation of standards rather than certification, training, or licensing of technicians.[50]

Because the mandate of the OTA was to provide Congress with policy options, not recommendations, the report concluded with the various courses of action that could be taken on the five major policy issues surrounding DNA typing. The suggestions ranged from taking no action, to encouraging voluntary action, to encouraging the use of a formal, open consensus committee, to enacting legislation. As I have shown thus far, each of these positions had advocates throughout the country. Up until now, though, debates over the best way to ensure the validity and reliability of DNA evidence were decentralized, taking place in various forums in California, Minnesota, and New York. There had not yet been a discussion at the national level about these issues.

While the 15 March Senate hearing was primarily informational, with Hicks presenting an introduction to the FBI's DNA analysis project and with various professors of forensic science outlining the basic issues surrounding the use of the technique, the House hearing a week later under the auspices of the Subcommittee on Civil and Constitutional Rights, Committee on Judiciary, elicited considerable debate over the role of the FBI in schemes to ensure the validity and reliability of this new technique. This hearing began with John Hicks reprising his testimony from the previous week, followed by some technical background on forensic DNA analysis from James J. Kearney, the assistant section chief of the FRSTC, who had participated in the development of the bureau's DNA technology. Hicks's testimony was largely unremarkable. He presented the FBI's work in the best possible light and echoed the claims made previously by Lifecodes and Cellmark that if there are any problems with forensic DNA analysis, one gets no result, not the wrong result. He further stated that the FBI was continuously improving its population frequency databases and that any substructure in their population groups was so minor as to be insignificant. Unlike the Senate hearing, however, Hicks (and Kearny) were asked difficult questions by Representative Don Edwards, who was chairman of the subcommittee at the time, as well as various Judiciary Committee staffers. Their views were also challenged by the other witnesses at the hearing—Barry Scheck, Eric Lander, and Philip J. Bereano, a professor of engineering and public policy at the University of Washington, and a member of the activist groups Council for Responsible Genetics and American Civil Liberties Union—as well as in a letter sent into the subcommittee after the hearing had taken place from Jeff Brown, San Francisco's public defender.

Of immediate concern to the three non-FBI witnesses, and Brown as well, was the issue of who had the authority, expertise, and credibility to regulate the production of DNA evidence. Underlying this question was a much broader disagreement about what constituted peer review in an arena in which there was significant disagreement among participants over which scientific communities were relevant to forensic DNA analysis. As we shall see, the FBI witnesses claimed that peer review of DNA identification regimes should be done by parties with an interest in the success of the technique. After all, who would be more concerned about the reliability of the technique than those who use it everyday to solve crimes and catch criminals? The non-FBI witnesses, on the other hand, believed that the hallmark of peer review was organized scrutiny by scientists and other individuals who had no interest in the success of the technique. In making this claim, the non-FBI witnesses sought not only to ensure equal protection under the law and to promote civil rights, but also to protect the peer-review process from the bias and vested interest of people who developed the technique and wanted to use it as quickly as possible. In their view, they were not just standing up for the rights of criminal defendants; they were also standing up for science.

Philip J. Bereano was most critical of the TWGDAM approach taken by the FBI. He chided the bureau for bringing together a group that consisted solely of technical specialists and forensic practitioners, without considering others like academic scientists, defense lawyers, civil libertarians, and policy makers. Invoking a form of the social construction of technology framework, Bereano testified before Congress that "one of the things that have concerned me about some of the comments today, particularly Mr. Hicks' response to an early question by counsel and indeed in some of the implications of the other technologies, is that somehow the scientific and technical concerns and the policy concerns are—exist in separate compartments. The reality is, of course, that scientific developments and technological developments occur in a social and political framework. As anthropologists will tell us, technologies are cultural artifacts."[51] For Bereano, the FBI's decision to convene a standards-setting body made up almost exclusively of people inside the realm of forensic science was simply wrong, because the very notion of standardization implied a combination of political, social, scientific, and technical decisions that could not be meaningfully separated. Later in his testimony, he said, "I would urge you every time he used the word 'scientists' to expand that to the scientists and other interested policy people. Again, the issues are not just going to be technical issues. They're going to be that mix of technical and policy issues and I want to reiterate what I said in my earlier remarks."[52] In other words, he believed that the relevant peer community should be extended to include nonscientists, a view reiterated by Scheck in his testimony.

In his letter to the subcommittee, San Francisco Public Defender Jeff Brown outlined what he believed to be an appropriate regulatory framework for the production of valid and reliable DNA evidence. At the heart of his system was the belief that, although the FBI should be commended for the leading role it has played in the development of forensic DNA analysis, it has a "legitimate law enforcement interest in the use of DNA evidence." As a result, Brown argued that it was important to keep the role of the FBI as a de facto regulator of the technique in perspective. The FBI, he wrote, "ought to be regarded as any other lab would be and held to the same standards and requirements."[53]

Instead of the FBI setting and maintaining these standards and requirements, he argued, it was necessary for an independent, federal agency such as the Food and Drug Administration (FDA), which already regulated other diagnostic tests as well as a host of medical devices, to perform this function. Under Brown's system, this agency, among other things, would be responsible for setting educational, training, and technical standards, such as which probes, reagents, controls, and protocol steps were approved for use by forensic labs, and for setting match criteria. It would also conduct regular proficiency tests, license DNA analysts, and certify DNA labs, as well as serve as a repository for information on laboratory error.[54]

In a subsequent letter to Edwards, Brown reiterated and reaffirmed his belief that the FBI could not be trusted to regulate either the forensic community or even itself. He wrote that there was a fundamental disconnect between FBI director William S. Sessions's recently stated emphasis on the need for scientific information to be openly shared and carried from place to place, "passing it around at the speed of light," and the FBI's actual practices.[55] Brown went on to describe the *Kiles* case and the principle of self-critical analysis and continued that "the case illuminates to what extent the FBI is willing to go in order to conceal its supposedly scientific work, which, by definition and by Judge Sessions's directive, ought to be an open process."[56] At the conclusion of his letter, Brown pointed out that his proposal for FDA regulation of forensic DNA identification laboratories was echoed in the New York State Panel's recommendations for an independent body to license, accredit, and perform proficiency testing for both private and public labs, as well as a Scientific Review Board to approve standards and debate other technical issues.

Although I have no record of the FBI's response to Brown's proposal for FDA regulation of forensic DNA analysis, it is clear from their reaction to the New York State Panel's recommendations that they vigorously disagreed with any program that placed forensic scientists, and the FBI in particular, under any form of outside regulation. In a letter to Evan A. Davis, counsel to New York Governor Mario Cuomo, Hicks wrote that such efforts would "be an unnecessary duplication of effort that would conflict with ongoing efforts within the forensic science community."[57] He went on to describe the FBI's sponsorship of TWGDAM, as well as their close working relationship with American Society of Crime Laboratory Directors (ASCLD), a national organization of crime laboratory directors that works with the forensic science community to set standards, create protocols, develop quality control and quality assurance programs, and train forensic analysts and laboratory managers, "to coordinate policy issues regarding the implementation of DNA testing in crime laboratories throughout the country."[58] He continued:

> The fact that these efforts to set standards and guidelines for DNA testing are voluntary should not be regarded as a weakness. The term "guideline" leaves the emphasis on what results are desired rather than specifying exactly how the results are to be achieved in each laboratory. Several forces work to ensure compliance with guidelines even though such compliance is voluntary. First, peer review and open discussion of technical guidelines is a powerful stimulent [sic] for rigorous thinking. The TWGDAM guidelines were not prepared in a vacuum, but rather as the result of several rounds of discussion and review by senior members of the forensic community having bench-level experience performing DNA tests. Second, the intense scrutiny of DNA testing procedures by defense attorneys and the courts virtually guarantees that any crime laboratory seeking to have its DNA test results admitted as evidence will necessarily comply with pertinent standards and guidelines. From an implementation standpoint, a voluntary approach preserves the prerogatives of a crime laboratory director to determine how the standards will be implemented with out changing the standards.[59]

In addition to responding to the specific policy that the FBI cede regulatory authority to the state on issues of scientific review, proficiency testing, licensing, or accreditation, Hicks also laid out his vision for the proper way to go about ensuring the validity and reliability of DNA evidence—that is, according to the TWGDAM voluntary guidelines. He essentially belittled any effort by a state to regulate a federal entity, calling it "novel, to say the least, and without precedent in Federal-state relationships."[60] He further stated that if such a provision would become law, it would be "untenable and insistence on applying it to the FBI laboratory would make it impossible for use to continue offering DNA testing services to New York law enforcement agencies." With respect to a better approach, he urged New York to "work within the established framework of the forensic science community to promulgate national standards for reliability of DNA testing and the use of DNA profiles in state-level, regional, and national databanks. Efforts by states towards these objectives should be carried out in concert with other states through organizations or groups having a national perspective, rather than in isolation and potential conflict with broader national initiatives."[61] According to Hicks, there was no need for the FBI to organize outside or impartial peer review, because the intense scrutiny of the defense community more than adequately served this purpose.

In a position that was an interesting blend of the FBI's actual practice and Brown, Scheck, and Bereano's calls for greater intervention into the FBI's activities, Eric Lander stated that he would be perfectly happy with the TWGDAM's standards if the FBI were to publish their proposed standards and allow the general scientific community to peer review them. He said:

> I have very warm feelings toward the intentions of the FBI and I think they want to see good things come of it. All of us know that the best things come through peer review, and in fact if the FBI prepares such a standard, they should do so as I think they intend to, in collaboration with the scientific community, should publish it, compare it with what the OTA panel feels, and something that we can all feel good about should come forward, because this is the sort of technology where in fact I think reasonable people sitting around the table will agree. We do have control experiments that verify things; we do have standards we can apply, and somehow I feel scientifically secure that we can agree upon these things.[62]

Thus, Lander advocated a dominant role for the academic scientific community in setting standards for forensic DNA analysis. In many respects, Lander was calling for a system that mimicked the way that scientists internally peer review one another's work for funding purposes and for publication in scholarly journals. The basic idea was to enroll scientists who did not have a direct interest in the FBI's forensic DNA analysis regime, but who were intimately familiar with the associated technical and scientific issues, to evaluate the bureau's protocols and standards. The only role of lawyers, policy makers, and others in his vision was to serve as a sounding board and peer-review system for the scientific and technical community.[63]

The FBI Forges Ahead

Ultimately, Lander's hope that the FBI would wait for the scientific community at large to peer review their protocols and TWGDAM's standards would go unrealized. Although debate surrounding the regulation of forensic DNA analysis would continue in Congress for the next five years, the FBI proceeded at full speed with its plans to set voluntary guidelines and develop a nationwide DNA database. Further, while the FBI's validation protocol explicitly stated that this work should be published quickly so it could undergo the scrutiny of the scientific community, it was only submitted to the *Journal of Forensic Sciences* in June 1990 and published in the September 1991 issue.[64] The FBI also did not publish until 1991 the criteria it used to determine whether similar bands in two DNA profiles match and its first report on the allele frequencies within various populations.[65] This lack of publication and peer review before beginning casework would eventually be challenged by defense attorneys and expert witnesses.

The FBI made no attempt to hide the fact that it was proceeding with the implementation of a DNA typing regime before all of its work had been published in peer-reviewed journals. They felt that there was too much demand in the forensic community for the technique to wait the nine months to a year that would be required for the publication process. As John Hicks admitted at the Banbury Conference in November 1988, while the FBI tried to approach the validation of DNA typing responsibly and systematically, they could not afford to wait until all fourteen steps of the initial process had been accomplished. "We are going to be moving forward with the technology without having it in publication," he said. "The papers are being prepared and will be published; but knowing the publication process, it will be months from now, a year perhaps, before the studies that we have carried out will be in print. In the meantime, we are going to be doing the work."[66]

The FBI's Experts: Scientists in Service of the Prosecution and the State

One of the main reasons put forth by FBI representatives for their decision not to wait for full publication of their protocols and validation experiments was that they felt that their work had already been subjected to adequate peer review. Not only did the FBI present their work at numerous conferences and meetings, they also quickly formed collaborations and close relationships with members of the human genetics and medical research communities. In addition to Ray White at the University of Utah, the FBI's closest academic affiliates were geneticists, medical researchers, and statisticians from Yale University and academic institutions in Houston, Texas; population geneticist Bruce Weir at North Carolina State University; and P. Michael Conneally from Indiana University Medical School. The Yale group included population geneticist Ken Kidd and biostatisticians Bernie Devlin, Neil Risch, and Kathryn Roeder. The Houston group included C. Thomas Caskey, who held an endowed chair at Baylor Medical College and was the director of its Institute for Molecular Genetics, and who was actively involved in developing spin-off companies that commercialized products and services for the DNA identification market; Ranajit Chakraborty, who was professor of population genetics, human ecology, and biometry at the University of Texas Graduate School of Biomedical Sciences; and Stephen Daiger, professor of human genetics and ophthalmology at the University of Texas Health Science Center.

Of these individuals and groups, the Houston contingent became especially helpful to the FBI as they continued to develop and refine their DNA typing program. In addition to receiving substantial fees for testifying on behalf of the prosecution,[67] members of this group also received significant grants from the National Institute of Justice, the parent organization of the FBI, to do work related to forensic DNA analysis. Tom Caskey eventually reaped significant profit from the licensing of various technologies he developed while at Baylor, including a significant portion of the system currently in use for forensic DNA analysis around the world: short tandem repeats.

Most notably, the National Institute of Justice gave $300,000 to Daiger, Chakraborty, and a colleague, Eric Boerwinkle, for their proposal "Analysis of DNA Typing Data for Forensic Applications." This grant was written in direct response to the challenges that forensic DNA analysis had been subjected to throughout 1989. In their abstract they wrote that "although the laboratory methods used in DNA typing are widely accepted, evaluation of typing results and the appropriateness of control data have been hotly contested. Controversies in the analysis of DNA typing data have recently led to several well-publicized court challenges to the admission of such evidence. If DNA typing is to fulfill its great promise to the criminal justice system, these criticisms must be addressed. That is the aim of this proposal."[68] According to their grant application, the proximate goal was to produce peer-reviewed publications—supported, of course, by funding from outside the normal peer-review mechanisms set up to determine which research the scientific community considered worth doing—that would answer the five following questions:

1. How should DNA data from controls and case subjects be recorded and disseminated for effective use by the forensic science community?
2. How different are the data sets generated in different laboratories and is it appropriate to pool data to establish a national database?
3. How different are the data from racial groups and is there evidence in these data to suggest substructure within supposedly homogeneous groups?
4. What is the extent of deviation from expected values in allele frequencies, both within loci and between loci, and how can these deviations be incorporated into case calculations?
5. Finally, what is the impact of new mutations and abundant variation on forensic applications of DNA typing?[69]

Responding to criticisms from various defense experts, they also hoped to create standardized population data sets that were machine readable and could be analyzed by anyone with the appropriate software programs.

Their ultimate goal, however, was to "assure the credibility and accuracy of DNA typing evidence for use by prosecutors and the courts."[70] Indeed, in a section titled "DNA typing controversies," Daiger, Chakraborty, and Boerwinkle noted that while most of the debates surrounding laboratory errors and artifacts had been resolved, challenges to the validity of the statistical inferences drawn from DNA typing had not. Noting that these issues were deeply embedded within both the scientific, legal, and public contexts, they wrote, "Such challenges have been reported widely in the lay press and scientific journals. Criticisms have been voiced by knowledgeable scientists, informed attorneys, and experienced jurists." They went on to state that most of the criticisms had been directed to the assumptions made by the FBI and other testing labs, including that molecular weights of bands can be objectively determined and are reproducible; that the population databases are representative of the suspect's racial or ethnic heritage; that the group to which the suspect belongs is in HWE; that the alleles used in DNA typing are inherited independently.[71] Then, in all capital letters, they wrote, "IT IS THESE ASSUMPTIONS AND CONTROL DATA SETS WHICH HAVE BORNE THE BRUNT OF COURTROOM CHALLENGES AND WHOSE VALIDITY IS THE SUBJECT OF THIS PROPOSAL."[72]

Several aspects of this grant award caused alarm within the criminal defense community. To begin with, they were surprised to learn that a representative of the FBI laboratory sat on the grant review committee that approved both C. Thomas Caskey's grant to develop PCR-based DNA typing methods, as well as Daiger and his colleagues' grant to produce scientific data to combat defense challenges to the technique. In this case, it was James J. Kearney, the same person who not only played a crucial role in the development of DNA typing at the bureau, but also signed the letter from the FBI to Daiger that states the bureau's intentions to cooperate with the Houston team on the project proposed in the grant. Even more troubling was the fact that the grant proposal essentially outlined the task given to Daiger by Bruce Budowle, the FBI's lead man in developing their DNA profiling regime, in connection with his participation in the FBI's statistical standards committee.[73]

In an effort to convince the court of the impropriety of this arrangement, Scheck and Neufeld requested and received an affidavit from Sheldon Krimsky, a philosophy professor at Tufts University who was well known for his work on the scientific, social, ethical, and financial issues surrounding biological research in the academic setting. In his affidavit, Krimsky implied that the Daiger grant was unethical.[74] He also wrote that "particular attention should be paid to the fact that Dr. Daiger's grant appears to be an 'advocacy grant' designed to produce research publications and presentations at national meetings to address courtroom criticisms of DNA evidence," rather than to further scientific knowledge in the field.[75] Although Krimsky's statement was based on statements made by Daiger in an affidavit in *Yee*, rather than a detailed reading of the actual grant proposal, his point still has some merit, since the proposal was explicitly directed to the courtroom controversy and there appeared to be little doubt that the results would benefit the prosecution and not the defense. Whether the awarding of a substantial grant to advocates of the FBI by the bureau's parent agency was ethical or not, it demonstrated that the scientific information being produced to combat the claims of the defense was not the kind of independent, value-neutral work that is the hallmark of the traditional conception of science. While they were not explicitly banned from submitting proposals, doubters of the FBI did not stand a chance in securing funding and did not play a role in determining which scientists were in the best position to produce the research that could help resolve the disputes over DNA evidence. Acting as a consumer, the National Institute of Justice and the FBI purchased the information it needed to gain the upper hand on defense experts in court.

Early Developments at the FBI's New DNA Analysis Unit

By the fall of 1988, the FBI was in the final stages of equipping the purpose-built DNA Analysis Unit at its Washington, D.C., headquarters, which was initially staffed by Special Agent Adams and three technicians. In establishing the laboratory culture, particular attention was paid to developing a standardized protocol that, if followed, would ensure a correct result. As DNA examiner Lawrence Presley explained in a 1990 trial:

> We have an established protocol which is followed, literally, to the letter. There is no deviation from that protocol. Protocol is a recipe, like a cookbook. . . . If you do these things correctly, you receive this particular product. What we have is a series of hands who help us go through that particular protocol. . . . At no time does that person make any interpretation regarding the analysis.[76]

From this statement, it is clear that one of the main objectives of the FBI's DNA typing program was to eliminate the potential for human error and for defense challenge in its test. Technicians were conceived of, at least rhetorically, as a pair of disembodied, unbiased hands that make no decisions on their own. They simply follow the protocol created by the bureau's scientists. As recent scandals in the FBI laboratory, the Houston Police Crime Laboratory, and elsewhere have shown, however, DNA technicians frequently make decisions about deviating from the protocol, often for perfectly legitimate scientific reasons, but sometimes in order to save time or to deliberately produce false results.

Once this protocol was developed, but before beginning casework, however, the FBI needed to demonstrate that actual results produced by the DNA Analysis Unit were valid and reliable. The initial step in this process was the so-called developmental validation that was carried out during the spring and summer of 1988. The second type of validation that needed to occur was what TWGDAM would eventually call "in-house validation" of established procedures, following the approach set forth by the American Society of Crime Lab Directors in 1986.

As TWGDAM's 1989 quality assurance guidelines stated, prior to beginning case-work with a new procedure the forensic laboratory must first validate the procedure in its own laboratory by testing the method on known samples and then by undertaking individual proficiency tests, which can be administered internally, externally, or collaboratively.

During this process, the laboratory also had the responsibility to set legitimate DNA band-matching criteria based upon the measurement imprecision determined from repetitive analyses of the same known sample. As sociologist Linda Derksen explains in her work on the construction of matching criteria, the FBI chose a significantly different matching rule from Lifecodes and Cellmark.[77] Whereas Lifecodes determined that two DNA bands (or alleles) matched if they were within 1.8 percent of one another, and Cellmark did so if they were within 1 mm of one another, the FBI chose a matching rule of ±2.5 percent. This meant that two bands could differ by as much as 5 percent and still be declared a match.[78] (It should be noted that Cellmark's method cannot be compared numerically to either Lifecodes' or the FBI's because it does not vary with the size of the DNA band on the gel. The fact that defense witnesses did not criticize Cellmark for their decision, but did criticize the FBI, suggests that it was fairly conservative.)

The FBI also chose a different method for determining the size of band for the purpose of establishing its frequency within a given population. While Lifecodes and Cellmark used the "floating-bin" approach, the FBI chose to use a "fixed-bin" approach, which they felt would lead to more conservative probability statistics. Put simply, the two companies believed it was possible to determine the nearly exact sizes of alleles at a given locus and calculate their frequency within relevant population subgroups. When a match was determined in forensic case-work, the band size in the reference population that fell within 1.8 percent of the band in the forensic profile was used to calculate the probability of that band showing up by chance in the forensic profile. The FBI, on the other hand, argued that it was impossible to know the exact sizes of various alleles at a given locus due to the resolution issues associated with electrophoresis, as well as measurement imprecision.[79] Instead of calculating the size of the reference band in each case, the FBI arbitrarily assigned a fixed range of sizes, which were called bins, at each locus. They then determined how many alleles within each particular range of sizes were detected in a particular reference population. After a visual match was declared in forensic casework, a semiautomated computer system determined the size using a special camera.[80] This size would then be "placed" into a particular bin, and the frequency of alleles in that bin in the reference population would be used to calculate the probability of a random match. In cases where the allele fell in between two bins, or was at the outer edge of a particular bin, the largest frequency possible was used to make probability calculations.[81]

There were several advantages and disadvantages to each of the approaches taken by the FBI and the two private companies. Overall, on the issue of declaring a match between bands (technically called alleles) in two DNA profiles, the two private companies decided to be more conservative by deciding that two bands could only differ by a relatively small amount in order to be considered the same (3.6 percent or 2mm). The FBI, on the other hand, created a large match window that allowed two bands to differ by as much as 5 percent and still be considered a match. When it came to declaring the probability of a random match, the FBI was more conservative than the two private companies, creating relatively large bins into which a wide range of alleles were grouped. The two private companies, however, calculated the frequency of the alleles that fell into the range of ±1.8 percent or 1 mm, which led to a smaller bin with fewer alleles. (Remember, a match between two alleles that fall into a bin with a lot of other alleles provides less identification power than a match between two alleles that fall into a bin with very few other alleles.) In the end, it is difficult to say which system was objectively better, since both were conservative in some operations and not in others.

The third aspect of the validation procedure was ongoing proficiency testing. In many ways, this issue would come to be one of the most contentious aspects of the debates surrounding forensic DNA analysis. While the FBI claimed that it was consistently engaged in carrying out a series of both blind (meaning that mock case materials are created and submitted to the examiner as if it were coming from a law enforcement agency) and open proficiency testing (meaning that the examiner was aware that he was being tested), defense attorneys and experts would quickly come to question both the veracity of this statement and the legitimacy of the actual results. Responding to claims made by advocates for criminal defendants that the FBI was engaged in an unscientific form of "self-validation," and that it sought at all costs to avoid external proficiency testing, the FBI pointed out that, at least in the late 1980s, there was no organization or company that was engaged in carrying out blind proficiency tests.[82] Responding to questions about why the FBI chose not to participate in a second CACLD proficiency test (which was supposed to be a follow-up to the first test described in chapter 3 that Cellmark made numerous errors on), Budowle argued that the FBI felt that the test was not well designed.[83]

Yet another debate that emerged surrounding proficiency testing was whether the defense needed, and had the right to demand, raw data from the FBI's proficiency tests in order to fully evaluate the work done in a particular case. While defense lawyers and experts argued that an essential part of discovery was gaining access to the proficiency records of the examiner in the case so that the examiner's potential for error or technical mishaps could be taken into account, the FBI believed that the purpose of proficiency testing was to critically analyze each aspect of its procedure in order to make their DNA profiling system as good as it possibly could, which they called the privilege of self-critical analysis.[84] Having to openly share the results could potentially lead to DNA typing laboratories providing easier proficiency tests to its examiner, or to not doing them at all. As Kearney testified in *State of Iowa v. Smith*, an unreported 1989 case, "In my personal opinion if the court requires us to give up the raw data in our proficiency testing in all the DNA cases, it's going to provide a chilling effect on proficiency testing in DNA testing laboratories. . . . We would like to have the ability to have this free flow of information between those that are actually performing the tests and those that are administering the tests, and we would like to be able to criticize them openly without that being part of the court records all over this country."[85] Kearney then went to on tell the court that a major company involved in forensic proficiency testing for non-DNA tests, Collaborative Testing Services, keeps the results of the tests confidential and does not release them to outside parties. Kearney failed to note, however, that Collaborative Testing was engaged in external proficiency testing (i.e., it was an outside agency) rather than internal proficiency testing (i.e., by the DNA typing laboratory itself).

After being questioned for several minutes by defense attorney Cynthia Moisan on why the FBI was being so secretive about the results of its proficiency testing, he blurted out that the bureau was afraid that the defense would use the results of the proficiency tests to "pistol-whip" the FBI Laboratory.[86] Although he did not elaborate on this comment, it is clear that he disagreed with Moisan's contention that it was considered good scientific practice and procedure to share proficiency test results and raw data so they could be validated by outside laboratories. Without this openness, she argued, "the only way that the defense has in this case to confirm that Dr. [Harold] Deadman [the FBI analyst in the case] followed your protocol . . . would be your word and the word of Dr. Deadman; isn't that correct?" They then had the following exchange:

> KEARNEY: Yes, I'm here to testify under oath in this case, yes.
> MOISAN: And Dr. Ford wouldn't have the opportunity to make that evaluation for himself, would he?
> KEARNEY: He has a copy of the protocol. And as we have testified, we use the same protocol for both processes [casework and proficiency testing] in our laboratory.
> MOISAN: And the only way that Dr. Ford has of verifying that is through your testimony; isn't that correct?
> KEARNEY: That's correct.[87]

It is interesting to note that openness and free flow of information was crucial to Kearney's notion of scientific progress, but such exchange could only occur among a well-defined, friendly group of scientists. To leave the bureau open to too many people would hamper its ability to do good science.[88] This testimony also highlights the importance that the FBI placed on the sanctity of the protocol in their work. For Kearney, its existence is the only thing that nonforensic scientists need to know about when deciding whether or not a particular piece of DNA evidence is valid.

When Bill Thompson and Simon Ford gained access to the raw data behind the FBI's proficiency testing scheme in another case later in 1989,[89] they found several minor errors but no "apparent misclassifications."[90] They noted that

Deadman had mistyped a sample in one of his proficiency tests (he called some-thing a homozygote that actually had two separate loci). This mistake was later determined by the FBI to be the result of age-related degradation of high molec-ular weight DNA—that is, very large fragments—based on the fact that Dead-man carried out his analysis at a significantly later date than Adams and Presley. Although this was not technically an error, it demonstrated the real challenges of working with forensic samples. Thompson and Ford also noted that the FBI had discovered a clerical mistake in the report of Presley, who miscopied a set of frag-ment sizes from the computer readout to the FBI worksheet kept in the case folder. Although the clerical error did not affect the results of the test, because the computer made calculations based on data in its memory, FBI officials urged examiners to pay closer attention to their worksheets in order to avoid the mistake in the future.[91] While the defense community found these errors disconcerting, the FBI argued that the discovery of these minor mistakes during proficiency testing was exactly what the process was designed to do.

EARLY TRIALS AND TRIBULATIONS

As previously mentioned, by December 1988 the FBI officials felt that enough work had been done to begin accepting casework from state and local law enforcement agencies. Throughout the early winter months of 1989, as cases poured in from jurisdictions across the country, the FBI DNA analysis completed the population frequency databases they needed to interpret matches. They also analyzed the results of the first set of proficiency tests. The first trial in which FBI DNA evi-dence was introduced was *United States v. Two Bulls* (918 F.2d 56), which took place in South Dakota in March 1989. In this case, Matthew Sylvester Two Bulls, a Native American, was charged with aggravated sexual abuse of a fourteen-year-old girl on the Pine Ridge Indian Reservation. The girl's underwear was sent to the FBI laboratory for testing, and based upon DNA typing the bureau con-cluded that the semen recovered from the underwear matched Two Bulls's DNA profile, with a probability of random match calculated to be 1 in 177,000, based upon a Native American population database.[92] Apparently, word of the *Castro* and *Schwartz* cases had not yet made its way to South Dakota. Echoing the unquestioning judicial acceptance of DNA typing, the judge ruled that DNA evi-dence in the case was admissible after hearing the testimony of only one prose-cution witness and before hearing any defense testimony. Based on this decision, Two Bulls entered a conditional guilty plea and set about appealing the admis-sion of DNA evidence.[93] On appeal, the Eighth Circuit Appeals Court ruled that the trial judge should have carried out a more detailed review of the FBI's DNA typing procedure, since it was a case of first impression, and ordered Two Bulls's conditional plea set aside. The appeals court further granted Two Bulls a new trial, in which an expanded pretrial hearing had to be held. The appeal was dismissed, and has not been followed up on, however, upon Two Bulls's untimely death.

In several other early cases, the FBI's DNA evidence was admitted without serious judicial skepticism or defense challenge.[94] In many of these cases, convictions were reached and pleas were made on the basis of allelic matches at two or three loci, compared to five loci a year later and thirteen or more loci today.

By late 1989, however, the "Irvine Mafia" and their associates had become fixated on interrogating and challenging the FBI's DNA typing methodology. Indeed, as noted earlier, in several cases Thompson, Ford, and associates pressured the FBI to hand over all raw data and materials relating to its proficiency testing and population frequency databases. The defense also put on an extremely strong, but ultimately failed, challenge to the FBI's DNA typing regime in a Vermont rape case of *United States v. Jakobetz* (747 F. Supp. 250, 1990).

In order to supplant Cellmark and Lifecodes as the dominant purveyors of DNA evidence in the American criminal justice system, the bureau made simultaneous adjustments in the material, structural, and sociopolitical aspects of forensic DNA technology. Most obviously, the FBI's choice of HaeIII over the restriction enzymes used by Lifecodes or Cellmark rendered private laboratories' products, technological systems, and validation work less relevant in the new DNA identification economy. Further, with its decision to build the nascent national DNA database on their own techniques, standards, and procedural requirements, the FBI forced the private companies to adopt its technological system.

This technical choice alone did not reshape the DNA typing landscape. From a more structural perspective, in order to become the dominant source of standards in the field the FBI had to build a network of forensic DNA laboratories that would readily adopt its techniques and practices. They did so not by convincing Lifecodes and Cellmark that it was in their best interest to fall into line with the bureau's regime, but rather by actively excluding them from the inner circle of their nascent network. In order to do this, they had to actively construct, in both a literal and theoretical sense, new DNA typing facilities within existing public crime labs, which already looked to the FBI as a source of standards, material resources, and guidance, to perform DNA testing using their system. Because most of the scientists and technicians working within these crime labs had little experience working with DNA, the FBI had to provide a significant amount of education and training to make their nascent network function smoothly. Further, the bureau decided, largely by fiat, that standards in the new era of forensic DNA analysis should be set almost exclusively by the forensic science community. This meant that numerous interested parties, including Cellmark, Lifecodes, defense attorneys, defense experts, and other academic scientists, were left out of the process entirely. Thus, the FBI did not build a forensic DNA typing network by bringing together preexisting actors and material objects, but rather by constructing the network and the human, material, and social aspects of it at the same time.

Although it is impossible to know whether the FBI would have been equally successful had they not made simultaneous adjustments in the material and social aspects of DNA typing in the process of building a network of laboratories using standardized techniques and products, it is clear that the combination had a very powerful impact on the future of forensic DNA analysis. Indeed, by the middle of 1990 it was clear that the FBI's protocol and probes would soon become the industry standard. The promise of a nationwide DNA databank, combined with the bureau's traditional role as a provider of forensic services to the law enforcement community, meant that this situation was almost a foregone conclusion. Despite the private companies' head start in the legal system, they simply could not compete with the incentives offered by the FBI: free service and a uniform, standardized DNA typing regime. From 1990, Lifecodes, Cellmark, and the increasingly large number of companies involved in the DNA identification arena would be forced to shift their strategy from providing a service directly to law enforcement agencies to supplying emerging publicly funded DNA typing laboratories with reagents, probes, equipment, and technical support.

This is not to say, however, that the bureau had a lock on determining how to ensure the validity and reliability of the evidence produced in their lab in Washington, D.C., as well as the countless other state and local labs. While almost all interested parties conceded rather quickly, if not willingly, that the bureau had the authority and expertise to determine the rudimentary technical aspects of a nationwide DNA typing regime, such as which enzymes, probes, and reagents to use, the same could not be said for how the results of this test would be interpreted, what kinds of quality control and quality assurance plans would be instituted, or who would develop these measures and determine how they would be enforced. Despite several attempts to resolve them, these issues would remain contentious for the next several years.

CHAPTER 6

The DNA Wars

The FBI's success in gaining control of DNA profiling quickly attracted the attention of a few defense attorneys who were deeply skeptical that the bureau's testing regime was any better than those of Lifecodes or Cellmark. Because the tactics used against the private labs—that is, making a strong distinction between forensic and nonforensic uses of the technique—achieved only limited success in court, the defense decided to focus on the population genetics issues that had played a secondary role in early cases such as *Castro* and *Schwartz*.

In a series of trials culminating in the Toledo, Ohio, federal court case of *United States v. Yee, et al.* (134 F.R.D. 161, 1991), the defense community argued that the method used by the bureau to calculate the probability of a random match between a suspect's DNA profile and the DNA profile found at the crime scene was dangerously flawed. Their central claim was that the bureau did not adequately take into account a phenomenon called "population substructure" when estimating the rarity of a given genetic marker within the major racial groups—the reference population used in forensic casework to determine the rarity of a particular allele. Instead, they assumed that all major ethnicities within a given race, such as Swedes, Norwegians, Irish, Jewish, and Italian in the Caucasian racial group, tend to intermarry and have gene frequencies that do not diverge significantly from the group average. Although these subpopulations may have slight differences from the overall racial profile, for the FBI and their supporters, these deviations were only of academic interest and did not affect the kinds of probability calculations made in forensic casework.

Making this assumption allowed forensic scientists to carry out simple probability calculations that did not correct for the possibility that the donors of the two DNA profiles being compared belonged to a subtype more genetically similar than the Caucasian population as a whole. The defense and their experts argued that the probability of a random match would be significantly higher if two donors belonged to a subpopulation of individuals with a lower level of

genetic diversity than the larger reference population being used for the probability calculation, because they are more likely to share particular combinations
of alleles in common.

After *Yee*, race and population genetics became the dominant source of controversy surrounding DNA evidence in the American legal system, leaving many
of the issues that arose in *Castro* and other previous cases unresolved. It is important to note that this controversy did not emerge from the scientific community.
Instead, lawyers—most notably Barry Scheck and Peter Neufeld—generated disagreement about population genetics by constructing an alternative set of expert
witnesses with which to challenge claims made by the prosecution and their
witnesses.

In *Yee*, as in previous cases involving the private companies, prosecutors
sought to control the relevant scientific disciplines and legitimate experts
involved in the evaluation of DNA evidence, a process that sociologist of science
Thomas Gieyrn has labeled "boundary work."[1] In this case, prosecutors argued
that the human genetics community was the "relevant scientific community" to
testify in court about the validity and reliability of the FBI's interpretation of
DNA profiles. Scheck and Neufeld, on the other hand, sought to convince the judge
that the relevant scientific community was much broader, encompassing individuals whose work was not directly related to the study of human populations or forensic investigations. They used this same tactic with significant success in *Castro* when
they had Eric Lander qualified as an academic scientist who was knowledgeable
enough about forensic science to be considered an expert by the judge, even though
he was not considered to be part of the forensic science community itself. The prosecution and its allies sought to counter these moves by co-opting the strategy that
Scheck and Neufeld used in their early challenges to prosecution experts testifying
on behalf of the private companies' DNA typing regimes.

As important as the minutiae of population genetics were to the outcome of
Yee, there was much more at stake in this case and the controversy that emerged
in its wake. Most important, the prosecution and defense were doing battle over
the identity of the expert witnesses. Not only did the two sides disagree on
whether the FBI's were sound, they also disagreed on who had the authority and
expertise to make these decisions, how peer review of forensic science should
take place, as well as which forum the review should take place in. Underlying
these questions was the fundamental question of what counts as good science, or
alternatively, what science is good enough in the context of the legal system.

The federal magistrate who presided over the pretrial hearing in *Yee* was
forced to temporarily resolve these problems because he had to decide which
experts were most relevant to the issue of admissibility. In doing so, he also determined which information about the genetics of human populations would be
considered in the courtroom context. Closure in the courtroom, however, did
not necessarily ensure closure outside of its confines. Although expertise took
shape and was policed within the legal system, expert witnesses also became

engaged in debates about DNA evidence taking place in the less procedurally controlled world outside of the courtroom.

UNITED STATES v. YEE, et al.

On the evening of 27 February 1988 in Sandusky, Ohio, David Hartlaub was murdered in his van in a bank parking lot. He was about to make a cash deposit from the music store where he worked. After shooting him, the perpetrators threw Hartlaub's body out of the van and drove away. Police later found the van abandoned in the parking lot of a nearby motel, with the murder weapon stashed between the seats and a significant amount of blood splattered both on the gun and the van's upholstery. The motive of the killing did not appear to be robbery, since the bank bag with almost $4,000 in it was found on the front seat. Serological analysis subsequently revealed that the blood belonged both to Hartlaub and another person, most likely his killer.

Based on bystander testimony (there were no direct eyewitnesses, but several people reported unusual activities in the area), an investigation of the gun, and numerous pieces of corroborating evidence, three members of the Cleveland Hell's Angels motorcycle gang were arrested on charges of murdering Hartlaub. According to the U.S. District Attorney's Northern Ohio Organized Crime Strikeforce's theory, the three men, Wayne Yee, Mark Verdi, and John Ray Bonds, had planned to "hit" a member of the Sandusky Outlaws, a rival motorcycle gang, who drove a van almost identical to Hartlaub's. The trio mistook Hartlaub's van for their enemy's and accidentally shot him.

When the police arrested Bonds, he had a severe injury on his arm and was holding it in a sling. According to the court records, it was "later established that he had a serious ricochet wound." When investigators searched Yee's car, they found a significant amount of dried blood in the backseat. This evidence led police to conclude that Bonds had been the gunman. In their view, the most likely scenario was that he was dropped off by Yee and Verdi at the scene of the crime, shot Hartlaub, and then drove the van to the hotel, where he met Yee and Verdi in Yee's car. He then got in the backseat of the car, and they drove off. This theory was bolstered by the fact that the trio was pulled over by police, but not issued a citation, for making an illegal turn, shortly after the crime had been committed.

Upon analysis by the FBI laboratory, Bonds's DNA profile was shown to match the bloodstains found in Yee's backseat and Hartlaub's van. In its initial report, dated 7 April 1989, the FBI reported that the probability of a random match between the bloodstains and Bonds was 1 in 270,000. In a subsequent report, issued in May 1990, just before the trial began, the FBI modified the way it calculated the frequency of an allele in a given population and lowered the figure to 1 in 35,000.[2] This move was clearly an effort to make the bureau's calculations as conservative and impervious to defense challenge as possible.

As the pretrial admissibility hearing approached in late June 1990 under the auspices of Federal Magistrate James Carr, it quickly became clear that the challenge to DNA evidence in *Yee* would be nothing less than spectacular. The journal *Science*, echoing a *New York Times* headline, declared that the hearing was being billed as the "ultimate showdown" on the validity and reliability of DNA evidence.[3] By the time that the pretrial admissibility hearing actually began, nobody was surprised to learn that Bonds, Verdi, and Yee would be represented by Barry Scheck and Peter Neufeld. Bolstered by the national publicity and scientific expertise that they had obtained in *Castro*, Scheck and Neufeld set about bringing together a collection of the nation's top academic researchers to "examine the scientific basis for the FBI's methods, which were based upon unpublished studies, and render fully informed opinions to the court."[4]

The trial and subsequent legal wrangling lasted for nearly two years and was full of the drama and intrigue that make newspaper reporters and editors salivate. By the end of the *Yee* trial, the prosecution described Scheck and Neufeld's efforts as a "vicious, mean-spirited and baseless attack on the character, ethics, and actions of numerous FBI agents, prosecutors, and nationally prominent scientists,"[5] while the defense attacked the prosecution and the FBI for approaching the trial with a "win at any cost mindset."[6]

In order to understand the controversy that emerged out of the *Yee* case, it is important to first understand the basic issues addressed in the pretrial admissibility hearing. In his report and recommendation (federal magistrates' judgments are highly influential but not binding on federal courts), Carr wrote that "despite the complexity of much of the evidence, the issue about which the experts testified can be fairly easily described." The first of these issues related to the FBI's protocol and procedures for determining that DNA fragments match. In order for the bureau's DNA evidence to be admissible, the government must "show that there is general acceptance in the scientific community with regard to the FBI's ability reliably to declare matches over several loci." The second issue Carr highlighted was the ability of the FBI to make "a reliable and scientifically acceptable estimate of the probability that a match once observed, would be encountered within the American Caucasian population."[7]

The Defense Case: Law as a "Surrogate for Scientific Procedure"

In presenting the defense case, Scheck and Neufeld argued that the FBI's DNA typing regime had not been subjected to adequate peer review and scrutiny within the scientific community. Despite being mindful of the notion that the courtroom is not a research laboratory, it was nonetheless necessary "to use the legal process as a surrogate for scientific procedure so that critical predicate data could be obtained, and for the first time independently assessed."[8] They went on to make the argument that "science can only be considered reliable when there has been experimental validation that not only has been repeated, but repeated by others. . . . [I]t is simply not science unless the results are thoroughly empirically validated; those results are reproduced; and those results are reproduced by other laboratories."[9]

Scheck and Neufeld argued that the court must be focused on DNA technology as specifically applied to the forensic context. In their view, it was not enough for prosecution witnesses to state that the technology was accepted in other fields, or that proficiency tests and validation had been done on pristine, nonforensic samples. They wrote, "unless the government can prove that the FBI's DNA technology *as applied*—that the FBI's methods of declaring matches and calculating probabilities for forensic specimens—would be generally accepted as reliable, the evidence cannot be admitted even though DNA technology may be reliable for other purposes."[10]

For them, forensic DNA testing was significantly different from most other technological systems because it was very difficult to recognize failure: it could never be definitively proven that a mistake in the system would cause no result at all, rather than the wrong result. Unlike other systems, in which, for example, patients die, planes crash, or chemical compounds do not get synthesized, "mistakes in the transfer of DNA technologies to forensics will go undetected because complaints from innocent defendants cannot be distinguished from the protestations of the guilty."[11] Further, unlike in other scientific pursuits, where there is a continuous accretion of new knowledge and most errors or inconsistencies will eventually be detected, the result of a DNA test is a scientific end point. It would no longer be useful to any other experiment or test and would therefore never again be checked for accuracy; unless, of course, systematic problems with the technology were discovered down the road, such as in the recent example of the Houston Police Department, where hundreds of cases have had to be reanalyzed.

Ultimately, Scheck and Neufeld were not just making an argument about the validity and reliability of the FBI's DNA typing regime, they were also making a claim about what constitutes good science and legitimate peer review that they hoped Magistrate Carr would use to render his judgment.[12] Despite acknowledging that the FBI did consult a subset of the scientific community in developing their system, they argued that the bureau had not received input and criticism from all of the relevant scientific communities. They lambasted the FBI and other DNA typing laboratories for not seeking out critical voices, but rather "retain[ing] members of the scientific community whose reverence toward DNA technology was limitless—the gene mappers. Doctors like Caskey, Conneally, Daiger, and Kidd are either part of, or direct labs that have a major stake in, the government's as yet undecided underwriting of the big science human genome project."[13] In calling the witnesses they did, Scheck and Neufeld attempted to redress this perceived problem.

The defense witnesses attacked the validity and reliability of the FBI's forensic DNA analysis regime in two main ways. First, as previously mentioned, they argued that the FBI's procedures had not been subjected to adequate peer review. Second, they asked scientists who they believed were relevant, or rather, whose relevance could be constructed in the courtroom, but had been left out of the FBI's peer review process, to produce peer reviews of the bureau's DNA typing regime for the benefit of the magistrate. Thus, they constructed an alternative relevant scientific community in which to examine the validity and reliability of the FBI's work on DNA that was not based solely on membership in the human genetics community, who they labeled "the gene mappers."

The first of these witnesses was Peter D'Eustachio, a professor of biochemistry at New York University's School of Medicine, who received his Ph.D. in genetics from Rockefeller University. D'Eustachio's general complaint was that the FBI did use adequate controls when conducting validation experiments but did not run validation tests under the same conditions that actual casework was performed. He also argued that the FBI's initial tests produced several unexpected results that should have been followed up with additional experiments and took the FBI to task for using a match criterion that was significantly less conservative than the private DNA typing labs.[14] The bottom line from his testimony and expert's report was that the FBI had serious problems with reproducibility, that their methods and conclusions were scientifically flawed, and that many of these experiments needed to be done again if the FBI's methodology was to be considered scientifically acceptable.

Paul Hagerman, an M.D./Ph.D. who specialized in nucleic acid chemistry and molecular biology at the University of Colorado Health Sciences Center, also testified on behalf of the defense. He addressed the issue of how DNA loading variability and the use of ethidium bromide (EtBr) in the FBI's gels affected the quality and reliability of the bureau's test results. Ethidium bromide is a molecule that binds to DNA and emits an orange glow when subjected to UV radiation. It is often used in electrophoresis gels to allow easy visualization of DNA fragments as they travel down the electric gradient. Binding of EtBr slows the progress of DNA as it passes through the electrophoresis gel.

Based on his review of the bureau's DNA typing regime, Hagerman concluded that the bureau's inability to precisely measure the amount of DNA in a particular sample to be run on a gel, combined with the use of EtBr, meant that the reliability of their casework and population database analyses were both severely compromised. He pointed out that the problems of DNA mobility under conditions of high sample volume and EtBr could lead to errors in band identification in casework and incorrect band assignments in databases.[15] The bottom line for Scheck and Neufeld was that "we now have decisive evidence that the FBI's ability to match RFLP patterns is fatally impaired by EtBr loading dependent shifts. Until EtBr is removed from the FBI's analytic gels, its methods cannot be considered reliable."[16]

Scheck and Neufeld's third major witness to testify about the validity and reliability of the FBI's DNA typing protocol was Conrad Gilliam, who had previously testified in *Castro*. In addition to agreeing with and building upon the critique of the FBI made by D'Eustachio and Hagerman, Gilliam also specifically addressed the issue of expertise in the context of developing matching criteria. He argued that there was a fundamental difference between developing systems to detect discrete alleles—that is, when there are only a few well-known and characterized alleles at a given locus, such as in the medical diagnostics context— and the kind of quasi-continuous allele system that the FBI had developed, in which there could conceivably be alleles of any size.[17] In his view, this issue had never emerged in the medical genetics community and had "only come up in forensics communities."[18] Thus, he implied that the witnesses brought to the stand by the prosecution to testify to the general acceptability of the FBI's matching criteria were not representatives of the relevant scientific community.

Although his testimony was primarily focused on issues of population genetics, Daniel Hartl also testified about what he perceived to be the FBI's excessively large matching criteria, as well as the issues surrounding the validity and reliability of the FBI's protocol. At the time, Hartl was professor and chair of the Genetics Department at Washington University in St. Louis, considered one of the best in the country, and was a coauthor of the leading textbook of population genetics in the United States. A few years after *Yee*, he was wooed away by Harvard University and given a large laboratory there.

With respect to matching criteria, Hartl argued that the FBI failed to distinguish between inter-gel comparisons, which have a larger measurement error because conditions can never be identical for two gels, and intra-gel comparisons, which should have lower measurement errors because conditions should vary little between lanes of a single gel. As a result, Hartl claimed that the FBI should either use a match criterion more in line with other laboratories—for example, the Canadian police's use of a ± 2 percent window—or adjust it based on fragment size—that is, allow for more error when measuring larger fragments and less when measuring smaller ones.[19] Although lacking any empirical evidence, Hartl declared in his expert's report that "the problem with using the 5% window for within-gel comparisons is that too many false [i.e., incorrect] matches are declared."[20]

With respect to validity and reliability, Hartl's most damning testimony was with what he perceived to be severe problems related to the FBI's recent retest of its Caucasian population frequency database of 225 FBI affiliates. After comparing the results of the DNA profiles analyzed for the database in 1988 with those created when the biological samples were retested a little over a year later, in 1990, Hartl noticed that several of them did not match. In his view, the inescapable conclusion was that the FBI "is unable to identify its own agents as being themselves."[21] He went further, arguing that the discrepancies between the first and the second tests indicate "grossly inadequate control over the reproducibility of the sample preparations and electrophoresis protocols, inconsistent classification of the bands into bins, or inexcusable slovenliness in data storage and retrieval."[22]

Based on his courtroom reaction, Bruce Budowle was extremely upset by Hartl's claims. Although he was unable to persuasively explain the differences while on the stand,[23] in the July 1991 edition of *Crime Laboratory Digest*, Budowle and FBI legal counsel John Stafford attempted to set the record straight. In doing so, they attacked Hartl's credibility on several counts.

> At the outset, it should be emphasized that 1) Hartl did not contact any scientist at the FBI to discuss the basic assumptions set forth in his report, 2) he did not present any empirical experiments in the report to substantiate his criticisms, 3) the report does not reference any information from the many scientific meetings where the FBI's data regarding DNA typing in forensic applications has been presented, 4) the report has never been peer-reviewed, and 5) the tone of the report is decidedly adversarial, unlike the scientific tone of Hartl's peer-reviewed articles.[24]

Budowle and Stafford when on to point out that the purpose of the 1990 retest was not to test the reproducibility of the FBI's DNA protocol. Rather, it was to establish a new population frequency database based on an updated protocol. Not only had the FBI halved the amount of DNA that they used to run their test, which created less EtBr-related band shifting, they had also adopted a new DNA sizing ladder from Lifecodes that had twice as many fragments in it than the initial one they used, which made it easier to accurately size unknown fragments. Had Hartl bothered to talk to anyone at the FBI to find out what they were doing, they argued, he would not have been dismayed to find that some of the bands from the first and second tests, especially the larger ones, did not match perfectly.[25] Instead, he would have expected this phenomenon to occur.

In responding to the defense witnesses on the other issue, the FBI and prosecution witnesses did not deny that the use of EtBr, or variations in the amount of DNA loaded in a gel lane, could potentially lead to band shifting. Instead, prosecution witnesses Caskey, Conneally, Daiger, and Kidd argued that it was highly unlikely for these effects to occur in such a way that a match would result at multiple loci between two biological samples. Magistrate Carr found Kidd's testimony to be especially compelling on this issue. He summarized it by stating that

> the true underlying DNA pattern of the actually-different-but-apparent-identical suspect would be have to be proportionately distributed over several loci in a manner that the same degree of band shifting that moved one of the suspect's bands into a false match with the sample band would have to move all other of the suspect's bands into a false match with the equivalent sample band. In other words, all the bands in the forensic sample would have to be so uniquely located that the band shifting—which cased the alteration of all the band positions to an equal extent and in an equal direction—resulted in a false match.[26]

The defense attempted to discredit the prosecution witnesses' argument by labeling it the "multiple locus ploy." In their posthearing memorandum, Scheck and Neufeld dismissed the argument as a "seductively simple suggestion [but] not science."[27] As the court's witness, Eric Lander agreed with this assessment. (Lander repeatedly refused to testify in court after *Castro*, but was persuaded by Magistrate Carr to act as a neutral, court-appointed witness in *Yee*. For the most part, however, his statements tended to support the views of the defense.) However, Magistrate Carr was apparently persuaded by the prosecution witnesses' testimony. He further argued that any problems with band shifting caused by EtBr and DNA concentration went to the weight of the evidence rather than to its admissibility.[28] That is, it did not affect the general acceptance of the technique within the scientific community, but rather affected results on an individual basis.

In replying to the defense's claim that the FBI's matching criteria was not generally accepted within the scientific community, Budowle pointed out that his team had used actual forensic samples in determining the measurement error of their DNA typing system. He also noted that all of the other laboratories, including Caskey's, used pristine DNA that had not been subject to degrading effects of time or the environment to develop their match criteria. Therefore, it was only natural that the FBI's matching window would be slightly higher than other laboratories.' He also pointed out that 97 percent of all matches found by the FBI were within 3 percent of one another, not the maximum difference of 5 percent. Magistrate Carr seemed to be convinced by prosecution witnesses that the ±2.5 percent window was valid and reliable, and again concluded that any questions surrounding its scientific acceptability went to the weight of the scientific evidence rather than to its admissibility.

In his "Report and Recommendation," written at the conclusion of the summer's testimony and adopted by the trial court on 10 January 1991, Magistrate Carr began by noting that the *Frye* standard still held in this jurisdiction despite recent cases that either added a "requirement of conformity to a generally accepted explanatory theory" or made this the chief requirement. For Carr, the government's responsibility was to prove general acceptance within the relevant scientific community. In an effort to apply this standard, Carr next discussed three main issues: specification of the pertinent scientific community, standard of proof, and the meaning of the term "general acceptance."[29]

With respect to the issue of the pertinent scientific community, Carr agreed with the defense's contention that the government had to demonstrate the general acceptance of forensic DNA analysis outside of the forensic community that uses it on a regular basis. He wrote, "to the extent that the government seriously intended to contend that scientists from broader fields of molecular biology and population genetics, including theorists in those fields, were not credible if they had not had experience with the forensic application of DNA and genetic theories, I reject that contention."[30] He also chided the prosecution for the "conduct of its cross-examination of many of the defendants' witnesses and its emphasis on the frequency with which the FBI DNA test results have passed muster in other courts" and praised the defense for bringing a diverse array of witnesses to testify about the general acceptance of forensic DNA typing.[31]

Ironically, though, he essentially ignored the defense's contention about relevant expertise and ruled that forensic DNA analysis was generally accepted within all relevant scientific communities. "In making my determination," Carr wrote, "I take note of the relative professional standing of the prosecution witnesses and the defense witnesses regarding the band shift issue [i.e., the effects of DNA concentration and EtBr]."[32] He then pointed out that Conneally, Caskey, and Kidd were all invited by selection to be members of the Human Genome Organization and that Caskey was president of the American Society of Human Genetics. "In addition to the scientific stature and judgment that such election implies, participation in the activities of such organizations and affiliation with fellows of that rank gives such individuals, I believe, a somewhat better basis on which to gauge the views of those colleagues about the acceptability of new developments related to their discipline."[33] Carr, however, made no effort to justify why he thought membership in various human genetics societies, as opposed to the kinds of prestigious scientific societies to which defense witnesses belonged, gave individual scientists greater insight into the behavior of DNA during electrophoresis. After all, band shifting was not unique to human genetics but was rather a problem faced by all molecular biologists working with DNA.

Another interesting example of Carr's view of relevant expertise was his treatment of Caskey. Although the defense essentially wrote him off as a pawn of the FBI who could not be trusted, Carr interpreted his activities differently. While noting that Caskey had a personal, professional, and financial stake in the admissibility of the FBI's DNA typing regime, because he used it extensively in his laboratory, he concluded that Caskey's adoption of the FBI's protocol gave his testimony greater credibility rather than less. The logic behind this statement was that Caskey must really believe in a particular technology to adopt it and then come to court to defend it despite the serious challenges being made by defense attorneys.[34] With respect to Caskey's expertise, Carr noted that even though Caskey was "currently . . . within the community of forensic DNA scientists, he remains, as he was at the time he was making his decision to adopt the FBI protocol, a pre-eminent academic and clinician. His views, accordingly, reflect those of someone who may be viewed as being both 'inside' and 'outside' the forensic community."[35] Thus, based on his perception of the expertise of the prosecution's witnesses, Carr determined that the FBI's DNA typing regime produced results that were valid and reliable enough to be admitted into evidence.

ESTIMATING THE PROBABILITY OF A RANDOM MATCH

Yee was the first case involving DNA evidence to ensnare the "leading lights of the population genetics community" in the debate over population substructure and statistical calculations, most notably Daniel Hartl and Richard C. Lewontin.[36] Although these issues had been debated to a limited extent in other cases, most notably by Lander in *Castro*, they had never played a major role in any previous trial or defense case. Yet, the *Yee* case did not represent some sort of rational progression in debates over DNA typing. Had Scheck and Neufeld not actively pursued the participation of Hartl and Lewontin and convinced the judge that they had relevant information to present concerning the validity and reliability of DNA evidence, perhaps the legal and scientific history of the technique would have turned out differently.

Lewontin's participation in the trial was especially dramatic, since he was considered by many to be the founder of molecular population genetics in the 1960s and 1970s through his pioneering use of electrophoresis to study the evolutionary implications of enzyme polymorphisms. Additionally, a significant proportion of the academic population geneticists actively engaged in research at the time of *Yee* had either done their doctoral work with Lewontin or had spent time in his lab at the University of Chicago or Harvard after receiving their Ph.D. In a very real sense, he was a pillar of the population genetics community, both from a personal and intellectual standpoint.

His 1972 article "The Apportionment of Human Diversity," in which he argues that genetic variation is greater within races than between them, is considered a landmark paper in human genetics and was frequently cited by scientists on all sides of the debates surrounding the population genetic aspects of DNA typing. Further, his classic 1974 work, *The Genetic Basis of Evolutionary Change*, was, and still is, considered required reading for aspiring population geneticists. Lewontin's reputation, however, was not based simply on his many scientific and academic accomplishments. He was also well known for his left-leaning politics, as well as scathing critiques of sociobiology, the commercialization of agribusiness, and the use of biology to explain social and economic differences among racial groups. He achieved notoriety in the scientific community for resigning from the National Academy of Sciences in the early 1970s to protest the academy's willingness to engage in secret research on behalf of the military. Further, as Peter Neufeld told me in a recent interview, Lewontin had a unique ability to communicate the essence of arcane scientific issues to him and Scheck in a way that allowed them to incorporate his views into their legal arguments.[37]

The essential claim of the population geneticists who testified for the defense (Lewontin and Hartl), as well as the court's witness (Lander) was that the FBI's method for calculating the probability of a random match was not scientifically acceptable and was perhaps even fraudulent. They believed that many of the FBI's claims about the lack of substructure of the Caucasian racial group, the hereditary independence of the various VNTR loci used by the FBI, as well as the soundness of the FBI's Caucasian database, were scientifically untenable.

Before getting into the defense critique, however, it is necessary to review the basic population genetics claims made by the FBI. To begin with, the bureau and its allies argued that there was no relevant substructure within the Caucasian population, since Caucasians tended to mate randomly with one another, at least with respect to their VNTR genotypes. They justified this claim by testing the Caucasian database for deviations from the expectations of Hardy-Weinberg Equilibrium at the various VNTR loci used by the bureau.[38] That there were no significant deviations was confirmation to them that the Caucasian population indeed lacked significant substructure.[39] Prosecution witnesses in *Yee* claimed that the Hardy-Weinberg test was a valid way of identifying substructure, and, further, even if a minor deviation from Hardy-Weinberg expectations were found, this would still not invalidate the FBI's use of its Caucasian database as a reference population to calculate the rarity of a given genetic profile. Second, the FBI and prosecution witnesses claimed that since Caucasians mate at random, so long as different VNTR loci are located on different chromosomes, they will assort independently and be inherited in a random fashion. As a result, the bureau argued that it could use the part of the Hardy-Weinberg equation that deals with heterozygotes ($2pq$) to calculate the frequency of a particular allele pattern at a given locus.[40]

In their testimony, Lewontin and Hartl spent a great deal of time attacking both the scientific assertions made by the FBI, the assumptions underlying them, as well as the credibility of the prosecution's witnesses. To begin with, they argued that it was highly likely that there was significant substructure within the Caucasian population and hence with the FBI's database. Lewontin claimed that the FBI's assertion of no relevant substructure was scientifically untenable for three main reasons:

- First, Caucasian subgroups (such as Sicilians, Poles, Jews, Germans, and Norwegians) may have different allele frequencies at specific VNTR loci;
- Second, only a few generations have passed since these populations immigrated to the United States and began to mix with one another;
- Third, there is a significant body of scholarly research that suggests that people tend to marry individuals with similar religious, cultural, language, and geographic characteristics.[41]

In essence, the FBI assumed that the United States was a melting pot, where, at least within racial groups, individuals moved away from their insular ethnic enclaves and tended to intermarry with people who were different from them enough to mitigate any genetic differences that may have existed at the time that the various ethnic groups came to the country. Lewontin and Hartl, on the other hand, believed that racial and ethnic segregation, both voluntary and involuntary, were still very much alive in the United States and caused both social substructure and genetic substructure.

Unlike the prosecution and FBI, which made their assertions about the structure of the Caucasian population based on a test of Hardy-Weinberg Equilibrium, Lewontin cited extensive nonpopulation genetic evidence to support his case. He not only delved into the literature of sociology, demography, and anthropology in an effort to establish that American Caucasians did not mate at random, even if they associated together in places like work, entertainment venues, or public spaces, but also cited Census data that showed that most Caucasian immigrants came to America between 1905 and 1924, not in the nineteenth century, as is commonly believed.[42] In Lewontin's view, this short time did not provide enough generations to erase the unique nature of the various ancestral subpopulations. As we shall see, however, despite having a large body of sociological and demographic research against them, the FBI's assumptions proved to be relatively correct from a genetic standpoint.

Lewontin and Hartl also attacked the FBI's use of the Hardy-Weinberg principle both to determine whether there was significant substructure within the Caucasian population and to calculate the frequency of heterozygotes at a particular locus. In their expert reports, testimony, and a subsequent article in *Science*, Lewontin and Hartl argued that statistical tests for substructure, such as those used by the FBI and their allies, "are virtually useless as indicators of population substructure because, even for large genetic differences between subgroups, the resulting deviations from HWE are generally so small as to be undetectable by statistical tests."[43] They went on to argue that the lack of statistical deviation from HWE does not imply an absence of substructure and concluded that "statistical tests for HWE are so lacking in power that they are probably the worst way to look for genetic differentiation between subgroups in a population."[44]

Hartl further testified that he had serious reservations about the FBI's use of a statement made in his textbook *Basic Genetics*,[45] that for rare alleles it is possible to calculate the frequency of heterozygotes using the HW principle in certain cases, provided that the frequency of one of the homozygous genotypes was known. For instance, if about 1 in 1,700 Caucasian newborns are afflicted with cystic fibrosis, a recessive, single-gene trait, one can extrapolate that the frequency of the cystic fibrosis gene in the general Caucasian population is 0.024 (since $q^2 = 1/1700$, or 0.00059). Knowing that $1 - q = p$, one can calculate that the frequency of heterozygotes in the Caucasian population is $2pq$, or $2(0.024)(0.976)$, which equals 0.047, or about 1 in 21.

In Hartl's view, the example provided in the textbook was meant only to give beginning genetics students a rough idea of how to calculate gene frequencies.[46] Because it was a basic textbook, rather than an advanced one, he made little effort to warn students about the effects of even minor substructure on these calculations. He assumed, for the sake of pedagogy, that the frequency of the gene for cystic fibrosis was uniform across all subgroups of the Caucasian populations. If one wanted to use this method for diagnostic, prescriptive, or public health purposes, one would have to carry out studies to ensure that the conditions

Lewontin and Hartl used the following example to illustrate their point: Suppose that a population consists of two endogamous subpopulations in which each subpopulation has different allele frequencies at a diallelic locus (A and a). Group I has a ratio of 5:5, while group II has a ratio of 9:1. If there is random mating within these groups, then Hardy-Weinberg equilibrium will produce the following genotype frequencies:

	AA	Aa	aa
group I	0.25	0.50	0.25
group II	0.81	0.18	0.01

Suppose further that group I represents 90 percent of the population, while group II represents 10 percent. In the overall population, the overall genotype frequencies will be:

AA: $(0.25)(0.9) + (0.81)(0.1) = 0.306$
Aa: $(0.5)(0.9) + (0.18)(0.1) = 0.468$
aa: $(0.25)(0.9) + (0.01)(0.1) = 0.226$

Being unaware of the substructure, we would expect the average allele frequencies to be:

A $= (0.9)(0.5) + (0.1)(0.9) = 0.54$ and a $(0.9)(0.5) + (0.1)(0.1) = 0.46$

under conditions of HWE, which would correspond to genotype frequencies of:

AA: $(0.54)(0.54) = 0.2916$
Aa: $2(0.54)(0.46) = 0.4968$
aa: $(0.46)(0.46) = 0.2116$

Because these genotype frequencies found under known conditions of substructure (the first set of calculations) do not differ significantly from what one would assume from HWE without knowledge of substructure (the second set of calculations), using deviation from HWE is not a good test for substructure. According to the NRC report, one would require a sample population of nearly 1,200 people to show substructure using HWE, whereas sampling the DNA of individuals in various subgroups would require only 22 samples per group to determine the degree substructure found in a given racial group, because one would have direct access to the gene frequencies in each subgroup.

necessary for the HW principle to be valid actually existed in that particular instance. As he wrote in his expert's report:

To make the Hardy-Weinberg assumption without adequate validation in forensic application is like arguing that since we do not routinely test lenses in opera glasses before assembly, we need not test the lenses before we launch the Hubble telescope. Different standards are obviously called for in the two cases because the consequences of an error are so much greater. While the Hardy-Weinberg frequencies may be invoked in some contexts in human population genetics, in forensic applications the tolerance must be tightened. In some cases when Hardy-Weinberg is assumed without evidence, the contexts are ones in which errors have virtually no significant consequences—except possibly the embarrassment of the investigator. On the other hand, no physician in his right mind, or one who values his license to practice, would ever diagnose a condition or prescribe a drug based on the outcome of a Hardy-Weinberg calculation![47]

In their rebuttal to Hartl's report, Budowle and FBI lawyer John Stafford wrote that they found Hartl's position on the need for stricter standards in forensic science to be a social commentary rather than a scientific one, despite the fact that Hartl's two examples came from astronomy and medicine. They felt that imposing higher standards on forensic science was unfair and lacking support in legal precedent.[48]

Ultimately, because of the possibility that undetectable substructure existed, Lewontin, Hartl, and Lander all argued that the FBI could not use a unified Caucasian database to generate probability statistics that a match between two profiles occurred at random. Lewontin and Hartl argued that the only way to develop a scientifically reliable method for calculating probability statistics would be to go out into nature and take a large number of DNA samples from representative members of the ethnic subpopulations of the Caucasian racial group. Only if VNTR alleles proved to be similarly dispersed in all ethnic groups tested, would the FBI would have the necessary scientific justification to use their Caucasian database to calculate probability statistics.

During the trial, the FBI and prosecution witnesses never directly addressed Lewontin and Hartl's contention that Hardy-Weinberg calculations were not an adequate test for population substructure. Conneally, Daiger, and Kidd all testified that they *believed* that the Caucasian population was in HWE and that Americans intermarry outside their neighborhoods and across ethnic lines.[49] While there was a sizable body of literature suggesting that this was increasingly the case in postwar America, the prosecution witnesses did not cite it. They did not provide any significant demographic, sociological, or even genetic data to back up their assertions about the minimal substructure in the Caucasian population. They simply assumed, based on common knowledge, that there was no endogamy and justified their views of Hardy-Weinberg on visual inspection of printouts of various Caucasian subpopulation databases.[50]

They also spent a great deal of time convincing Magistrate Carr that any substructure in the Caucasian population would be exceedingly minor and that the FBI's conservative approach to creating bins and dealing with homozygotes was more than enough to compensate for any minor effects caused by substructuring.

Specifically, Kidd, Connelly, and Daiger all argued that the built-in conservatism of the FBI's binning procedure—in which rare alleles are grouped together in one large bin, so that each bin had a minimum of five alleles—the use of $2p$ instead of p^2 when calculating the frequency of an apparent homozygote,[51] as well as their efforts to use the largest possible allele frequency when an allele fell at the intersection of two bins, intrinsically corrected for any distortion that may result from substructuring. On a related note, they argued that this conservative approach was warranted by the fact that many of the apparent homozygotes were actually artifacts of the technical limitations of the electrophoresis process.[52]

They further noted that while the minimal substructure that existed was interesting at a theoretical level, it was irrelevant at the practical level of forensic science. In their view, the exact characteristics of the natural world were irrelevant to forensic practice, since the purpose of the product rule was to estimate the probability of a random match in a very large racial group. As Kidd testified, for a particular bin in the database the FBI's method of bin construction and allocation of alleles resulted in "an estimate of the frequency of that class of patterns where the alleles fall into the bins that were observed, and that is not the estimate of the frequency of any one pattern. It's a frequency of a class of similar patterns."[53]

For Lewontin, Hartl, and Lander, however, the prosecution witnesses' views about what was good enough with respect to population genetics in the forensic context was simply not sound science. Lander stated that he did not believe that there were adequate proofs to allow the use of the FBI's methodology.[54] The reason, according to both Lewontin and Hartl, was that the population genetics community had never before had to address the question of determining the rarity of a particular genetic profile in a population with unknown levels of population substructure.[55] In their view, the FBI's multiplication procedure would only become valid once they could be sure of the allele frequency distributions within the population that most represents that of the individual in question. The only scientifically valid way to do so, in their view, was to undertake detailed surveys of numerous subpopulations within a given race, to determine empirically which allele frequencies were most appropriate to use in determining the rarity of a particular profile.[56]

From the very beginning, Magistrate Carr made it clear that deciding whether the prosecution or defense witnesses most accurately represented the views of the general scientific community was an extraordinarily difficult task. He noted that he could not use individual reputation as he had with other issues, since "each principal witness on this issue is within the first rank of his profession; indeed, it is fair to say that each occupies a primary spot within that top rank."[57] In an effort to replace this standard, he constructed one that took into account the extent to which a given scientist's views "have been available for consideration by scientists attentive to the application of RFLP procedures to forensic uses."[58] This standard was based not on any kind of sociological, philosophical, or even scientific research about how science works; rather, he created it largely out of thin air. It seems that Carr decided that if prosecution witnesses were aware of the defense witnesses' views about the FBI's inability to accurately estimate the probability of a genetic profile, yet still believed that the FBI's methodology was generally accepted within the scientific community, then this was enough to rule in favor of the prosecution.[59] For instance, he noted that the views of Lander and Lewontin were expressed in Lander's 1989 *Nature* article, as well as in the *Castro* case, and were thus known to the scientific community. As such, Carr argued that Caskey was well aware of these views and had taken them into consideration when he decided that the FBI's methodology for calculating probabilities was generally accepted within the scientific community.[60] Thus, Carr decided between two seemingly equally credible scientific theories not by invoking scientific principles or logic, but by an unsubstantiated gut feeling about how scientists think and act. It is no less distressing to realize that he would have preferred to make the decision based solely on the reputation of the various experts testifying for each side, but he simply could not determine who had a higher standing in the scientific community.

In his decision, Carr made a rather awkward admission that showed just how tenuous the reasoning was behind his standard for general acceptance. He noted that Caskey's primary expertise was in molecular biology and medical genetics and that he was not as much of an expert in population genetics as Lewontin, Lander, and Hartl. "Nonetheless," he stated, "I remain persuaded that the views of Dr. Caskey, with reference to the issue of the degree of acceptance of the reliability of the FBI's probability estimates, more correctly describe the level of acceptance within the general scientific community."[61] Although he provided no sociological, psychological, or scientific justification for his conclusion, Carr argued that Caskey was more likely than Lander to reflect the views of the scientific community. He did, however, list three reasons for this continued faith: Caskey's stake in the outcome of the hearing, his awareness of the debate about the ability to calculate probabilities, and his continuing reliance on his own procedures that are similar to the FBI's.[62] Incredibly, Carr failed to admit that there was a chance that because Caskey was so dependent upon the validity and reliability of DNA profiling, he might be less likely than other witnesses to try to identify its shortcomings in a public setting. For Carr, Caskey's belief was further supported by the fact that the other prosecution witnesses agreed with Caskey even though they, too, knew about the objections of Lewontin, Lander, and Hartl.[63] Thus, he concluded that "it is more likely than not that the general scientific community accepts the reliability and scientific suitability of the FBI's protocol and practices."[64] Although it carried the weight of law behind it, Carr's conclusion was in reality little more than a leap of faith.

REACTION TO *YEE*

At the end of the pretrial hearing, Magistrate Carr ruled that the FBI's DNA profiling regime was indeed admissible. The judge in the actual case accepted this ruling, and DNA evidence was admitted into the murder trial of Bonds, Verdi, and Yee. All three men were subsequently convicted of murder and immediately, but unsuccessfully, appealed the decision.[65] Thus, despite the best efforts of Scheck, Neufeld, and their parade of eminent scientists, the FBI and the prosecution emerged from *Yee* victorious.

Although many judges around the country took note of the *Yee* decision, only one disagreed with the ruling. In 1991, the Superior Court of the District of Columbia determined in *United States v. Porter* (618 A.2d 629, 1992) that there was indeed a significant dispute in the scientific community over the validity of the FBI's population genetics assumptions. The court ruled that DNA evidence in the case was inadmissible because the FBI's methodology for calculating probability statistics was not generally accepted. Thus, the *Porter* court believed, contra Magistrate Carr, that probability statistics were an integral part of DNA evidence and therefore went to its admissibility, not just its weight.

While the case had only a minor impact on subsequent admissibility rulings, its presence was felt much more strongly within the scientific community at large.[66] In the wake of the case, scientific journals increasingly became sites of intense debate and disagreement among academic biologists, forensic scientists, lawyers, and corporate executives. In a series of letters to the editor responding to Eric Lander's 1991 invited commentary on *Yee* in the *American Journal of Human Genetics*,[67] individuals on all sides of the debate over population genetics issues in forensic DNA analysis presented their views about what counted as good science in the legal context. As *AJHG* editor Charles Epstein noted in a statement accompanying this correspondence:

> Several years ago the Letters to the Editor section of *American Journal of Human Genetics* was filled with relatively arcane discussions of the calculation of probabilities of paternity and non-paternity based on the use of blood groups and similar genetic markers. This interchange of letters gradually abated, but once again the *Journal* is being used to debate the forensic uses of the fruits of genetic research. This time the rhetoric has become somewhat more shrill, perhaps because the stakes are greater. . . . Because of the great scientific and societal importance of these issues and because the issues are clearly within the province of human genetics, I have felt that the *Journal* is an appropriate vehicle for both scientific communication and relevant commentary on the forensic applications of molecular genetics. . . . As editor, I have sought to make the *Journal* a forum for the discussion of many issues of concern to human geneticists, issues which may sometimes go beyond the purely scientific. The forensic applications of molecular genetics is clearly one of these issues.[68]

Ultimately, this lengthy set of letters to the editor would be seen as a significant landmark in the struggle to articulate the numerous views on forensic population genetics issues.

In his response to the numerous correspondents, including Harmon, Wooley, Caskey, Chakraborty, Daiger, as well as a group of scientists from Cellmark, Lander developed a three-part classification to describe the various perspectives in this debate.[69] He called the first category the "keep-it-simple school," which was made up of people like *Yee* prosecutor James Wooley, who argued that the issue of population substructure should be resolved "in the context of the courtroom setting."[70] Although there was much to be learned by carrying out detailed studies of individual population subgroups, this information was primarily of interest to academic population geneticists and not to judges. In his letter, Wooley asked, "if we were to develop subpopulation databases, who would we use them in U.S. courts? . . . Do we use a Neapolitan or Sicilian data base for a fourth-generation Italian defendant? Are we compelled to use a Columbian data base on a Columbian immigrant defendant? Do we use a selected African data base on a U.S. Afro-American defendant? . . . Where does population gathering stop, and how is it used?"[71] In his view it was better to stick to standard racial groups than to get mired in the problem of determining exactly to which ethnic group the suspect belongs.

In their *Crime Laboratory Digest* article, Budowle and Stafford argued that Lewontin and Hartl did not understand what the relevant subgroup was in the context of forensic science. Although Lewontin and Hartl both insisted in their testimony, and in their subsequent *Science* article, that the crucial level of racial subdivision was ethnic, or ancestral, Budowle and Stafford argued that it was geographic. In this view, the issue was not whether there was ethnic substructure within a broad racial group, but whether there was regional variation of this race within the United States. Thus, the FBI believed that it was important to ensure that the distribution of VNTR alleles of representatives of the major racial groups in specific geographic locations (e.g., Massachusetts, Texas, California, and Florida) did not differ significantly from the national reference database used in calculating probability statistics. Indeed, this was exactly the kind of population genetic work that the FBI had done and reported in an article that was published shortly after the *Yee* case.[72]

Geneticists who testified on behalf of the prosecution and FBI in *Yee* were heavily represented in Lander's second group, the "statistical school." Proponents of the statistical approach believed that the population substructure issue could be resolved by collecting random DNA samples from individuals in broadly defined racial categories and testing the resulting allele frequency databases for Hardy-Weinberg equilibrium and linkage equilibrium. As long as there were no major deviations, one could use the frequencies in these databases to calculate the rarity of a particular profile using the product rule.

The third school was most associated with Lewontin, Hartl, and Lander. Proponents of the "empirical school" argued that the only way to resolve the population substructure problem was to ascertain the actual level of variation among human subpopulations of interest. In their view, there was no statistical test that could replace the act of going out into the real world and collecting biological samples of people from a wide range of ethnic backgrounds.[73] Only if these excursions showed that there was no substantial substructuring within the "racial" categories could the product rule be used without some sort of correction factor.

THE *SCIENCE* AFFAIR

Although Lander would articulate the basic views of the empirical school in his reply to the various letters to the editors,[74] it would receive its most complete articulation when Lewontin and Hartl decided to publish their critique of the FBI's population genetics assumptions in *Science*. This article led to one of the fiercest controversies in the recent history of science, with numerous accusations of impropriety and unethical behavior being slung by both defense and prosecution advocates. Hartl even accused Wooley (the prosecutor in *Yee*) of threatening and intimidating him in an effort to persuade him not to publish the *Science* article.[75] Hartl and Lewontin, however, could not be dissuaded, and they continued with the process of revising their article, which had already been accepted for publication. In the meantime, though, *Science*'s editor-in-chief, Daniel Koshland, began receiving a great deal of pressure from the FBI, prosecutors, and certain members of the scientific community, most notably Caskey and Kidd, to do something about Lewontin and Hartl's article. Because he could not simply pull an article already in galley proofs, Koshland instead decided to delay the publication of the peer-reviewed article until a more pro–DNA typing perspective piece could be commissioned that would rebut many of the claims made by Lewontin and Hartl.

In the end, *Yee* witness Ken Kidd and longtime FBI collaborator Ranajit Chakraborty wrote an article that actually preceded Lewontin and Hartl's in the 20 December 1991 issue of *Science*. This article, entitled "The Utility of DNA Typing in Forensic Work," argued that, contra the claims of Lewontin and Hartl, the FBI's methodology was fundamentally sound and generally accepted within the scientific community.[76] Kidd said that he was motivated to write the rebuttal because "some aspects of the Lewontin and Hartl calculations were really out in left field, and in my opinion, the article was clearly motivated by Dick Lewontin's philosophical bent of always being for the defendant and against the state."[77] He and Chakraborty, as well as several other scientists who testified for the prosecution, believed that it was a "gross perversion of science" and that something needed to be done about it to prevent it from having an adverse affect in the legal system.[78]

The crux of their argument was that the issue of substructure had to be understood "within the context of the courtroom applications of DNA typing."[79] Although Lewontin and Hartl's claims were potentially interesting from an academic standpoint, Chakraborty and Kidd believed that it was "necessary to draw the distinction between exact values and valid estimates. The issue under debate is whether, when a match occurs, a meaningful estimate can be obtained for the frequency of the DNA pattern."[80]

Although argued more eloquently than by Budowle and Stafford, their conclusion was essentially the same. In their view, the purpose of reference databases was not to capture the exact structure of the reference population, but rather to generate a database that provides "conservative approximations of the relative frequencies" of particular alleles in the region where the perpetrator of a crime is likely to be found. Thus, one need not prove that there is no substructure within the given reference population, only that "its effect on deviation from HWE and LE is so small that its effect cannot be detected in practice."[81] Interestingly, Chakraborty and Kidd did not directly address Lewontin and Hartl's contention that Hardy-Weinberg equilibrium was not a valid test for substructure. Instead, they repeated the claim that it was an appropriate approximation and that the FBI's conservative methodologies more than amply corrected for any substructure that may exist.

They did, however, contest Lewontin and Hartl's claim that mating was not random in the Caucasian population. They then launched into an extended critique of Lewontin and Hartl's claim that human populations form small, isolated inbreeding groups at both a genetic and demographic level, also known as endogamy. They began by noting that Lewontin and Hartl cited demographic studies that suggest different U.S. ethnic groups are "largely endogamous" and "the American tends to marry the boy or girl next door." They then wrote that "the qualifications 'tend to' and 'largely,' however, have significant implications in genotypic probability calculations because population genetic theory shows that even a small amount of gene migration across ethnic and religious boundaries will quickly homogenize populations. Both the portion of mixed marriage and mixed ethnicity (20%) and that of marriages outside the 10-mile radius (67.6%) per generation are high."[82] Chakraborty and Kidd also criticized the demographic and sociological studies used by Lewontin and Kidd to formulate their generalizations about American marriage patterns as being outdated. They argued that since the time that the studies Lewontin and Hartl quoted were carried out, there had been a sea-change in the increased mobility and mixing of groups in the general U.S. population in the postwar "baby boom" era. Summarizing their arguments against Lewontin and Hartl, Chakraborty and Kidd wrote that "American demography for descendants of Caucasians is closer to a 'melting pot' than to a rigid subdivision."[83]

Chakraborty and Kidd concluded their perspective article by arguing that Lewontin and Hartl's three potential remedies for the population genetics problems they believed existed were unnecessary. Specifically, these remedies were:

1. Don't multiply the frequencies at each locus using the product rule.
2. Employ "ethnic ceilings" based on the study of separate databases constructed for major subpopulations within each ethnic group. Although a

particular defendant's subpopulation may not be represented, the range of variation within the racial group will probably be accurately captured anyway. Using this method, "the estimated frequency of the VNTR phenotype is taken as the maximum value observed among the relevant ethnic subpopulations, and these locus-specific estimates are multiplied together to obtain the composite estimate of the probability of a multi-locus match."

3. Fix the current method by undertaking detailed investigations of VNTR frequencies in a wide variety of ethnic subgroups. Once these frequencies are compiled they can then be used in conjunction with the product rule to gain an accurate representation of the likelihood of a random match within the relevant population.[84]

Chakraborty and Kidd wrote that Lewontin and Hartl's suggestions were misguided because they mistakenly believed that it was necessary to know the exact frequency of VNTR alleles within the subgroup to which the suspect or defendant belongs. They believed, on the other hand, that it was only necessary to obtain a general frequency in the total population from which the suspect or defendant could have come. In the end, they concluded not only that the methodology currently used by the FBI "does not require 'fixing,'" but also that if Lewontin and Hartl's critique led to the inadmissibility of DNA evidence in the courtroom, then "the prospect of convicting true criminals, as well as exonerating the falsely accused, will be substantially diminished."[85]

In the same issue in which the articles by Lewontin and Hartl and Chakraborty and Kidd could be found, *Science* ran a news article that described the contentious events surrounding this unusual dual publication. The article noted that the debate over DNA typing had become "decidedly nasty" and that the stakes were much higher than in a typical scientific disagreement.[86] A sidebar to the story also asked whether *Science* was fair to its authors in commissioning a response to the article. The answer, according to the story, although not stated explicitly, was yes.[87]

News accounts of this episode appeared in countless media outlets. *U.S. News and World Report* noted in a story entitled "Courtroom Genetics" that "a flap over DNA evidence raises questions about the relationship of science to the law."[88] The *Sacramento Bee*'s lead editorial on 28 December 1991 lamented the increasing pressure that prosecutors were exerting on scientists and editors in an effort to manipulate the content of scientific journals.[89] Perhaps the most breathless account took place in the august pages of the *New York Times*, in which reporter Gina Kolata, who delighted in documenting every setback of the technique, focused on the harassment of "dissenting scientists" by cruel prosecutors and government officials.[90] Increasingly, media stories referred to the debates over forensic DNA typing using the language of battle and attack.[91] Indeed, many participants and observers began taking these words literally and began referring to these disputes as the "DNA Wars."[92]

WHO WERE THE DNA WARRIORS?

Although most of these battles were framed in terms of technical issues, under-lying all of them was a serious discussion over which scientific community, or communities, had the expertise necessary to determine whether or not the prac-tices and methodologies surrounding forensic DNA analysis were valid and reli-able. Unlike the prosecution witnesses, whose research was almost exclusively dedicated to human genetics, most of the academic scientists who testified on behalf of the defense studied nonhuman model organisms like fruit flies. These defense witnesses generally resided in departments within faculties of arts and sciences, rather than medical schools. Of the witnesses who testified on behalf of the defense, or primarily in their interest, only Lander was involved in research on human genetics.[93]

Like their prosecution counterparts, defense witnesses were not immune from allegations of unethical behavior and financial impropriety. Almost imme-diately after the defense community found scientists willing to testify that certain aspects of forensic DNA analysis were not yet generally accepted within the scientific community, the prosecution derided these individuals as hired guns who were testifying for dollars, not in service of the truth. Among the most vociferous proponents of this view was Rockne Harmon, from the Alameda County District Attorney's office in California. According to *Science*, Harmon was "arguably the most aggressive prosecutor defending DNA finger-printing against court challenges."[94] In addition to devoting the vast majority of his time to such cases, he was considered by many commentators to be the "linchpin of an unofficial network of prosecutors and the FBI" and had devel-oped a close friendship with Bruce Budowle.[95] His dealings with several aca-demic scientists who testified on behalf of the defense, most notably Laurence Mueller, were so acerbic that *Science* devoted an entire article to them, entitled "Prosecutor v. Scientist: A Cat-and-Mouse Relationship."[96] Harmon not only rigorously cross-examined Mueller but also sent him numerous letters asking him to renounce his criticisms of DNA typing for the good of society. He even went so far as to send letters to the editors of *Science* and *Genetics* alerting them that they should subject two articles submitted by Mueller to extra-stringent peer review because he had presented erroneous and low-quality testimony in numerous court proceedings. Harmon even suggested four potential scientists whom he believed were especially well qualified to peer review Mueller's work: Bruce Budowle, Bernie Devlin, Ranajit Chakraborty, and Bruce Weir.[97]

By early 1992, the relationship between prosecutors and representatives of the FBI, especially Budowle, and members of the defense community had grown so nasty that defense attorneys Peter Neufeld and William C. Thompson began a letter-writing campaign to various high-ranking government officials, including William Sessions, the director of the FBI, and James X. Dempsey, the legal counsel of the U.S. House of Representative's Subcommittee on Civil and Constitutional Rights. In one such letter, Thompson wrote to Dempsey:

> My impression is that harassment and intimidation of scientific critics of DNA testing has been a serious problem. Ugly rumors have been spread about virtually every scientist who has had the temerity to come into court to express concerns about forensic procedures. I stay in touch with defense lawyers around the country who handle forensic DNA cases. Constantly I hear about rumors, spread by prosecutors, concerning ethical, financial, and even sexual improprieties of defense experts. Common themes are that the defense experts engage in fraudulent billing practices, make unethical use of materials turned over to them in discovery, and are homosexual. On those occasions when I have been able to trace this toxic stream of lies to its source, it has always been either someone at the FBI or . . . Rockne Harmon. . . . The effects of this systematic smear campaign should not be underestimated. I have talked to several scientists who are reluctant to express their views in court because they simply, and quite understandably, do not want to expose themselves to this sort of harassment.[98]

Although Harmon characterized himself as an assertive and aggressive, but ultimately fair, prosecutor, Mueller described him to me as a "bully."[99]

In addition to sending Mueller letters attacking his personal and professional credibility,[100] Harmon regularly spoke to the press about the fees that various defense witnesses had collected for their testimony against the validity and reliability of DNA evidence. He was especially critical of Mueller, who made $60,000 in consulting fees in 1991, and Simon Ford, who by 1991 had left academia to consult for the defense full time, as well as their associate Randall Libby, who made about $80,000 in court fees in 1991. He also pointed out that Dan Hartl made close to $28,000 for his testimony in Yee. When I asked several of the individuals who served as defense witnesses why they accepted such large payouts, they said that it allowed them to justify to themselves and their families the difficulty, stress, and tremendous amount of time required to prepare a case.[101]

For some scientists who testified, however, the only way to maintain their credibility in the courtroom was to refuse to be compensated. This was the position taken by such well-known critics of DNA evidence as Eric Lander, Richard Lewontin, as well as Lewontin's former student, Jerry Coyne, among others. In an interview with Science, Lewontin said, "One appears [in court] as a matter of principle. It is our obligation as scientists to do it. I don't think we should be paid."[102] Ultimately, however, Lewontin believed that all efforts to maintain the guise of objectivity in the courtroom were rather limited. He concluded, "I think your credibility is damaged no matter what you do. If the attorney asks if you are paid, you answer yes, they [call you] a hired gun. If you say no, the question is, are you a zealot with a political agenda?"[103]

Why particular scientists tend to testify on behalf of one side of a legal dispute as opposed to the other is a longstanding question surrounding the relationship between science and law. There is no single, easy answer to this question. Rather, explanations can only be constructed on an ad hoc basis. Some general trends, however, can be discerned. With the exception of Bruce Weir, who was well known as a pillar of the theoretical population genetics community, all of the scientists who collaborated with and advised the FBI studied the genetics of humans and were generally eminent scientists and vocal boosters of the Human Genome Project. Although these individuals never explicitly linked their work on forensic DNA analysis with their desire to see the human genome mapped, it is conceivable, at least, that they worked with the FBI partly to raise the profile of genetic research in the minds of taxpayers, politicians, and funding agencies.

More important, FBI researchers said that the bureau went to the human genetics community for guidance and support because they felt that they were dealing with a technology that was already widely used in the context of medical research and diagnostics. Once they made contact with the human genetics community, initially through Ray White, it was only natural that they would continue to work within that community. White told me that the community was quite close-knit at the time and often consulted one another on problems of mutual concern.[104]

Others, however, think the issue has to do with differing philosophies of science. Among the most sophisticated representatives of this camp was William C. Thompson, who believed that there may be an element of different scientific models:

> [Academic researchers] who are more theoretical scientists take a hypothesis testing approach to science. They ask "is there data that exists that can disprove the hypothesis of independence"? And they basically won't accept the hypothesis until they adequately test the hypothesis. So, it's just a more rigorous approach. Whereas, people in [medical research], I think are not quite so rigorous about [proof]. They're more accustomed to saying, "does HIV cause AIDS? Yeah, well, it probably does, we don't have to rule out every other possibility, but ..." So, I think it may not be a coincidence that the prosecutors were finding supporters for their methods from the medical community, and the defense lawyers were finding critics from the theoretical community.[105]

Richard Lewontin, on the other hand, suggests that intellectual background had little to do with the positions that various scientists took. Instead, it was simply a matter of who requested the scientist's help first, the prosecution or the defense.[106] In his view, most scientists held no opinion about the statistical and population genetics debates surrounding forensic DNA analysis until they were induced to do so by an adversary in a criminal proceeding. This explanation works well for at least some of the people who testified in court about DNA typing. Dan Hartl, for instance, became involved in the *Yee* case when the defense contacted him and asked if he approved of the way that certain statements in his seminal population genetics textbooks were being used by the FBI. He did not and agreed to testify in court.[107] In a similar fashion, Bruce Weir began testifying on behalf of the prosecution when he was contacted by Bruce Budowle and asked whether he approved of the way that Lawrence Mueller was using his work on population genetics, and a computer program that he had developed, to argue that there was a significant divergence from HWE in the FBI's population database. Weir did not and agreed to testify in court on behalf of the bureau. That said, it seems highly unlikely that defense attorneys as skilled as Neufeld, Scheck, Thompson, and Sullivan would contact a scientist for help without having some inclination that he or she would provide testimony that would be useful for their cause.

What is interesting to note, however, is that only in a few cases did experts testify for both prosecution and defense, or even publicly state that their position had changed. The three most obvious examples of this phenomenon were Dan Hartl, who did so largely because he wanted to lend support to a forensic laboratory that he felt was doing probability corrections carefully and with due diligence;[108] Eric Lander; and William Shields, a professor of biology at the State University of New York College of Environmental Science and Forestry in Syracuse. Other than these few individuals, most academic scientists maintained their allegiances to a particular viewpoint throughout time. One explanation for this phenomenon, which was suggested to me by Lewontin but cannot be verified one way or the other, was that once scientists stated their initial opinion for the record they were forced to maintain that view and defend it because opposing lawyers would be able to use their original statement against them if their views changed.

Ultimately, it is difficult for the historian to demonstrate why particular groups of scientists became partisans for one or another side in the disputes over DNA typing. What can be more easily analyzed, however, is the extent to which the divergent collective biographies of defense and prosecution witnesses would have a profound impact in the disputes over what counted as the relevant scientific communities in which forensic DNA analysis had to be generally accepted. One of the great ironies of the history of DNA typing is that the very critique that was used by the defense community to disparage early prosecution witnesses would quickly be turned back onto their witnesses by the prosecution. Specifically, just as Scheck and Neufeld had so skillfully argued that there was a fundamental difference between the forensic and medical uses of DNA typing, by the autumn of 1990 prosecutors would begin arguing that there was a fundamental theoretical and practical difference between the use of DNA technology in the forensic context and the kind of work done by most academic scientists who testified on behalf of the defense.

The Debate in Washington

While all participants in the debates over DNA evidence conceded that the FBI's technical standards had clearly become dominant over Lifecodes' and Cellmark's by 1990, the fate of the bureau's population genetics and statistical methods used to calculate the probability of a random match, as well as the social arrangements that it had created to ensure the validity and reliability of its evidence, were hotly contested. Witnesses from both sides of the debate believed that a nonpartisan, extralegal mediator needed to step in. It was in this context that the U.S. Congress carried out hearings on how to regulate DNA identification in the criminal justice system. Many people involved with DNA profiling hoped that the National Research Council's forthcoming report on *DNA Technology in Forensic Science* would provide a definitive set of recommendations to Congress on how to ensure the validity and reliability of DNA evidence and would help resolve the population genetics dispute that had emerged in *Yee*. In the end, such hopes proved to be spectacularly unfounded.

In understanding the failure of the first National Research Council (NRC) report to resolve the "DNA Wars," it is important to realize that the NRC committee explicitly sought to blend scientific facts, technological standards, concepts of conservatism and regulation, as well as notions of justice in their solutions to the problems that had been plaguing DNA profiling in court. One of the major reasons why the NRC committee ultimately failed to achieve closure was that the FBI, and many members of the scientific community on both sides of the debate, rapidly recognized the hybrid nature of the report's recommendations and disagreed with the particular orderings of science and law that the committee had constructed. Specifically, they believed that the "ceiling principle," the solution proposed by the committee to solve the problem of population substructure, was unnecessarily conservative and was not based on sound science. In many ways this was true, since the basic purpose of the ceiling principle was to limit the frequency of a particular allele in a DNA profile from dipping below a purposefully conservative level

(5 percent). This action ensured that the probability of a random match between two profiles remained relatively high compared to what would be calculated using actual allele frequencies (e.g., 1 percent).

Using its political power and control over the forensic science community, the bureau actively sought a new report that would reconfigure the relationships between the natural and the social in ways that were more in line with their views about the world. Specifically, they sought to maintain the apparent boundary between scientific and legal issues, leaving them to be resolved independently of one another. The second report issued by the NRC was much more to the FBI's liking, thanks in part to the tremendous political pressure that the bureau placed on the NRC and the exclusion of committee members that may have been sympathetic to the ambitions of the first report.

DEBATE OVER THE DNA ACT(S) OF 1991

Congressional debate over the validity and reliability of forensic DNA evidence continued in June 1991 with a joint hearing before the House Subcommittee on Civil and Constitutional Rights and the Senate Subcommittee on the Constitution.[1] Much of this discussion revolved around the boundaries of forensic science in relation to the other realms of science and who possessed the authority to regulate forensic laboratories. While the FBI and its supporters argued that forensic science was a unique branch of science, because of the conditions under which it operates, and could therefore only be regulated from within, others believed that forensic science should be treated like any other science and be open to scrutiny from outsiders.

The hearing opened with brief messages from Representatives Don Edwards and John Conyers (D-Michigan) hinting at various problems that had emerged surrounding forensic DNA evidence since the first hearings in June 1989 and highlighting the need for appropriate standards and proficiency testing to ensure that DNA typing lived up to its enormous potential. Interestingly, despite the recent events of *Yee*, population genetics was only tangentially mentioned in the hearing. Central to this hearing was a discussion of recently proposed bills in the House and Senate that would provide formal frameworks for the regulation of DNA typing.

The first was H.R. 339, the "DNA Proficiency Act of 1991," which Representative Frank Horton (D-New York), had recently introduced in the House. Horton hoped his bill would address two major issues that were plaguing DNA typing at the time: the lack of an effective proficiency testing program for American crime labs and the lack of funding necessary to buy equipment for DNA typing at the local level. The backbone of what came to be known as the "Horton Bill" was a $5 million appropriation by the federal government to help state crime labs acquire DNA typing equipment. In exchange for this money, however, these state labs had to meet or exceed the guidelines established by TWGDAM in July 1989, and they had to agree to participate in at least two proficiency tests a year. Further, the FBI was given the responsibility of developing proficiency testing standards, as well as approving testing programs offered by private companies if the director decided that they were efficient and effective.[2]

The second bill, the "DNA Identification Act of 1991" (S. 1355) was introduced into Senate in June 1991 by Senator Paul Simon (D-Illinois). In addition to covering the areas of funding and proficiency testing addressed by the Horton Bill, the "Simon Bill" went further to enable the creation of a nationwide DNA databank by the FBI, set requirements for participation by state and local labs in the genetic index, mandate the development of privacy standards, and create a criminal penalty for law enforcement officials who violated these standards. Additionally, the Simon Bill mandated that the population databases used by labs to calculate probability statistics be made publicly available, and it required the FBI director to submit the results of proficiency tests to Congress so they could be continuously monitored. It also called for the creation of a permanent interdisciplinary advisory board to help decision makers work through complex issues surrounding DNA typing.

In some respects, the substance of the Simon Bill built upon and echoed the Horton Bill. It too provided funding for the development of state and local labs on the condition that these labs partake in FBI issued proficiency tests at least twice a year. There were major differences, however. To begin with, while the Horton Bill gave the FBI significant leeway to develop and approve proficiency testing standards and regimes, the Simon Bill vested this authority within the National Institute of Standards and Technology (NIST), which had been involved with the development of standards for various reagents, sizing ladders, and other chemical products used in forensic DNA analysis. Under the Simon Bill, the director of NIST would not only be responsible for giving ultimate approval to the FBI director's DNA proficiency testing program, but also would appoint the advisory board that would provide recommendations to the FBI director. The bill further mandated that the board be made up of "molecular geneticists, population geneticists, experts in law, and experts in privacy [because of concerns about the storage of genetic information in the database]."[3] The bottom line of this requirement was that the FBI could not create its own scientific community, as it had with TWG-DAM, to approve its vision of how the validity and reliability of DNA should be ensured. (NIST was not contacted by Simon's staff to discuss this provision, however, and proved to be reluctant to take on an explicitly regulatory role.)

The Simon Bill met with immediate criticism from the forensic science community. In his initial analysis, Jay Miller from the FBI noted that one of the bill's major objectives was to "diffuse authority for developing DNA testing standards." He also pointed out that in the FBI's experience, external control of proficiency testing was not any more effective than an internally run program.[4] The FBI also received letters from nineteen directors of forensic crime laboratories in response to the FBI's request for comments on the proposed legislation. Although there was significant variation in their views on how the validity and reliability of DNA evidence should be ensured, the overwhelming response was that any effort to do so must come directly from the forensic science community.

Representing one strand of the community's reaction was Paul Ferrera, who directed Virginia's Division of Forensic Science and spearheaded its effort to establish a DNA typing laboratory and databank in the state. In his letter to John Hicks, Ferrera wrote the following:

> As you well know, forensic science laboratories have a unique responsibility and role unlike that of any other type of analytical laboratory. Certainly in no other field are examiners and their results of examinations so subject to scrutiny and review by independent experts as in the forensic setting. For this basic reason, and for so many more which are intuitively obvious to those of us in the field of forensic science, I believe it is critical that practitioners in the field of forensic science be charged with the responsibility of regulating their own profession. . . . Those outside the forensic science field do not recognize that historically the forensic scientist has successfully drawn on and adapted many sophisticated technologies from the various physical sciences and applied them to the criminal justice system.
>
> Therefore, I have steadfastly maintained that the leadership role relating to the DNA technology should remain with the Forensic Science community. Within that community, the FBI, through its FSRTC, Quantico and sponsorship of TWGDAM has assumed an effective leadership role in the implementation of DNA technologies.
>
> Therefore, with regard to the proposed legislation (S. 1355), it seems inappropriate to place now the most important responsibility of appointing an advisory board on DNA forensic analysis methods or approving a DNA proficiency testing program to the director of any agency not directly involved providing forensic laboratory services. With all due respect to NIST and NRC, one must question the experience and knowledge of those agencies regarding the experts actually engaged in the forensic applications of this technology.[5]

Representing the other major strain of thinking within the forensic science community was Rod Caswell, director of Florida's Forensic Laboratory System. Like Ferrera, he believed that representatives of law enforcement and forensic laboratories must appoint any oversight committee. If not, the hands of forensic scientists would be tied with "unworkable rules."[6] Unlike Ferrara, however, he believed that the FBI should not serve as the prime mover in the development and implementation of a proficiency-testing scheme or the creation of accreditation procedures, unless it was mandated that a federal agency must do so. He wrote, "the task of oversight is so critical . . . that I don't think it should be left to one laboratory alone, regardless of that laboratory's qualifications."[7] In his view, the body that was most equipped to handle such a task was the American Society of Crime Laboratory Director's Laboratory Accreditation Board (this board is abbreviated ASCLD/LAB, while its parent organization is abbreviated ASCLD).

ASCLD/LAB had been formed a decade earlier with the purpose of serving as a body that would design and administer a crime laboratory accreditation program.

This program consisted primarily of onsite inspections, proficiency testing, as well as various requirements for laboratory personnel. During his testimony before the Joint Subcommittee hearing, ASCLD's president Richard L. Tanton made the case for the involvement of the ASCLD/LAB in any nationwide quality assurance program. He also cautioned against overzealous attempts by nonforensic scientists to intervene in the regulation of forensic DNA typing. In his view, the main goal of any legislation should be to provide funding for state and local laboratories for the implementation of the technology, as well as for participation in quality control programs. He also testified that his organization supported legislation aimed at setting quality assurance standards, "provided that these standards and regulations are developed from within the forensic science community and are administered by a professional regulatory body whose purpose it is to administer and regulate these standards."[8]

ASCLD leadership absolutely abhorred any legislation that sought to diminish the authority of the forensic science community to regulate itself. In his prepared statement, Tanton contrasted the positive legislation described above with the kind of bills being considered in the New York State Assembly.[9] He stated that the New York legislation was highly problematic because it removed oversight responsibility from the forensic science community and assigned it to the state. In ASCLD's view, "the most viable and effective DNA Quality Assurance Programs will come from within forensic science using external expertise where appropriate. On the other hand, to place the responsibility for such programs with a well-meaning, technically competent organization unfamiliar with the forensic science venue would not only be unwise, but probably unsuccessful."[10]

Ultimately, Tanton believed that ASCLD/LAB should play as central a role in the regulation of forensic DNA analysis as the FBI did. Indeed, he believed that doing so would lend the technology more credibility than allowing the FBI to go it alone. In a letter to Don Edwards, he suggested that one way to solve various criticisms of the Simon Bill would be to maintain the proposed role for the FBI, but give ASCLD/LAB the responsibilities assigned to NIST. This change would open up the standard-setting process to a broad spectrum of the forensic science community; give ASCLD/LAB "important recognition outside the forensic science community"; aid ASCLD/LAB in obtaining the resources necessary to implement their DNA accreditation program; and "avoid creating the impression that Congress feels the forensic science community is incompetent to regulate itself."[11] Interestingly, two of the four outcomes of giving ASCLD/LAB responsibility for partnering with the FBI to regulate DNA profiling directly addressed the organization's reputation and funding. As such, it seems that Tanton's rationale for making ASCLD/LAB a prominent player may have been more focused on raising the prominence of his organization than making sure that DNA evidence was as valid and reliable as it could possibly be.

During the question-and-answer period after Tanton's presentation, Senator Simon made it clear that he was skeptical of the forensic science community's ability to regulate itself. He asked Tanton why less than a third of all crime labs in the country had been accredited by ASCLD/LAB in its ten years of existence. When neither Hicks nor Tanton had a ready explanation for this situation, Simon then expressed his worry that under a voluntary system, labs could simply pop up and start offering DNA testing without any oversight. In an attempt to assuage him, Tanton responded to this anxiety by stating that although the ASCLD/LAB accreditation program was voluntary, he believed the tremendous attention being paid to DNA typing would lead to its becoming the "de facto standard" in courts for the admissibility of forensic DNA evidence.[12] For Simon, however this was not a good enough answer. In response to subsequent discussion, he stated that it was unacceptable that forensic laboratories were not subjected to outside scrutiny. "I don't mean any disrespect to the fact that you have a lab and you're doing something," he said to a representative of the forensic laboratory for Illinois, "but it does seem to me that somebody on the outside has to come in and say: 'Are we getting quality work?'"[13]

This sentiment was the main thrust of Barry Scheck's testimony before the joint committee hearing. At this hearing, Scheck reiterated the kind of testimony that he had given on previous occasions, focusing specifically on the need to institute a system of blind external proficiency testing that modeled actual forensic casework as much as possible (against the wishes of the FBI), and the fact that the FBI should not be trusted to regulate itself. In his view, the only requirement for the agency that carried out the testing was that it was independent and did not have a vested interest in the results. He suggested ASCLD or the College of American Pathologists. Scheck further argued that a regulatory scheme similar to that set up by the recently passed Clinical Laboratory Improvement Act (CLIA) of 1988 should be created for forensic DNA typing. As he noted, quarterly blind external proficiency testing was the linchpin of the CLIA, which also dealt with the accreditation and licensing of medical diagnostic laboratories. Interestingly, and foreshadowing future developments, Scheck concluded his testimony by describing how DNA evidence could be useful in getting the innocent out of jail. As such, he argued that both the prosecution and the defense have a vested interest in making DNA evidence as accurate and reliable as possible.[14]

In a letter to Don Edwards written shortly after the hearing, John Hicks sharply rebutted Scheck's testimony. He began by pointing out the great difficulty associated with undertaking truly blind proficiency testing, in which realistic crime scene evidence must be created and a law enforcement agency must be found that is willing to submit the case and then answer any subsequent questions a forensic laboratory may have about it. He also noted that CLIA guidelines had recently been amended to remove the requirement for blind proficiency testing, and that the FBI was already engaged in a significant proficiency testing scheme that was as good as, if not better, than anything that Scheck could come up with. Perhaps ironically, Hicks concluded by invoking the same kind of distinction between public and private science that Scheck had initially used to criticize Cellmark and Lifecodes when he first testified in Congress in 1989. Hicks wrote that a CLIA model was inappropriate for the FBI because it was directed at commercial clinical laboratories.

> Crime laboratories are government, not commercial activities. They are typically adjunct to a law enforcement agency and therefore subject to public oversight and scrutiny. Virtually every probative test result from a crime laboratory is subject to exhaustive scrutiny in the courts. . . . In contrast, the commercial clinical laboratories are subject to minimal outside scrutiny and the test results are typically seen only by the requesting physician. There are many other distinctions between commercial and clinical laboratories and law enforcement crime laboratories, including the nature of the samples received for testing, the nature of the tests utilized, as well as the nature of the client communities served by these laboratories. While formal regulations may have been warranted for the commercial clinical laboratories, it is not clear that the same is appropriate for public crime laboratories.[15]

Thus, Hicks attempted to co-opt the defense's claim that there was a fundamental difference between public and private science, as well as forensic and diagnostic testing. While the defense used these dichotomies to argue for more oversight of crime laboratories, Hicks used them to argue for less.

Based on this kind of rhetorical jousting, then, it is not surprising that very little consensus came out of the 1991 joint hearing. If anything, the process served to lay bare just how much disagreement still remained between prosecution/law enforcement communities and the defense/civil liberties communities over who had the authority and expertise to ensure the validity and reliability of forensic DNA analysis. While the latter group continuously advocated for outside oversight of the technique, based largely on their discoveries of defects in FBI practice in *Yee* and other trials, the former communities resisted these efforts with equal passion and conviction.

Toward the end of 1991, in response to the failure of either the House or Senate bills to win the unqualified approval of the forensic science community and the FBI, Representative Edwards introduced a late-session replacement, H.R. 3371, which was also called the DNA Identification Act of 1991. The major difference between this bill and the Horton and Simon Bills was that it did not give the FBI an explicitly regulatory role, but made it the institution that issued standards that were to be followed by the forensic community. The result of this decision was that the bureau was given the ability to constitute the expert panels that would set rules for the field. In order to ensure that these rules and standards were followed around the country, Congress tied funding of state laboratories to compliance with FBI-issued standards.[16] The Edwards Bill ultimately served as the basis for the DNA Identification Act of 1994, which passed both houses and became the major federal law regulating forensic DNA analysis in the United States. It also served as the enabling legislation for the national DNA database.

Among other things, the DNA Identification Act of 1994 mandated that a DNA Advisory Board (DAB) be established by the FBI to provide advice on quality assurance and the setting of procedural standards. The FBI director was given the authority to choose the board's members based on the nominations of the National Academy of Sciences and professional societies of crime laboratory directors (i.e., ASCLD). In a strong boundary-setting move, Congress mandated that the board be composed of scientists from public and private laboratories, academic molecular geneticists and population geneticists, and a representative from NIST. Like TWG-DAM, lawyers, policy experts, and nonscientists in general were not welcomed. In another nod to the bureau, Congress stated that the FBI director merely had to take into consideration the recommendations of the DAB when issuing standards on quality assurance and proficiency testing—they were not binding in any way. While this was the ideal arrangement for people who believed that the FBI was the fittest agency to regulate the production of DNA evidence, some members of the defense community were deeply troubled by Congress's decision. Upon learning of this decision, William C. Thompson lamented that "having the FBI Director appoint the DNA Advisory Board is a bit like having the president of the Oscar Mayer Company appoint a board charged with regulating the content of bologna."[17]

NRC I

While DNA profiling was being debated in Congress, however, the importance of figuring out how to regulate the technique took a backseat to the debate over population genetics caused by the Yee trial and the NRC's first report. On 14 April 1992, the National Research Council released its long-awaited report on DNA Technology in Forensic Science. The report (hereafter referred to as "NRC I") came out more than three and a half years after it was initially proposed, took more than two years to produce, and was seven months overdue.[18] The report's scope was broad, covering everything from technical and statistical issues to legal admissibility to the social and ethical issues associated with development DNA databanks. The primary purpose of NRC I was to help resolve the numerous disagreements and debates that had emerged in the wake of Castro and to provide guidance to forensic scientists, policy makers, and judges who were concerned with ensuring the quality, validity, and reliability of DNA evidence and associated probability statistics. Its effect, however, was quite the opposite.

Although the majority of the committee's recommendations met with little resistance in the forensic, academic, and legal communities, the most contentious ones became the subject of severe criticism from all directions. The bulk of this disapproval was leveled at the NRC committee's creation of the "ceiling principle," an extremely conservative method for determining the probability of a random match between two DNA profiles. Critics charged that the ceiling principle was both "unscientific" and "irrational" and unnecessarily limited the strength of DNA evidence. Underlying these claims, however, was a disagreement about which legal and scientific values would be incorporated into the solution to the population genetics debate. This situation is unsurprising when one realizes that the ceiling principle was neither scientific nor legal, but rather both at the same time.

Soon after DNA typing was introduced into the American legal system, a few scientists (most notably Eric Lander) and defense lawyers began to call for an NRC study on the uses of the technique in the criminal justice system. They argued that the technique was so radically new that it needed serious study before being considered valid and reliable for use in convicting individuals suspected of violent crime.

In many ways the National Academy of Sciences was an ideal venue in which to begin resolving the issues that emerged as forensic DNA analysis was deconstructed in the courtroom. Long considered to be one of the primary sources of scientific advice to the U.S. government, the academy had significant experience bringing together a diverse array of knowledgeable individuals to study policy-relevant scientific matters. The academy is a "private, non-profit, self-perpetuating society" of distinguished scientists, engineers, and medical professionals created by an 1863 Act of Congress with a mandate to "advise the federal government on scientific and technical matters." It is an extremely influential, trusted, and powerful body that has provided advice to policy makers on a wide range of issues, including nutrition, defense strategy, transportation, public health, environment, and science education, as well as various scientific, technical, and medical questions of significant importance to the public. The academy's high credibility stems partly from the fact that membership is restricted to the most accomplished scientists, engineers, and medical professionals in the country, as well as a few from around the world. Indeed, election to the academy is generally considered one of the highest honors a scientist can achieve. It is important to note, however, that members of the academy do not engage in the policy studies themselves. This work is carried out by the academy's operating agency, the National Research Council, which was created in 1916 to serve this function. It should also be noted that neither the NAS nor the NRC are federal agencies and therefore receive no line-item federal funding. Their revenue is generated almost entirely through contracts from various governmental organizations to perform specific studies.

As a result of these early efforts, in August 1988, well before there was any serious defense challenge to DNA evidence, the NRC requested $310,000 from the FBI to conduct an eighteen-month study entitled "An Evaluation of the Application of DNA Technology in Forensic Science." Initially, the FBI declined to fund the entire project. At least in the early stages the introduction of DNA typing into the legal system seemed to be going quite well for the law enforcement community. Indeed, as Eric Lander drolly noted in his 1989 *Nature* article, a government official had recently told him that "the study was unwelcome: scientists had done their part by discovering DNA; it was not their job to tell forensic labs how to use it."[19]

By mid-1989, however, the bureau's views had changed significantly. They now needed an authoritative report in order to close down the debates that had emerged in the wake of the *Castro* case and to prevent their DNA profiling system from becoming embroiled in similar controversies. On 31 May 1989, at least in part through the efforts of Eric Lander, James K. Stewart, director of the National Institute of Justice, and Roger T. Castonguay, director of the Laboratory Division of the FBI, asked John Burris, executive director of the NRC's Commission on Life Sciences, to submit a grant application to convene a panel of experts to study the emerging controversy over DNA typing. In their letter, they estimated the cost of this study to be approximately $100,000.[20] Although such a figure was unrealistically low to the NRC, they decided to ask for significantly less than the $381,000 they had initially projected for the study. In the summer of 1989, Burris sent letters to various institutions (including the State Judicial Institute, the National Science Foundation, the National Institutes of Health, the Sloan Foundation, and the National Institute of Justice) requesting support for a scaled-back fourteen-month study estimated to cost only $251,865. All of the organizations solicited agreed to provide some assistance for the study, which began in January 1990.[21]

According to one participant, George Sensabaugh, many of the people involved in the NRC committee thought that it would do a great deal to limit the controversy that emerged in the wake of *Castro*. As he told me in a recent interview,

> Actually, the FBI, as well as a number of other people, and myself included, thought that the first NRC report was going to be a closure, not closure in the same sort of way because there hadn't been so many questions, really I think more of a ratification. John Hicks stated at one of the public hearings that he really saw the first NRC panel as coming forth with the statement rather like what you find on the tube of toothpaste, that with proper use, diligence and application, that this would be conducive to good health.... I was, even though I had a sense that Eric Lander had raised a number of questions, I didn't think that the questions that had been raised were going to present the kind of challenge to the technology that they ultimately did.[22]

As the deliberative process unfolded during four meetings in 1990 and dragged on through almost continuous debate, dissention, and negotiation throughout 1991 and into 1992, it became apparent to all involved that the final report would only be a temporary solution until more work had been done on the increasingly contentious issues relating to population genetics and statistics. According to a news article in *Science*, NRC I emerged only "after much strife, a threatened minority opinion, and countless leaks of confidential drafts," and was "described by NAS staff and committee members as one of the most contentious reports in recent years."[23]

Unfortunately, information about the process by which members of the NRC committee came to their conclusions will remain unknown for some time because documents are only released twenty-five years after their creation, beginning in 2015 in this case. Based on the composition of the panel, however, it can be inferred that every effort was made to bring in a roughly equal mix of academic scientists, forensic scientists, legal scholars, and judges with opposing viewpoints who could still work together to craft a consensus solution to the problems that had emerged in *Castro*.[24] As several commentators have noted, however, the committee was handicapped by the lack of a statistician or population geneticist.[25] While the decision not to include a population geneticist or biostatistician seems unwise in retrospect, it was not surprising at the time, since population genetics and statistics were only a minor issue before the *Yee* case, which took place in the summer of 1990.

Indeed, at least in the NRC's initial proposal, statistics and population genetics were mentioned only once in passing. Instead, the proposal stated that the study would address the following general areas:

- The general applicability and appropriateness of the use of DNA technology in forensics
- The need to develop acceptable standards in data collection and analysis
- An assessment of instrumentation and technology needs as well as the basic science needed to support forensic uses of DNA technology
- Management of DNA typing data
- Societal and ethical issues surrounding DNA typing
- Legal issues and the impact of DNA typing on the judicial system[26]

By the time that the report was issued, however, statistics and population genetics would become the key issues in the debate over the committee's findings.

Based on my interviews with committee members, the negotiations over the issues of standards, methods, and quality control were somewhat less contentious than the population genetics and statistical issues. Overall, the report was more in line with what the defense community had been advocating for. On most issues, it erred on the side of caution, acknowledging uncertainties where they existed and highlighting many problems that still existed surrounding DNA typing. At least according to one participant, George Sensabaugh, many of the forensic scientists and advocates of the use of DNA evidence went along with certain recommendations knowing that they would probably not be carried out in the real world. He told me that many of the NRC committee's recommendations were meant to accentuate the handwriting already on the wall about the need for improved standards, quality control regimes, and oversight mechanisms in a forensic DNA typing market increasingly dominated by the FBI.[27] Although the report was cautious and qualified, its ultimate message was that DNA typing is an extraordinarily powerful forensic technique that is reliable when DNA samples are collected and analyzed properly and testing is carried out under sufficient quality control and assurance mechanisms.[28]

With respect to the issue of governance, the report began in an uncontroversial manner by noting that "critics and supporters of the forensic use of DNA typing agree that there has been a lack of standardization of practices and uniformly accepted methods for quality assurance. The lack is due largely to the rapid emergence of DNA typing and its introduction in the United States through the private sector."[29] Further, the lack of standardization was also caused, at least in part, by the unique nature of forensic casework, in which there is "little or no control over the nature, condition, form, or amount of sample with which laboratories must work."[30] Also, they believed it was important to keep in mind that forensic DNA typing was a relatively young technology and that better technologies and practices were constantly being developed. As such, it was imperative for the committee that any recommendations they offered should not have the effect of locking forensic scientists into a particular technology prematurely. That said, the committee believed that the time had come for enough standardization and regulation to occur to "assure the courts and the public that results of DNA typing by a given laboratory are reliable, reproducible and accurate."[31]

Ultimately, the committee's recommendations were based on those found in the Report of the New York State Panel on Forensic DNA Analysis and the Horton Bill on DNA Identification, which had recently been introduced in the U.S. House of Representatives.[32] The most notable aspect of the committee's recommendation was the important role to be played by federal agencies. While the forensic community largely shunned governmental intervention, the committee saw it as the best way of providing the authority to regulators to make sure that DNA typing laboratories complied with standards.

The committee felt that while TWGDAM guidelines were an excellent starting point for establishing an overall quality assurance regime for DNA identification laboratories, they were not entirely sufficient because there was no mechanism to ensure that forensic laboratories complied with them. In an effort to ameliorate this concern, as well as the problem that TWGDAM was rather insular, the committee proposed that the responsibility for developing quality standards be shared by an additional body, the National Committee on Forensic DNA Typing (NCFDT). In the committee's view, the NCFDT would be administered by an appropriate federal agency (either NIST or NIH) and would be convened to provide expert advice on scientific and technical issues for the forensic community as they arose. In justifying its decision to place the NCFDT in the federal government, rather than in the law enforcement community, the committee wrote that "because its task is fundamentally scientific, we feel that this agency should be one whose primary mission is scientific rather than related to law enforcement. To avoid any appearance of conflict of interest, an agency that uses forensic DNA evidence itself would be unsuitable."[33]

The committee also concluded that "private professional organizations [such as ASCLD/LAB] lack the regulatory authority to require accreditation." This meant that a mandatory accreditation and licensing program needed to be developed by the Department of Health and Human Services, in consultation with the Department of Justice. In the committee's opinion, DHHS was the appropriate body to undertake this task because of its extensive experience with the Clinical Laboratory Improvement Act, as well as significant expertise in molecular genetics through the National Institutes of Health. Further, such a system was the most efficient because laboratories needed to be licensed and accredited only once, and not in numerous states. Interestingly, although the committee implied that external proficiency testing was an important part of ensuring the validity and quality of DNA evidence, it made no recommendations about whether this testing should be open or blind, who it should be performed by, or how the tests should be analyzed.[34] Finally, the report noted that the National Institute of Justice did not receive enough money to undertake adequate education, training, and research in the field of forensic DNA analysis.[35]

In addition to making several structural and organizational recommendations to improve the governance of forensic DNA analysis, the committee also made numerous technical recommendations about laboratory procedures and match criteria. Few of the committee's recommendations were new or caused any substantial controversy or debate. For instance, the committee recommended that ethidium bromide only be added to gels after electrophoresis; that in cases where a only a single band is visualized, the interpretation should always include the possibility that a second band was missed—either because it was so small that it ran off the gel, because it cannot be distinguished from the visualized band due to similar size, or because there was degradation; that at least one blank lane should separate evidence and suspect DNA samples on the gel so that leakage from one lane to the other can be detected; and that all anomalous results should be stated in writing on the final report. With respect to the measurement of fragments, the report recommended the use of monomorphic probes in order to detect band shifting; computer-aided systems to minimize bias and subjectivity; and an unknown control in each test to ensure that the analyst and his or her equipment are generating accurate sizing results.[36] The committee made no judgments about which DNA typing laboratory's match criteria were the best. Instead, it merely stated that this standard should be based on extensive replication studies using forensic-quality samples. Finally, the committee stated that new DNA typing methods and techniques should be published and peer reviewed before being introduced into court rather than after.

THE CEILING PRINCIPLE

While regulatory and technical issues had been the primary focus of the debates over DNA typing from late 1988 to mid-1990, the *Yee* trial would bring the issue of population genetics and statistics to the fore. The discussion of the *Yee* controversy and the Lewontin-Hartl/Chakraborty-Kidd *Science* affair noted that some scientists had expressed concern about the possibility of significant substructure within racial groups, while others considered the degree of substructure to be irrelevant to forensic statistics. The report boiled the issue down to two fundamental statements. In short, it noted, population geneticists from the first camp believed that the absence of substructure must be proven and cannot be assumed, while the latter camp believed that enough evidence had already been collected to assume that any substructure would not have a significant impact on statistical calculations.

According to the *Science* news article about the report, this schism was replicated within the committee, with Lander taking the cautious view and Caskey strongly advocating the latter view. In the end, the report took a precautionary approach to dealing with the substructure issue. In both scope and tone, it resembled to a remarkable degree the argument put forth by Lewontin and Hartl in their *Science* article and by Lewontin in his recommendations to the committee.[37] The report pointed out that there was a high probability that significant population substructure existed within American racial groups and that examining divergences from Hardy-Weinberg equilibrium was not a scientifically reliable method to uncover this substructure. It went on to state that "population differentiation must be assessed through direct studies of allele frequencies in ethnic groups."[38]

The report also stated that "the committee has chosen to assume for the sake of discussion that population substructure may exist and provide a method for estimating population frequencies in a matter that adequately accounts for it." The decision was justified in pragmatic terms and sought to incorporate conservatism and justice into probability statistics. The committee wrote the following:

- "it is appropriate to prefer somewhat conservative numbers for forensic DNA typing;"
- "it is important to have a general approach that is applicable to any loci used for forensic typing," and not just the ones already in current use;
- "it is desirable to provide a method for calculating population frequencies that is independent of the ethnic group of the subject."[39]

In this formulation, one can clearly see the simultaneous production of what is considered fair when probability statistics are used in the criminal justice system and the statistical techniques themselves. This precautionary tone is unsurprising when one realizes that Eric Lander largely controlled the final discussions of population genetics during the panel's deliberations, since Caskey was forced to resign from the committee in December 1991, shortly after Scheck and Neufeld highlighted his potential conflicts of interest in their appeal in the *Yee* case.[40] Caskey was in the process of patenting and commercializing the method of DNA profiling currently used today (short tandem repeat analysis), and therefore had a financial and professional interest in the admissibility of DNA evidence. Many defense attorneys and other commentators believed that this situation would make it difficult for him to objectively recommend standards and policies for conducting DNA testing.

After a considerable amount of debate and dissension, the committee ultimately decided to approve the method for accounting for possible substructure that Lander had proposed in a recent publication and that Lewontin and Hartl had reiterated in their *Science* article.[41] Called the "ceiling principle," the method would enable forensic laboratories to use the product rule (i.e., multiply the genotype frequencies at each locus to get an overall probability of random match for a DNA profile) in a way that would be independent of the conditional probability associated with ethnicity or subpopulation. This goal would be accomplished by determining a so-called ceiling frequency that would serve as the minimum frequency for each allele used in forensic casework. Thus, while waiting for detailed studies of the genetic makeup of numerous ethnic groups, which would take many months of hard work to investigate, one could use the ceiling frequency in place of the general allele frequency in the racial reference databases.

One of the many issues debated within the committee was the best way to calculate this ceiling. In its report, the committee argued that it was necessary to balance rigor with practicality. As such, it would be impossible to calculate the allele frequencies of all alleles used in forensic casework in every conceivable subpopulation. Thus, the committee decided that the best approach would be to take random samples of 100 individuals from fifteen to twenty relatively homogeneous subpopulations. Then, allele frequencies at each forensically relevant locus would be determined. Finally, the largest frequency in these populations, or 5 percent, whichever was larger, would become the ceiling frequency.[42] The 5 percent figure was chosen, somewhat arbitrarily, because the committee concluded that "allele frequency estimates that were substantially lower would not provide sufficiently reliable predictors for other unsampled subgroups. Our reasoning was based on population genetic theory and computational results, and we aimed at accounting for the effects of sampling error and for genetic drift."[43]

Until such time that the appropriate sampling studies could take place, the committee went on to advocate for a so-called interim ceiling principle, which stated, "In applying the multiplication rule, the 95% upper confidence limit of the frequency of each allele should be calculated for separate U.S. 'racial' groups and the highest of those values or 10% (whichever is larger) should be used. Data on at least three major 'races' . . . should be analyzed."[44] In simple English, the interim ceiling principle called for the setting of the minimum allele frequency at 10 percent.

Although the ceiling principle was only one of many recommendations issued by the NRC committee, it quickly became the most controversial. Scientists and population geneticists on both sides of the debate weighed in with opinions about why the method was "unscientific," "illogical," or "just simply wrong." The bulk of the criticisms centered on two issues: first that the recommendations were ad hoc and not based on sound science and, second, that they were unnecessary.

In addition to criticizing the NRC's reliance on Lewontin and Hartl's *Science* paper, and asserting (again without sociological, demographic, or anthropological data) that American ethnic groups show considerable interbreeding, statisticians Bernie Devlin, Neil Risch, and Kathryn Roeder also attacked the NRC's ceiling principle. They began by calling the 5 percent and 10 percent ceiling values arbitrary and stated that the overall methodology was lacking scientific justification.[45] They also argued that the proposal to examine the degree of substructure by sampling 100 individuals within fifteen to twenty ethnic groups was fraught with serious flaws. Specifically, because there were more than twenty allelic bins within each locus, the number of individuals would need to be much larger than 100 to capture their true distributions. This undersampling would lead to dramatically exaggerated maximum allele frequencies and would potentially undermine the statistical power, and admissibility, of forensic DNA analysis.[46] In a December 1992 *Proceedings of the National Academy of Sciences* article, Bruce Weir made similar criticisms, noting, for instance, that "the merits of the ceiling principle were somewhat diluted by the NRC report suggesting ad hoc frequencies of 5 percent or 10 percent." British geneticist Newton Morton went even further, calling the ceiling principle "illogical" and "absurdly conservative." He sarcastically noted that the report could just have easily urged forensic laboratories to "move the decimal point a couple of places."[47]

The FBI based its official, albeit subtle, criticism of the ceiling principle on the notion that its methods and procedures were already conservative enough, and that the only major difference between its methodology and the NRC committee's was the degree of conservatism.[48] At least in the bureau's view, "the degree of conservativeness is a legal, rather than technical, question regarding the weight given to DNA evidence, not its admissibility."[49] In making such a claim, the FBI was arguing that the decision to use the ceiling principle was a choice to be made, but not a fundamental requirement for valid results. If individual prosecutors decided that it was in their best interest to use the ceiling principle, or if the court demanded it, the FBI made it clear that they would comply with this request.[50]

In addition, the bureau noted that there might not be sufficient scientific evidence to warrant the kinds of subpopulation studies and subsequent ceiling frequencies suggested by the committee. The FBI's official response to the NRC report, which took credit for initiating it back in 1989, states only that the proposal to study allele frequencies in fifteen to twenty homogeneous populations must be reviewed to determine whether

1. the presumed problem of potential substructure affecting the statistical estimate of a DNA profile in relevant U.S. forensic populations really exists [the exact definition of such populations is not provided],
2. the Committee's proposed approach can achieve the desired result, and
3. the proposed study is practical and cost-effective.[51]

The FBI response also noted that it was in the process of analyzing variable number tandem repeat (VNTR) data from several subpopulations collected by various foreign forensic science laboratories and TWGDAM member laboratories in the United States. It noted that the results from this large study would be available by the summer of 1992 and that any significant population substructure would become apparent. The FBI decided to take "no position on the advisability of the population studies recommended by the Committee" until that time.[52]

The FBI and its allies were not the only critics of the ceiling principle. Mathematician and population biologist Joel Cohen, for instance, pointed out in a letter to the editor of *American Journal of Human Genetics* that the ceiling principle was truly conservative if and only if a population contained no subpopulations with linkage disequilibrium between loci or divergences from HWE at one or more of the loci being used.[53] Thus, in his analysis, the very problem that led to the formulation of the ceiling principle was not adequately addressed by it. As a result, he concluded, the principle should not be used until research had been done to assess the conditions under which it failed to be conservative. In the meantime, he suggested that forensic scientists use the counting method for calculating the frequency of a given genotype. In this method, a large database of DNA profiles is amassed and the frequency of a particular genotype is calculated by dividing the number of times it appears in the database by the number of individuals in that database. Similarly, Richard Lewontin felt that the ceiling principle did not adequately address the problems with substructure that he and Hartl had raised in their article. In the *Science* news article, he was quoted as saying that the arbitrary choice of a figure "out of the air," without detailed population sampling was "irrational."[54]

That said, Lewontin and Hartl did not feel that Devlin, Risch, and Roeder's criticism of the NRC report was entirely justified. In a letter to the editor of *Science* written a few months after Devlin and his colleagues' article was published, Hartl and Lewontin charged that their critique was "scientifically . . . a rehash of old arguments and inadequate data" that had already been discussed and refuted in their December 1991 article.[55] While Hartl and Lewontin criticized the ceiling principle as being "not excessively conservative," they reserved the bulk of their displeasure for the Devlin group's claim that the additional research advocated by NRC (and Lewontin and Hartl) to obtain data relevant to population substructure would do nothing to resolve the population genetics debate. In their view, "the call for 'no new data' will only guarantee more contentiousness and controversy."[56]

Indeed, Hartl and Lewontin cited a recently published study by Hartl, along with Dan Krane, a biologist at Wright State University in Ohio,[57] and others, that seemed to support the conclusions of Lewontin and Hartl (1991) as well as the NRC report.[58] In this study, Krane and his colleagues took data from two Caucasian subpopulations, ethnic Finns and Italians, and ran a series of analyses to determine whether the product rule still gave adequate approximations of the probability of a random match, even when substantial substructure existed within the reference population (i.e., the claim made by the FBI and their supporters). Krane and his coauthors concluded that a significant portion of the estimated probabilities were artificially small, both when the overall Caucasian database was used (i.e., when data from the ethnic populations were merged into a single database) and when the opposite ethnic database was used to calculate the probability of a random match.[59] The inescapable conclusion for them, and for Hartl and Lewontin, was that substructure did indeed have a profound effect on the probability statistics that emerged from the use of the product rule.[60]

Proponents of the ceiling principle, most notably NRC committee members Eric Lander and Richard Lempert, shot back that it was never meant to be a strictly scientific solution to the problem of population genetics in the forensic context. Thus, they believed the concerted attacks on the ceiling principle as "arbitrary," "illogical," "irrational," and "unscientific," although unsurprising, were misguided.[61] As

Lempert wrote in a 1993 *Jurimetrics* article, in which he expressed both his support for and misgivings about the NRC report, "The ceiling principle blends scientific and value considerations; indeed the blend of the two is in large measure what makes the ceiling principle vulnerable to scientific criticism. It is also what makes it an arguably more attractive forensic approach than current product rule procedures."[62]

In his view, the core value was to always err on behalf of the presumption that the defendant is innocent when making probability calculations. Although there was "no scientific basis for this value," it was at the heart of the FBI's fixed bins and the use of $2p$ instead of p^2 in cases of homozygotes. Driving home his point, he wrote, "The NRC Committee's central task was arguably to determine what scientifically sound procedures best implemented this legal value. This task is different from the task of determining how to secure the most probable point estimate of the likelihood that some random person left [DNA evidence at the crime scene], and the NRC Committee cannot be criticized for suggesting a method that does not yield the 'most probable' estimate. Science alone cannot provide a yardstick with which to measure the Committee's recommendations."[63]

Other committee members expressed similar sentiments. Lander, for instance, argued in his *Science* letter that the ceiling principle had two main purposes, neither of which had anything to do with arriving at the best solution to the substructure problem on strictly scientific grounds. From the perspective of the legal system, it was devised to facilitate the future admissibility of DNA evidence in courts of law by developing a standard of practice that was "so conservative as to ensure that there would be no serious scientific argument that the evidence could be said to overstate the case against the defendant."[64]

The second purpose of the ceiling principle, as Lander conceived of it, had more to do with governance of forensic DNA technology than its validity or admissibility. As he told me in a recent interview, "the whole point of the ceiling principle was ridiculously simple and nobody could just get their head around it, which is: don't attach too much weight to any one locus. If you want bigger numbers, do more loci." In his view, the ceiling principle was "merely a bit of engineering conservatism" that was designed to counteract the many things that could go wrong in the test, ranging from laboratory error to the effects of substructure. By limiting the weight that forensic scientists could attach to each locus to 5 percent or 10 percent, he hoped to coerce them into typing more loci. In this way, forensic analysis would still be able to generate the low probabilities of random match that prosecutors like to bring to court while at the same time significantly reducing the possibility that such a random match could occur. In his view, it was this aspect of the ceiling principle that angered the forensic science community. However, because they could not attack it on these grounds (for fear that they would come across as not being appropriately conservative), they decided to fight it from a population genetics standpoint, even though it wasn't meant to be strictly scientific at all.[65]

REACTION TO THE NRC REPORT

As science studies scholar Stephen Hilgartner has shown, through time, the National Academy of Sciences has developed an elaborate set of administrative procedures, as well as rhetorical devices and dramatic performance techniques, that seek not only to produce the most objective, independent, and high-quality science advice possible, but also to portray it as such to decision makers and the public.[66] These procedures include efforts to prevent conflicts of interest among panel members and to prevent sponsors from influencing the results of studies, as well as the publication of reports that have not been adequately reviewed by people not involved in the study itself. Further, although academy studies do solicit input from stakeholders through public meetings, the actual deliberations, draft reports, and internal documents are tightly guarded and kept private for a period of twenty-five years. Committee members are not allowed to comment on the negotiation process that leads to consensus. Although this makes the work of critics of the academy (as well as historians) difficult, it is meant to ensure that panel members can engage in frank and open discussion without fear of outside pressure or subsequent retribution.[67] At least in Hilgartner's view, it also allows the academy to maintain a tight boundary between the "backstage" politics that go into the production of a report (i.e., intense disagreement among experts; threats to leave the committee, etc.), and the "front stage" presentation of the final report as well reasoned, objective, and worthy of trust.[68]

These safeguards, and the strictly guarded boundary between front and backstage, sometimes break down in the face of the hotly contested issues that the NRC investigates. This was certainly the case with the first report on DNA typing. Two committee members, Tom Caskey and Michael Hunkapiller from Applied Biosystems, resigned during the deliberations as a result of their ties to private companies; supposedly confidential drafts were leaked to interested parties by committee members; and the dissensus that characterized the negotiations surrounding the ceiling principle was made very public.

Making matters worse for the academy, an erroneous representation of the report's main conclusions was leaked, presumably by someone close to the defense community, to *New York Times* science reporter Gina Kolata several days before the report was due to be released at a press conference on Thursday, 16 April.[69] Kolata's story, "U.S. Panel Seeking Restriction on the Use of DNA in Courts," ran on Tuesday, 14 April, and stated that the NRC committee had recommended that "courts should cease to admit DNA evidence until laboratory standards have been tightened and the technique has been established on a stronger scientific basis."[70] She went on to write that the committee believed that "the method is potentially too powerful and too important for its development to be left solely in the hands of prosecutors and law enforcement officials. Instead, the report says, it must be regulated and controlled by scientists and federal agencies that have no stake in the method's success or failure."[71]

Kolata's story forced the NRC not only to release the report two days before it had planned to, but also to schedule an impromptu press conference that cleared up her erroneous account of the committee's recommendations. Committee chair McCusick began the press conference by noting that he was very upset when he read the article, which was picked up by the *Baltimore Sun*. He continued, "It seriously misrepresents our findings. Its lead paragraphs are wrong. The impression it gives about our conclusions is misleading." He went on to say that the committee definitively did not call for a moratorium on DNA evidence and that it confirmed the general reliability of using DNA typing in forensic casework. Further, although the report called for more standardization, federal regulation, and accreditation, it never stated that courts should cease to admit DNA evidence until the recommended programs were put into place.[72]

Although most of the questions at the press conference dealt with the committee's recommendations or clarification of technical details, a few reporters asked the more fundamental question of why the committee would call for significant changes to a technique and a technological system that they claimed was already valid and reliable. Craig Fischer from Pace Publications, for instance, asked, "It seems that you're saying that the current system is not broken, but you want to fix it in half a dozen pretty serious ways. If the current system is adequate to send someone to prison for 20 years or more, why do we need to bother with all these steps that you recommend? These are not trivial suggestions."[73] Curiously, although the report was commissioned primarily to resolve debates surrounding the technique that had emerged in the first few years after its introduction, none of the committee members mentioned the problems of the past. Indeed, they all praised the DNA typing industry for regulating itself effectively. Their concern, they said, was that DNA technology was evolving very rapidly and would soon spread to laboratories around the country that might not have the same level of skill and expertise as the FBI, Cellmark, and Lifecodes possessed. Lander, after reiterating this basic line, continued:

> What the report is premised on is the notion that DNA typing is such a powerful technology that not only do we want it to be very good, we want it, in the long run, to always be assured to be perfect. Because it is such a powerful technology that it can lead to absolute identification, there's no need for, and no excuse for, errors to ever happen. We're looking to a future when there are more laboratories practicing DNA technology than there are today, many more technicians in the lab doing it, and what we have to do is make sure that as we move from the very good into the future that we ensure that it be perfect. And for that reason, it's the appropriate time to regularize the quality control, to regularize the evaluation of new technologies. No, we don't have to say something is bad today to say that we want to ensure that it's absolutely perfect in the future.[74]

Although they did not explicitly say so, it seems that all of the committee members who spoke were doing their best to make it known that courts should not use the NRC report as the basis for ruling pending DNA evidence inadmissible, or to reopen previously decided cases that rested on DNA evidence.

This was the interpretation that dominated Gina Kolata's article on the press conference that was published on the front page of the next day's *New York Times*.[75] In the story, headlined "Chief Says Panel Backs Courts' Use of a Genetic Test: *Times* Account in Error," Kolata recounted the statements made by McCusick and the rest of the committee members. She went on to report that the *Times* had based its previous interpretation on "the views of legal experts" and had erred in stating that the panel called for a moratorium. Kolata, however, seemed unwilling to back down from her previous stance that the report had the potential to cause serious problems with the legal admissibility of DNA evidence. The article highlighted significant confusion in the legal and scientific community about what the actual impact of the NRC report would be. For most of the people interviewed for the story, including Peter Neufeld, as well as legal scholars Paul Gianelli, Edward Immwinkelreid, and Randy Jonakait, the report was tantamount to calling for a moratorium because no laboratory currently met the standards recommended by the committee. Indeed, Jonakait was quoted as saying, "Here you have a disinterested scientific panel, in some sense the only people who have looked at this in a disinterested way, and they are saying that until the quality control systems are in place you shouldn't be sending people to jail based on DNA evidence. . . . Certainly, defense lawyers are going to argue that this should be the standard."[76] Curiously, and perhaps suggesting Kolata's partiality on the subject, no opposing viewpoints were included in the 15 April article, even though prosecutor Rockne Harmon was quoted in the previous day's story as saying that the report was irresponsible for not explicitly addressing this issue.

Despite the insistence of committee members that the NRC report was a forward-looking document that said nothing about the way that forensic DNA analysis was carried out in the past, defense lawyers did indeed quickly argue that the recommendations of the NRC report were, for all intents and purposes, the same as standards of general acceptance within the legal system. In several 1992 appellate cases throughout the country, the NRC report, as well as the media circus that surrounded the publication of Lewontin and Hartl's 1991 *Science* article, were cited as evidence that a significant controversy existed within the scientific community about appropriate standards and population genetics issues. Indeed, legal commentator David Kaye went so far as to describe these cases as a "third wave of cases . . . crashing down upon [the] battered legal shoreline" of DNA evidence.[77] In some cases, most notably *United States v. Porter* (618 A.2d 629, 1992), which involved a claim of statutory rape, appellate courts decided to remand cases in an effort to determine whether or not the recommendations of the NRC committee constituted the generally accepted position in the scientific community.[78] In others, the use of probability calculations not incorporating the NRC's ceiling principle was ruled inadmissible based on the apparent controversy.[79]

In the consolidated California appeals case of *People v. Barney and People v. Howard* (10 Cal.Rptr.2d 731, 1992), defense attorneys Linda Robertson and Victor Blumenkrantz argued that there was a lack of generally accepted standards, guidelines, and controls pertaining to numerous aspects of DNA testing laboratories' analysis. Although the First Appellate District Court rejected their arguments on most issues—including standardization as a precursor for admissibility and the lack of scientific consensus regarding match criteria or band shifting—it ruled that two Alameda County judges had erred in admitting probability statistics produced by the FBI (in *Howard*) and Cellmark (in *Barney*). The court based its decision primarily on "the *Science* articles of December 1991, [which] vividly demonstrate not merely a current absence of general acceptance, but the presence of a 'bitter' and 'raging' disagreement among population geneticists," as well as the discussion of this dispute in the NRC report.[80]

The court went on to note that Lewontin and Hartl and Chakraborty and Kidd significantly disagreed with one another about the existence of substructure as well as the effect that this phenomenon has on probability calculations. Because both sides had significant support within the scientific community, both in number and expertise, the court could only conclude that there was no generally accepted position on probability statistics in the relevant scientific community.[81] Cognizant of the hundreds of previous cases in which all aspects of forensic DNA evidence had been ruled admissible, the court argued that "the debate that erupted in *Science* in December 1991 changes the scientific landscape considerably, and demonstrates indisputably that there is no general acceptance of the current process."[82]

Although the court determined that the statistics in *Barney* and *Howard* were inadmissible, it was optimistic that the ceiling approach described in the NRC report would help point the way to some common ground between the two factions in the dispute. The court concluded, "the question now at hand is whether the interim and future methods of statistical calculation proposed by the NRC report will be generally accepted by the population genetics community. If, as it appears likely, this question is answered in the affirmative in a future *Kelly-Frye* hearing, then DNA analysis evidence will be admissible in California."[83]

When the California court revisited the issue eight months later in *Wallace*, it rendered the same decision, noting that "not enough time to confirm our speculation that the new methods of statistical calculation proposed by the NRC report will likely receive general acceptance resulting in future admissibility of DNA analysis evidence." This time, however, the court was less optimistic that the NRC report would solve the controversy that emerged in the wake of the 1991 *Science* articles. "Recent developments have shown that general acceptance may not be easily achieved," the court said dishearteningly. Instead of "attempting to come to terms with the NRC report or some other compromise on statistical calculation," the court continued, proponents of the FBI's methodology have "taken the offensive and attacked the report's proposed new method . . . as unsound."

In *Porter*, the District of Columbia Court of Appeals considered the government's appeal of a September 1991 trial court decision to exclude the FBI's DNA evidence in twelve consolidated cases. Judge Henry H. Kennedy's initial ruling was based on the belief that there was "a controversy within the scientific community [about population genetics] which has generated further study, the results of which will soon be available for study." The court declared, "it is *after* these studies and others, such as the study which is being prepared by the National Academy of Sciences, have been completed when the court should be called upon to admit DNA evidence, not before."[84]

Upon reviewing the facts of the original case, the recently released NRC report, as well as the prosecution's argument that the views of the prosecution witnesses were more representative of the general scientific community's than the defense's witnesses, the appeals court upheld Judge Kennedy's ruling, remanding the case back to the trial court for further consideration.[85] This trend toward ruling existing DNA evidence inadmissible, or remanding the case back to the trial court for further study, continued throughout the remainder of 1992 and into the first part of 1993.[86]

The FBI Demands a New Report

The prosecutorial, forensic science, and law enforcement communities were, needless to say, unhappy about the legal developments surrounding the NRC report and the *Science* controversy. They believed that the committee's efforts to integrate scientific and legal concerns in the ceiling principle were wrongheaded and rejected the mindset of members like Lander and Lempert. Advocates of DNA typing argued that science was a separate domain from law and that the two should be kept apart as much as possible. According to prosecutors James Wooley and Rockne Harmon, the "forensic DNA brouhaha" was based not on sound science, but rather on the willingness of defense witnesses to raise hypothetical problems that may result in unreliable DNA evidence in the courtroom.[87] Further, they believed that objections to the use of the product rule raised by Lewontin, Hartl, and others had been resoundingly discredited by the work of Chakraborty, Kidd, Devlin, Risch, Roeder, and others.

In a group letter to Burton Singer, chairman of the National Academy of Sciences' Committee on Statistics, more than one hundred attendees of the FBI-sponsored Second International Symposium on the Forensic Aspects of DNA told him the statistical and population genetic recommendations of the report "have not been well received by the scientific community."[88] They went on to argue that "published technical articles have demonstrated the validity of current DNA technology and of appropriate statistical calculations." The NRC report, however, had caused confusion over this fact in several recent court decisions. The group concluded by asking that the Committee on Statistics address these issues in a new report. Implicit in this letter was the idea that the statistical community was the audience and the proper evaluator of the ceiling principle.

Pressure on the National Academy of Sciences to convene a new panel on forensic DNA analysis would grow throughout the spring of 1993. In early April, John Hicks met with NRC staff members at least twice and expressed the FBI's and its allies' extreme criticism of the report and the "discredited Lewontin theory" upon which it relied.[89] Although one NRC staffer suggested to Hicks that "the fact that appellate courts are asking lower courts to use the ceiling principle was an indication that we had established the consensus of the relevant scientific community in the *Frye* sense," Hicks, apparently, was not persuaded by this line of reasoning. In his view, the FBI already had a perfectly good statistical method and the relevant scientific communities thought that the ceiling principle was completely unacceptable.[90]

After a couple of weeks of informal communications between the FBI and the National Academy of Sciences, William Sessions, director of the FBI, sent an official letter to Frank Press, president of the National Academy of Sciences requesting NRC's help in resolving the confusion and debate surrounding NRC I. Sessions began his letter by noting that before the report was issued, DNA evidence was ruled inadmissible in only two of fifty-eight appellate court decisions. Since the release of the report, that number had skyrocketed to eleven of thirty. Sessions pointed out that this crisis was largely caused by the ceiling principle and had occurred despite the fact that the report explicitly stated that DNA should continue to be used in the courts. He also complained that the principle was ambiguous and that numerous qualified scientists believed that it was less scientifically valid compared to other well-established methods, such as the FBI's, that barely received mention in the first report. Further, recent scientific work, such as that of Devlin, Risch, and Roeder, "may obviate the need for its use."

Sessions requested that the academy undertake an immediate review of the issues and "empanel a group with strong representation from the statistical community, as well as population genetics, molecular biologists, and the appropriate experts to prepare a new report." In addition to this report, Sessions stated that it would be helpful if the academy issued a "clarifying statement by the previous NRC committee" acknowledging that the ceiling principle was only "intended to serve a perceived public policy need." The statement should also address the "ambiguity in the wording of the report concerning statistical approaches to the use of DNA typing" that have allowed experts to arrive at vastly different probabilities of a random match. He concluded, "The NRC may have other ideas about the most appropriate approach to take in dealing with this matter. However it proceeds, since DNA technology is such a powerful tool in the criminal justice system, and one that can resolve issues of fact not possible by other means, I ask the NRC to act quickly to resolve the controversy it created over its report."[91]

The NAS reacted remarkably quickly to the FBI's request for a new report. By 18 May 1993, the academy had already put together a preliminary proposal, *DNA Forensic Science: An Update*, which it sent to the Department of Justice and various other funding bodies.[92] This proposal noted that the FBI had requested the study and that the NRC was considering forming a committee to review the merits of the ceiling principle as well as other alternative statistical approaches, as well as to "access and describe the degree of certainty of DNA evidence in ways useful to the courts; consider sources of error such as differences among laboratories or techniques; and lay out alternative perspectives about the assumptions and assessments of uncertainty and describe their implications in court cases."[93] Finally, the proposal acknowledged the controversy swirling around the first report and promised to complete this study in an unusually short timeline of six months.

The FBI was not happy with the academy's proposal. In the opinion of the bureau's top management, the NRC's job was not to reexamine the wide range of issues discussed in the first report, but rather to study the "scientific reasonableness of current or proposed statistical approaches and whether these various approaches accomplish what they were intended to achieve."[94] In a letter to Richard Rau from the National Institute of Justice (which was also sent to Eric Fischer at the NRC),[95] John Hicks wrote that the FBI's view "must be impressed upon the NRC before the DOJ grants funds for further study."[96] Presenting his view of the relationship between science and law, Hicks went on to write that the courts "do not need theoretical discussions on assumptions and assessments of uncertainty. As clearly indicated in dozens of legal decisions, the courts are seeking definitive statements from the scientific community on the 'general acceptance' of the validity of various statistical methods." Hicks also wanted to make sure that the NRC clearly understood what questions the FBI wanted the new panel to clear up. He wrote, "In assessing the significance of the association, the question is, 'What's the probability that another individual (unrelated to the suspect) could have contributed the material at the crime scene?' The question is not how many American Indians or Italian Americans or Irish Catholics or Karitiana Indians could have deposited the crime scene material. The next question should be, 'Can a statistical construct be applied which provides a reasonable estimate of the significance of the association and is that estimate sufficiently conservative to accommodate variations in ethnic diversity?'"[97] Hicks concluded by noting that the committee's success (at least in the eyes of the FBI) would depend on the "self-discipline of its members to confine their efforts to the questions central to the existing controversy."[98]

By the time that the final proposal had been prepared, the NAS had made almost all of the changes requested by the FBI regarding which issues were salient and the relationship between science and law. In the final August 1993 proposal, the NRC pointed out several times that the proposed study "will emphasize statistical and population genetics issues in the use of DNA evidence" and "will focus on scientific rather than legal issues." According to the proposal, the committee would "review relevant studies and data, especially those that have accumulated since the previous report" in the process of formulating its recommendations. The proposal continued:

Among the key issues examined will be the extent of population subdivision and the degree to which this information can or should be taken into account in the calculation of probabilities or likelihood ratios. The committee will review, and in its report explain the major alternative approaches to the statistical evaluation of DNA evidence, along with their assumptions, merits and limitations. The report will include specific, concrete, and realistic examples. It will also specifically rectify those statements regarding statistical and population genetics issues in the previous NRC report that have been seriously misinterpreted or let to unintended procedures.[99]

Shortly after this final draft was complete, the proposal was sent to various funding bodies for review, including the National Institute of Justice, National Science Foundation, and the National Institutes of Health.

NRC II

On 30 August 1993, James Crow, a well-known population geneticist who had never before been involved in the forensic DNA debate, was asked to chair the new NRC committee. Despite the perceived importance of the follow-up report, the NAS found it so difficult to secure funding for the committee that nothing happened for nearly a year.[100] Finally, in August 1994, the NRC named the people who would make up the committee. Unlike the first committee, which was made of members with extremely diverse views, the second committee was stacked with individuals less likely to produce a report that angered the FBI and the forensic science community. No outright critics of the FBI or members of the defense community were included in the committee. Five geneticists, including Chairman Crow, and two statisticians were included in the committee, as were three law professors. Haig Kazazian, chair of the Department of Genetics at University of Pennsylvania School of Medicine, and George Sensabaugh, a professor of forensic science at the University of California-Berkeley who was generally considered to be neutral in the debates, were the only members who also served on the first committee. The committee met in September 1994 and then twice in 1995.

When the second NRC report was finally released in mid-1996, the defense community's concerns were largely ignored and the FBI got almost everything it wanted when it asked for a new report. Of particular concern to Thompson and other critics of the DNA Identification Act and NRC II was the extent to which the FBI and the forensic community were left to police themselves. While the first NRC report broadly recognized the need for external scrutiny of forensic DNA testing, the act and NRC II both back off of this argument in a major way. In fact, while NRC I praised the work of the FBI's TWGDAM, it argued for the creation of a National Committee on Forensic DNA Typing (NCFDT) that would be housed in a government agency that had no vested interest in forensic science or law enforcement—for example, NIH or NIST.[101] While the membership of NCFDT would not differ substantially from TWGDAM or other existing groups in terms of expertise, it would not be subject to direct oversight by the very institution it was advising, and it would most likely consist of different people (there was a significant amount of overlap in the membership of the DAB and TWGDAM). The call for such an organization was gone in NRC II, replaced with the somewhat bland statement that "the 1992 NRC Report recommended that a [NCFDT] be formed to oversee the setting of DNA-Analysis Standards. The DNA Identification Act of 1994 gives this responsibility to a DNA Advisory Board appointed by the FBI. We recognize the need for guidelines and standards and for accreditation by appropriate organizations."[102]

In demanding a new report, the FBI insisted that science and law be kept separate and that legal issues, such as what is fair to the defendant, should be left to the judge and jury for resolution. The bureau further sought to control the identity of the expert in the resolution of the DNA typing controversy. In place of the first NRC report's heterogeneous community, the FBI insisted that the second NRC committee be comprised primarily of statisticians and population geneticists—the two major scientific communities that were not represented on the first committee. The FBI never suggested that members of the defense community should be on the second committee, and it left no room for that community to suggest alternative forms of relevant expertise. By accommodating the FBI's wishes, the NRC further solidified the bureau as the predominant source of regulation and standards for DNA typing. As we shall see in the next chapter, instead of taking an active role in the debates, as the first NRC committee had, the second NCR committee did little more than affirm policies and practices already in place when it belatedly released its report in 1996.

CHAPTER 8

The DNA Wars Are Over

When Americans think about controversies over DNA evidence that have taken place in this country, *Castro*, *Yee*, and the National Research Council probably do not figure too heavily in our collective memory. Rather, visions of white Ford Broncos, bloody gloves that don't fit, and footprints from "ugly-ass" Bruno Maglia shoes dominate our perceptions of the use of DNA in the criminal justice system. Who can forget the low-speed highway chase, Judge Ito, nonstop news coverage of the "Trial of the Century," the theatrics of the highly paid legal "Dream Team," and the racially charged reaction to the not-guilty verdict? Justified or not, the criminal trial of ex-football star and actor Orenthal James Simpson looms large over the history of DNA profiling.

Despite the publicity that the DNA Wars received in the run-up to the trial, the *Simpson* case was in many ways the denouement rather than the climax of the disputes that have been described in this book. Although the second NRC report had not yet been written, the debates over DNA profiling were essentially over when Simpson's criminal trial began in January 1995. The validity of DNA profiling as a technology was never actually questioned by Simpson's defense team, even though he had hired Barry Scheck, Peter Neufeld, and William C. Thompson to attack the forensic evidence in the case. Instead, the defense focused on the potential for accidental contamination and police malfeasance during the collection, processing, and storage of biological samples at the crime scene.

There was no one event or document that caused the debates to end in the months leading up to the trial. Rather, the legal and scientific debates over DNA evidence were extinguished slowly through a combination of FBI initiatives, judicial decisions, federal legislation, a very shrewd public relations move by former defense witness Eric Lander and the FBI's Bruce Budowle, technical changes in RFLP-based DNA profiling, as well as the development of an entirely new method for creating DNA profiles called short tandem repeat analysis, or STR for short.

But not everybody involved in the debates over DNA typing believed that all of the problems that had emerged over the previous five years had been resolved. One of the most important factors in the end of the disputes was that the issue of population substructure received so much attention in the wake of the *Yee* trial and the problematic ceiling principle, that it came to represent almost the entire debate in the popular press, the academic community, and the courtroom. Additionally, in the case of the population genetics debate, defense witnesses provided a clear challenge to the FBI: they called for the examination of racial groups for population substructure. While meeting this challenge, the FBI succeeded in making population genetics the dominant issue to be resolved. In their demands for a second NRC report, the FBI did not ask the National Academies to resolve questions from the first report that it did not think were legitimate (e.g., how to set up a blind proficiency testing scheme or how to estimate laboratory error). The FBI wanted a document that courts could point to as evidence that no controversy existed over DNA profiling, and this was largely what they got.

When the second NRC report was finally published in 1996, the kinds of social and political compromises made by the first committee were gone. As its executive summary noted, the recommended procedures for calculating the probability of a random match were based solely on population genetics and statistics—the two disciplines that critics of the first report complained were missing. The most important conclusion of the second NRC report was that recent empirical studies of allele frequencies in subpopulations conducted by the FBI had rendered unnecessary the use of the ceiling principle to limit allele frequencies.[1] They recommended that when the subpopulation of the evidence sample is known, the product rule should be used with the allele frequencies for that specific subpopulation.[2] If the racial group of the donor is known, but not the subpopulation, then allele frequencies from the reference database for that racial group should be applied along with correction factors for the amount of overall inbreeding (i.e., nonrandom mating) within that racial group. If the racial group and subpopulation of the donor cannot be determined, or if data does not exist for a particular subpopulation, then the probability of a random match should be made using several reference databases from several closely related or otherwise similar populations. The results from these groups could be reported individually in court, they could be averaged, or only the most conservative probability calculation could be reported.[3] On other issues, the report was generally deferential to the FBI and its two main expert commissions, TWGDAM and the newly formed DNA Advisory Board.

Once the empirical examinations called for by defense experts were carried out in 1992 and early 1993, it became clear that the population genetics assumptions made by the FBI were not nearly as problematic as the NRC report suggested and there was less cause for concern in the scientific community. While many of the early critics remained unsatisfied, arguing that issues like proficiency testing, laboratory error, and an unwillingness to allow the defense community full access to DNA technologies remained unresolved, by late 1992 and into 1993 enough scientists who testified on behalf of the defense became convinced that the forensic community, under the leadership of the FBI, had set the standards and laboratory protocols necessary to ensure valid and reliable DNA evidence. As a result, many courts determined that any lingering issues went to the weight of the DNA evidence and not its admissibility.

By the middle of 1993, almost all of the trial court decisions denying the admissibility of DNA evidence because of the population genetics controversy were overturned at the appellate level, including the *Yee* case. These courts argued that DNA profiling as a technological system could finally be considered fundamentally valid and reliable, because population genetic and statistical issues were no longer a major concern and protocols existed to prevent problems arising from the damaged and degraded nature of test samples. Further, they argued that it was up to the jury to decide whether the particular results in the case where produced following scientifically sound protocols and procedures. Thus, the burden of evaluating DNA evidence in individual cases was removed from the judge's purview and was placed squarely on the shoulders of the jury. An Ohio appellate judge in *State v. Pierce* (597 N.E.2d 107, 1992), for instance, held that the trial court did not abuse its discretion in admitting calculations of the probability of random match because "the jury was free to reject the DNA evidence if it concluded that the evidence was unreliable or misleading and it was for the jury to determine what weight, if any, to give such evidence."[4]

DAUBERT AND DNA

One major reason for this shift in judicial perspective was the Supreme Court's 1993 decision in *Daubert v. Merrell Dow Pharmaceuticals* (509 U.S. 579) to replace the *Frye* standard of general acceptance with criteria based on relevancy and reliability.[5] As a result of *Daubert*, federal judges, as well as those in states that adopted *Daubert*-like rules of evidence, became gatekeepers of scientific evidence and experts rather than merely deciding whether enough of the right scientists accepted the evidence as valid and reliable. Under *Daubert*, judges were required to first decide whether the evidence being proffered was relevant (by asking if it would assist the trier-of-fact). Once this basic criterion was met, these judges then had to make their own determination of whether the evidence was reliable enough to be heard by the jury. Although they did not give a precise set of qualities that good science possessed, the Supreme Court did offer a set of general observations about the validity and reliability of scientific evidence that judges could apply flexibly to the proffered evidence at hand.[6] These factors included the following:

- Whether the theories and techniques involved in the case had been tested[7]
- Whether the theory or technique has been subjected to peer review or publication (the court pointed out, however, that mere publication was not

necessarily indicative of validity, but rather only an important factor in judging science)[8]

- Whether the technique at issue has a known error rate and standards controlling the technique's operation[9]
- Whether, based on the old *Frye* standard, the theory or technique in question was generally accepted within the relevant scientific communities[10]

In essence, the court ruled that federal judges should think and act like scientists when evaluating scientific evidence. Instead of deferring to the relevant scientific communities, as *Frye* had dictated, judges were put in the position of determining what counts as scientific and what does not.

After more than a decade, the verdict is still out on *Daubert*. Although few evaluations of the quantitative and qualitative impact of *Daubert* on the admissibility of scientific evidence have been completed, a 2002 study from the RAND Corporation suggested that the percentage of scientific evidence excluded from the courtroom increased significantly after *Daubert* was decided.[11] It is important to realize that this study looked only at civil cases and not criminal ones. (It should be noted that various efforts are under way to empirically evaluate the impact of *Daubert* on forensic evidence.) Thus, the bulk of the evidence rejected was from plaintiff's experts who wished to introduce evidence of toxicity or harmfulness of product offered by the defendant (i.e., a large corporation or industry). Naturally, then, advocates of tort reform and limited liability laud *Daubert* as a decision that has rationalized the admissibility of scientific evidence, while plaintiff advocates and activist groups argue that it has prevented high-quality evidence from entering the legal system.[12]

Either way, *Daubert* has had a generally positive effect on the legal status of DNA evidence, with almost all rulings going in favor of admissibility. One of the most important evaluations of DNA evidence using the *Daubert* criteria was the appeal of *United States v. Yee, et al.* (134 F.R.D. 161, 1990), which created the controversy over population genetics and statistics. In this case (the name was changed to reflect a change in the lead defendant), *United States v. Bonds* (12 F.3d 540, 1993), the Sixth Circuit Court of Appeals upheld the trial court's decision to admit DNA evidence because it was both relevant and could withstand the four questions suggested by the U.S. Supreme Court. It should be noted that Barry Scheck and Peter Neufeld did not participate in the appeal of *Yee*.

In *Bonds*, the Sixth Circuit decided that, although there were "serious deficiencies" with the FBI's internal proficiency testing scheme, it was adequate to allow for the testing of the bureau's methods and techniques. As such, it passed the scrutiny of the first *Daubert* criterion. The court also determined that the FBI's DNA evidence met the second and third criteria, despite being troubled by certain aspects of the bureau's testing regime.

The most interesting aspect of the Sixth Circuit's *Daubert* review was its analysis of the fourth criterion—the old general acceptance standard. After determining that both the theories underlying DNA typing and the FBI's particular method-ology need to be generally accepted in the relevant scientific community, the Sixth Circuit went on to describe how it viewed "general acceptance." Instead of the old notion of consensus in the scientific community, the court argued in *Bonds* that "only when a theory or procedure does not have the acceptance of most of the pertinent scientific community, and in fact a substantial part of the scientific com-munity disfavors the principal or procedure, will it not be generally accepted."[13] In other words, the remaining holdouts, such as Dan Hartl, Dick Lewontin, and Lawrence Mueller, were not enough to indicate that DNA profiling lacked gen-eral acceptance in the scientific community. While the Sixth Circuit did not deny that population genetics remained a contentious issue for some individuals, they argued that the controversy should go to the weight of the evidence rather than its admissibility:

> Because the DNA results were based on scientifically valid principles and derived from scientifically valid procedures, it is not dispositive that there are scientists who vigorously argue that the probability estimates are not accurate or reliable because of the possibility of ethnic substructure. The potential of ethnic sub-structure does not mean that the theory and procedure used by the FBI are not generally accepted; it means only that there is a dispute over whether the results are as accurate as they might be and what, if any, weight the jury should give those results.[14]

Interestingly, the Sixth Circuit refused to consider the contents of the first NRC report in reaching its conclusions. The court stated:

> There is no dispute that the NRC report exists, but there is considerable dis-pute over the significance of its contents. We acknowledge that several appel-late courts have considered the NRC report retroactively, asked the parties to brief the significance of the report, or remanded for consideration of it. However, we do not agree with those courts that have considered the NRC report retroactively or remanded for consideration of it, and we decline to take judicial notice of an article published a year after defendant's convictions were handed down.[15]

Thus, unlike previous courts that had remanded cases back to the trial court in the wake of the NRC report, the Sixth Circuit was disinclined to hold scientific evidence to standards that did not exist at the time it was created.[16] The *Bonds* decision highlighted the extent to which *Daubert*, although it held the potential for a very rigorous examination of evidence in theory, made possible less strin-gent reviews than the old *Frye* standard.[17]

This trend was not just limited to the federal level, where the product rule was ruled admissible in all cases heard. By the end of 1994, it was accepted in almost all state and local jurisdictions around the country, with the only exceptions being Arizona; Washington, D.C.; New Hampshire; and Washington State.[18] In all of these jurisdictions, the ceiling principle was ultimately ruled admissible in place of the product rule. It was in this context of broad judicial acceptance of DNA evidence that the Simpson trial took place.

The Simpson Trial

On the evening of 12 June 1994, Nicole Brown Simpson and Ronald Goldman were found brutally stabbed to death outside Brown's Brentwood, Los Angeles, condominium. Los Angeles Police Department (LAPD) detectives who arrived on the scene in the early hours of 13 June described a particularly bloody scene, but one that provided a great deal of forensic evidence—including a hat and a single leather glove that may have belonged to the murderer, bloody shoe prints leading away from the crime scene, and five drops of fresh blood more than 120 feet away from where the two bodies lay. At 4:30 A.M., detectives left the crime scene and drove to Simpson's Rockingham, Los Angeles, mansion to inform him that his ex-wife, who was the mother of his two young children, had been killed along with an acquaintance. Upon arrival, Simpson was nowhere to be found. After looking around for a while, one of the detectives, Mark Fuhrman, noticed what seemed to be blood on the door of Simpson's white Ford Bronco. According to later testimony, Fuhrman stated that he saw additional bloodstains on the car, as well as items in the car that he found to be indicative of a crime still in commission.

Based on these observations, the police team entered Simpson's property without first obtaining a warrant. Once inside the gates of his estate, they managed to wake up Simpson's houseguest, Kato Kaelin, and Simpson's twenty-six-year-old daughter, Arnelle, who were both living in guest bungalows on the property. The detectives learned that Simpson had left for Chicago shortly after the time the murders took place and, later, that the limousine driver got no response from Simpson for nearly forty minutes after he arrived at the house to take him to the airport. The driver also testified that he saw someone entering the darkened house shortly before Simpson finally answered his calls.

During the course of a discussion, Kaelin told police that he heard noises behind his guest bungalow shortly before Simpson left for the airport. When Fuhrman went alone to investigate this area, he claimed to have found a right-handed glove still wet with blood that appeared to match the left-handed glove found at the crime scene. At trial, the defense would make much of the fact that Fuhrman went to investigate the rear of the bungalow without any other witnesses. The defense team would go to great lengths to paint Fuhrman as an unrepentant racist who detested interracial relationships (Nicole Brown Simpson was white), and called several witnesses to the bench who corroborated this claim. Of particular interest to trial watchers were two witnesses who testified that Fuhrman repeatedly used the word "nigger" in conversation, despite his testimony under oath that he had never in his life used such a word. In his closing arguments, defense attorney Johnnie Cochran went so far as to liken Fuhrman to Adolf Hitler and argue that he had to be stopped before he could cause serious harm to the black community of Los Angeles. "There was another man not too long ago in the world," he told the jury, "who had the same [racist] view and who wanted to burn people."[19]

As the sun rose, detectives reported finding blood inside Simpson's Ford Bronco in addition to blood drops on the sidewalk and in the foyer of the Rockingham house. Based on these findings, the LAPD was granted a warrant to search the rest of the house. The only physical evidence seized from the house under the warrant was a pair of socks found in Simpson's bedroom. Although criminalists made no note of the socks being bloody at the time they were recovered, several weeks later the LAPD crime lab reported that one of the socks contained a thick blood stain that matched his ex-wife's DNA profile. While the defense argued that the blood was planted by a racist and incompetent police force, the prosecution argued that the bloodstain was proof that Simpson was guilty of murder.

Simpson was finally contacted in his Chicago hotel room in the early hours of the morning and returned to Los Angeles on the first available flight. After a brief stop at home, Simpson went to police headquarters for an interview. There, he claimed that he was sleeping when the limousine arrived and only woke up around 11 P.M. During the course of questioning, detectives noticed a cut on one of his left-hand fingers. Although Simpson offered several explanations for the cut, police believed it was a crucial piece of evidence, since the positioning of the blood drops next to the footprints indicated that the murderer was bleeding from his left hand.

After the interview, the LAPD nurse on duty, Thano Peratis, took a blood sample from Simpson and placed it in a tube with a preservative called EDTA. Peratis handed the tube directly to LAPD detective Phillip Vannater, who promptly put it into his pocket. Because the tube was not entered into the evidence log until much later in the afternoon, there was no official record of the contents of the tube when it left Peratis's custody. Vannater then drove back to Simpson's estate to give it directly to LAPD criminalist Dennis Fung, who was collecting samples from the stains found at the crime scene and on Simpson's property, with the help of a trainee named Andrea Mazolla, who had little crime scene experience. At trial, a dispute erupted over exactly how much blood was drawn. While Peratis initially stated in a deposition that he drew approximately 8 cubic centimeters (cc), the LAPD could only account for 6.5 cc. The defense argued that the missing 1.5 cc was used to manufacture incriminating evidence against Simpson. The presence of EDTA in some of the crime scene samples attributed to Simpson provided further fuel for the defense's claims that Simpson's blood was planted by police at the scene of the crime. The prosecution, on the other hand, argued that Peratis did not accurately measure or remember how much blood he drew from Simpson.

Samples collected by Fung and his assistant were sent to the LAPD's Scientific Investigation Division (SID), where criminalist Collin Yamauchi conducted various tests. From the very beginning of the Simpson drama, most legal experts and commentators believed that DNA evidence would hold the key to Simpson's innocence or guilt. At least early on, the evidence did not look good for him. On 15 June, preliminary DNA results from the materials sent to the SID by Fung became available. Using a low-resolution, PCR-based DNA typing technique called HLA DQ-alpha, Yamauchi determined that Simpson's blood was consistent with stains recovered from the crime scene and that the glove found by Fuhrman at Simpson's house contained a mixture of the blood of Simpson and the two murder victims. Although this test was far from conclusive—Simpson's blood type was shared by about 7 percent of the overall population—it was enough evidence for a judge to issue a warrant for his arrest.

The same day that the initial results became available, Simpson, sensing he was in trouble, began to assemble a crew of defense lawyers who would come to be known as the "Dream Team." In addition to Robert Shapiro, F. Lee Bailey, Johnnie Cochran, and Alan Dershewitz, who were all famous for their work as defense attorneys in high-profile criminal cases, Simpson hired two lesser-known attorneys who had significant experience challenging forensic evidence: Barry Scheck and Peter Neufeld. Scheck and Neufeld later invited William C. Thompson, of "Irvine Mafia" fame, to join the team. The Los Angeles County District Attorney's Office also began to assemble an impressive team of California prosecutors to litigate the case, including several who specialized in forensic evidence. Most notable among these was Alameda County (Oakland) Deputy District Attorney Rockne Harmon and San Diego County Prosecutor George "Woody" Clarke.

On the morning that he was supposed to surrender, Simpson was nowhere to be found. As ninety-five million television viewers would come to learn, Simpson had written what seemed to be a suicide note and was picked up by a friend driving his now infamous white Ford Bronco sport utility vehicle. Simpson was sitting in the backseat with a gun pointed to his head, apparently ready to kill himself. Over the course of the afternoon of 17 June, Simpson's white Bronco was joined on the freeway by a convoy of twelve police cars and seven news helicopters. When his friend finally returned Simpson to his Brentwood Estate, he was greeted by dozens of LAPD personnel, including more than two dozen SWAT team members and a vehicle assault team. According to a Court TV chronology of the situation, the slow speed chase became the second most watched event in the history of television, with the first being the moon landing.[20] This publicity ensured that the trial, which was televised on all major networks, would garner a huge audience. It would put DNA evidence into the public eye as never before. Indeed, the acronym "DNA" would be uttered more than ten thousand times during the course of the trial.[21]

Simpson ultimately surrendered and was arrested by police. The LAPD Scientific Investigation Division spent the next few days building their case against Simpson in preparation for a hearing to determine whether or not sufficient evidence occurred to undertake a full criminal trial against him. Because DNA tests took several weeks to process in 1994, the prosecution presented a significant amount of blood typing evidence that they believed supported their position that Simpson was guilty of double homicide. The municipal judge in the case agreed, and the trial of the century was set in motion.

In between the preliminary hearing and the start of the trial in early 1995, DNA testing results came back from the laboratories doing the work. Because of the celebrity status of the defendant and the publicity the case was receiving from around the world, the Los Angeles County District Attorney's Office decided that it would be best to split up the crime scene evidence and send biological samples to more than one laboratory. As such, they sent samples to Cellmark Diagnostics in Maryland and the California Department of Justice's DNA laboratory in Berkeley (the central laboratory in the state's forensic network) for analysis. All told, more than forty-five bloodstains were subjected to DNA analysis, almost all of which provided incriminating evidence against Simpson. These results showed that the blood drops found alongside the footprints leading away from the crime scene matched O.J. Simpson's DNA profile, that blood found in Simpson's Bronco matched Ronald Goldman's profile, that blood found on the gloves matched Goldman's, Nicole Brown Simpson's, and O.J. Simpson's profiles, and that blood on the sock found in O.J.'s bedroom matched Nicole's profile.

PCR in the Forensic Context

Eleven of these stains were analyzed using RFLP technology by Cellmark and the state's DNA lab, while the remaining thirty-four were analyzed using PCR-based techniques called "DQ-alpha" and "Polymarker" at the LAPD's SID laboratory.[22] Unlike RFLP tests, the PCR-based techniques performed by the LAPD SID were not based on length polymorphisms. Rather, they depended upon slight differences in nucleotide sequences. In these tests, a particular region of DNA whose sequence is known to vary in humans is amplified using the polymerase chain reaction (PCR). PCR, which will be described in greater detail later, is a chemical process by which specific regions of DNA can be selectively replicated over and over again, yielding millions of copies of that region. The DNA produced by PCR is then placed on cards that have probes for the specific sequences of interest embedded within them. When a DNA fragment attaches to a probe, a color change can be visualized on the spot where the reaction takes place. To give an example, if an individual has one copy of DQ-alpha variant 1 and one copy of DQ-alpha variant 3, the DNA fragment from these genes will bind to the probes for variant 1 and variant 3 on the card. Because the no-PCR product will bind to the other variants, they will not be visualized. The advantage of PCR-based tests is that they are relatively quick and require only minute amounts of DNA, since the first step is the exponential amplification of the region in question.

But there are two major disadvantages to these tests. First, they are low resolution. Since there are only six DQ-alpha variants, this test is no more individualizing than traditional blood typing. In an effort to ameliorate this problem, five additional markers were chosen to accompany DQ-alpha, which are collectively called "Polymarker," but these genes have even less variability than DQ-alpha.[23] Because of this problem, these marker systems are almost never used in forensic testing anymore. More important, PCR is incredibly sensitive to contamination. Because the process starts from just a few strands of DNA and creates millions of copies from it over the course of thirty-plus replication cycles, even the smallest amount of DNA present in the reaction mix has the potential to become the major product of the reaction. Although contamination by spurious DNA is still a problem with RFLP-based testing, a significant amount of DNA must get into the reaction mixture for an incorrect result to be visualizable, because very small amounts of DNA generally cannot be visualized on a Southern blot. PCR, then, has the potential to magnify the slightest mistake in evidence collection and laboratory procedure. Even contamination of a single cell or strand of DNA has the potential to generate false or misleading results.

DNA's Prime-Time Debut

As became apparent over the course of the nine-month trial, there was more at stake than the guilt or innocence of a celebrity athlete and actor. Many supporters of DNA evidence believed that its legal validity, and public credibility, would also be in the balance. Scheck, Neufeld, and Thompson were well known to all who had participated in the disputes over DNA profiling. Many of these participants feared that their legal tactics would irrevocably alter the American public's view of DNA evidence, and even its collective understanding of how science works.

Such concerns led one of the most vocal early opponents of DNA evidence, Eric Lander, to declare a public truce with the principal architect of the FBI's DNA profiling program, Bruce Budowle, in the journal *Nature*. In their coauthored article, they declared that there was no longer any dispute about the validity and reliability of DNA profiling.[24] All major issues that had led to the outbreak of war had been resolved, and there was no longer any need to worry about false or misleading results. Not surprisingly to those who derided the article as opportunist propaganda, the coauthors gave the FBI, one of the chief combatants, the bulk of the credit for resolving the problems highlighted by the defense attorneys and their witnesses in previous legal battles.[25]

In his letter to the editor of *Nature* about the Lander-Budowle truce, Richard Lewontin did not hide his view that there was some sort of conspiracy at work directly linking the O.J. Simpson trial to the assassination of John F. Kennedy. He wrote, "Why did Lander and Budowle choose to embrace in the pages of a leading journal of science, just before Budowle is scheduled to appear before tens of millions on television as a witness for the prosecution in what is surely the most publicized crime since the assassination of John Kennedy? As the French say, it gives one to think."[26] Of course, nobody questioned why Budowle would agree to

author such an article, though many people were puzzled, but not totally surprised, by Lander's decision to do so. In numerous conversations, people I spoke
to described Lander as a careerist who was primarily concerned with advancing
his own agenda—which at the time was starting up a center at MIT to map the
human genome. Many also stated that they believed that he did not want to be
seen as being anti–U.S. government at a time when genome sciences were receiving increased funding and attention from Congress. Although nobody was willing to go on the record with such a sentiment, it was widely shared by scientists
and lawyers alike.

When I asked Lander what he thought about such accusations, he laughed,
pointing out that he was invited by James Watson, the Nobel Prize–winning
geneticist who was the first head of the Human Genome Project, to apply for a grant
to start a genome center, and that his proposal was peer reviewed by fellow scientists, not politicians. In his view, if there was any reaction to his article with
Budowle, it would be negative, since his peers were inclined to be suspicious
of the government themselves.[27] Lander did, however, admit to me that he was
troubled by the way that people were invoking his name when arguing against
the validity and reliability of DNA evidence—he felt that the statements he made
in 1989 in the context of the *Castro* case had become "talismanic," invoked whenever a lawyer wanted to criticize DNA evidence, irrespective of the context in
which DNA evidence was produced. He felt that he needed to make a strong statement that his old views were no longer relevant, before his name was invoked by
the Simpson defense before millions of television viewers. And what better way
to do so than in a joint statement with Budowle?

Whatever Lander's motivation, he and Budowle made no effort to hide the
wellspring of their new friendship. They began the article by pointing out that
the upcoming Simpson trial had made the U.S. public obsessed with DNA evidence, thanks in large part to an endless stream of media reports on the subject.[28]
They then noted that the trial would "probably feature the most detailed course
in molecular genetics ever taught to the US people." While such an education
would normally be welcomed, Lander and Budowle lamented that "the catch is
that the syllabus is being prepared by attorneys whose primary roles are as adversaries; the likely result is confusion."[29]

In their view, this confusion would most likely be caused by the media reporting to the American people that the technique remained contentious and controversial in the scientific community. According to Lander and Budowle, however,
nothing could be further from the truth. The problems had little to do with the
technology itself, they argued, but rather its implementation by private companies rushing into court without appropriate quality controls, standards, or forensic experience. "There is broad agreement today," they wrote, "that many of these
early practices were unacceptable, and some indefensible. For its part, the [FBI]
moved much more deliberately in developing procedures, sought public comment and opted for conservative procedures."[30] Among these procedures included

the adoption of "precise guidelines for implementing the ceiling principle" in cases where it was required by the court for the admissibility of DNA evidence. Although they believed that the ceiling principle was scientifically problematic and unnecessarily conservative, they were willing to accept it on pragmatic grounds (i.e., to appease courts in those jurisdictions where the product rule was declared inadmissible because of the controversy that emerged after *Yee*).[31]

The article also highlighted the completion of the empirical investigations of allele frequencies in subpopulations demanded by Lewontin, Hartl, and Lander.[32] In a multivolume study, entitled *VNTR Population Data: A Worldwide Study,* and in related publications, the FBI reported on the allele frequencies of their major RFLP markers in more than twenty-five distinct subpopulations spread across more than fifty geographic locations.[33] This report represented a tremendous amount of work and worldwide cooperation among academic scientists, forensic scientists, research institutions, and crime laboratories around the world. Ultimately, the FBI concluded that while there was some statistically significant substructure in the major racial group reference databases that they used, any differences in allele frequency among subgroups were "modest" and did not have "forensically significant effects."[34] In other words, at least according to the FBI's interpretation of the data they had collected, Lewontin, Hartl, and Lander's fears about the detrimental effects of population substructure, which emerged in *Yee* and were published in 1991, proved to be unsubstantiated. The FBI argued that the cases of substructure that did occur in the major racial groups were trivial and solely of academic interest.[35]

In a letter to the editor of *Nature*, Hartl expressed his shock at the claims made by Lander and Budowle in their article.[36] He took them to task for marginalizing or dismissing many aspects of the controversies over DNA profiling as "purely academic," when in his view they were real and deeply relevant to forensic practice. He pointed out that statistical significance was "an objective, unambiguous, universally accepted standard of scientific proof. When differences in allele frequencies among ethnic groups are statistically significant, it means they are real— the hypothesis that genetic differences among ethnic groups are negligible cannot be supported."[37] Hartl argued that the FBI and its supporters could not simply write off such "real" differences as being irrelevant in the forensic context; in his view, real substructure demanded scientifically and legally appropriate corrections to the product rule. To do anything else, he implied, was dishonest and disingenuous.

Taking a dig at the FBI, he asked, "When does a statistically significant difference become 'forensically significant'? Well, [quoting an article written by FBI scientists],[38] 'when the likelihood of occurrences of the DNA profile would be meaningfully different.' And guess who decides whether differences are 'meaningfully different' . . . ?"[39] The answer, of course, was the FBI, which had been granted authority to set standards for DNA typing, including those on population genetics and statistical issues, by the recently enacted DNA Identification

Act of 1994. In his view, the FBI would be more honest in stating that the two methods for calculating the probability of a random match sat on a spectrum of conservatism, with the product rule at the nonconservative end and the ceiling principle at the more conservative end. By providing the jury with the results of both methods, it was likely that the "true value" would be bracketed somewhere in between.[40]

While one can argue with Hartl's overall perspective and claim that he was predisposed to disagreeing with the Lander-Budowle article, he was correct in taking them to task for minimizing the importance of many issues that had not yet been resolved. Nowhere in their article did Lander and Budowle mention that the initially poor implementation of DNA profiling was only exposed through the concerted effort of the very defense attorneys who were now busy creating the syllabus for the nation's upcoming lesson on molecular biology. Nor did they mention that the FBI played no role in performing the DNA analyses in the Simpson investigation (Budowle testified about non-DNA issues), or that the LAPD SID did not follow the guidelines produced by TWGDAM. In fact, one of the major revelations of the trial was that some personnel within the laboratory did not rigorously and routinely follow any guidelines at all.

Lander and Budowle also failed to mention that much of the DNA evidence in the Simpson investigation was produced through new PCR methods that had been subjected to limited scrutiny in court. Although they were correct that many of the problems associated with RFLP DNA analysis had been uncovered and corrected through the adversarial legal process, the same could not be said about the DQ-Alpha and Polymarker tests that were central to the prosecutor's case against Simpson. Thus, their conclusion that "the scientific debates served their purpose . . . [b]ut now it is time to move on"[41] was highly disingenuous in that it failed to distinguish among various forms of DNA profiling. In fact, such a claim blatantly disregarded one of the most important messages found in the 1992 NRC report, that DNA typing is a "'catch-all' term for a wide range of methods for studying genetic variations."[42] The NRC report went further, stating that, while there is no scientific dispute about the validity of the general principles underlying DNA typing, "a given DNA typing method might or might not be appropriate for forensic use. Before a method can be accepted as valid for forensic use, it must be rigorously characterized in both research and forensic settings to determine the circumstances under which it will and will not yield reliable results. It is meaningless to speak of the reliability of DNA typing in general—i.e., without specifying the methods intended to be covered."[43]

Despite his critical letter to the editor about Lander and Budowle's article, Hartl's views about DNA evidence had also evolved since he testified in Yee. Although he clearly still disagreed with many of the FBI's policies on DNA evidence, in early 1994 he testified on behalf of the prosecution in Minnesota v. Bloom (516 N.W.2d 159, 1994). In an interview, he told me that from the beginning he believed that DNA typing was an incredibly powerful and useful tool for forensic science. He testified in *Yee* only because he believed that the FBI's use of the product rule was fundamentally unscientific—he harbored no ill will against the law enforcement community in general. He was distressed, therefore, when he was quoted by defense lawyers around the country as being against DNA typing, and so he started looking for a case in which he could testify on behalf of the prosecution in which the DNA typing methods and statistical calculations were done correctly.[44]

His chance came when Steve Redding, the Minneapolis prosecutor, asked if he would testify about the population genetics and statistical issues surrounding DNA evidence in the case against a man who was accused of forcibly abducting a woman from her home and raping her in the back of her car. Minnesota was unique among the fifty states in that its state supreme court had barred the use of statistical probability evidence to prove identity in criminal trials in a trilogy of cases.[45] The rationale for this decision was that "there is a real danger that the jury will use the [statistical population frequency] evidence as a measure of the probability of the defendant's guilt or innocence, and that the evidence will thereby undermine the presumption of innocence, erode the values served by the reasonable doubt standard, and dehumanize our system of justice."[46] Prosecutors in *Bloom* sought an exemption from this rule for DNA evidence and needed expert testimony that the NRC's ceiling principle provided the necessary degree of conservatism to ensure that the probability of random match between suspect and crime scene samples would not be overly prejudicial.

When Hartl began examining the information sent to him by the prosecutor, he was extremely impressed by the work done by the Minnesota Bureau of Criminal Apprehension (BCA), which began doing DNA testing after Cellmark's evidence was ruled inadmissible in the *Schwartz* case. It is important to note, however, that Hartl's testimony was restricted specifically to the issue of population substructure and statistical calculations. The additional concerns that he had addressed in his 1991 *Science* article with Richard Lewontin—most notably the inadequacy of the FBI's laboratory procedures and the lack of an effective quality control regime for DNA profiling—were actively and intentionally avoided during his initial testimony in the case. When Hartl stated to the court what kinds of documents he had received about the BCA's DNA profiling regime, there was no mention of quality control protocols, proficiency testing results, or information about the overall performance of the laboratory and its personnel.[47]

In his testimony, Hartl argued that the issue of population substructure was only acute when very few loci were examined—as Lifecodes, Cellmark, and the FBI all did in the early days of the technique. With each additional loci, the probability that members of a particular subgroup will share the exact same DNA profile decreases. Although the product rule would have still been invalid on strictly scientific grounds, its forensic effect would have been much less troubling.[48] He stated that "there is a point where statistics simply don't matter anymore . . . that the probability of a random match . . . is so small that identity can be inferred."[49] In his view, the nine VNTR loci used by the BCA constituted overwhelming evidence that DNA from the victim's vaginal swab belonged to the defendant in the case: "in my best judgment, a nine-locus VNTR match is tantamount to identity."[50] Curiously, Hartl was making the same kind of forensic relevance determinations that he chided Lander and Budowle for in his letter to the editor of *Nature*.

In his cross-examination in *Bloom*, Public Defender Patrick Sullivan pointed out that Hartl had not taken potential laboratory error into account in making such a statement.[51] Sullivan argued that if a simple human error such as a sample switch was made, then a match at 9 loci or 100 loci would be meaningless. Following this logic, he went on to argue that in order to fully understand the probative value of a match at any number of loci, one must know how often the laboratory in question makes mistakes. Although Hartl acknowledged this point in general, he still maintained that the BCA performed their analysis perfectly in this case. He stated that the special circumstances of the case meant that it was highly unlikely that the BCA made any serious mistakes: Troy Bloom became a suspect only after his genetic profile in Minnesota's DNA database (he was a convicted felon) matched the rapist's profile. The police then took another blood sample from him, which matched both his profile in the database and the semen found at the crime scene. In the end, Hartl qualified his claim that a nine-locus match is tantamount to identity with the following statement: "the implicit claim in my affidavit is that it is my best judgment, to a reasonable degree of scientific certitude, that the crime scene sample matches the blood that is purported to have come from the defendant."[52]

During his testimony, Hartl also advocated the use of the ceiling principle. He testified that it was designed explicitly to allow for the most conservative presentation of DNA evidence possible, and that it sufficiently guarded against the possibility of overstating the probability of a random match. Even when an expert witness believed that the probability of a random match was much lower than the ceiling principle calculation suggested, he argued that the jury should hear the most conservative calculation possible so that the expert's higher calculation would not be prejudicial.[53] This was the same idea that he advanced in his letter to the editor about the Lander and Budowle article.

Once on the record in favor of what he considered to be the proper way of calculating the probability of a random match, Hartl stopped testifying in court on this issue and bowed out of the disputes over DNA evidence. With Lander and Hartl out of the picture, Richard Lewontin remained the only eminent geneticist who had not renounced or modified his earlier views on the technology. While several other scientists continued to testify on behalf of the defense, by the time the Simpson trial started, their complaints were increasingly drowned out by the chorus of voices arguing that DNA evidence was no longer the problematic technology that it was just a few years ago. This reality would ultimately shape the defense's challenge to DNA evidence in the Simpson trial.

The Dream Team Prepares for Battle

In the early preparations for the trial, the defense team planned to rely on the notion that each new method of forensic DNA analysis must be subjected to scientific scrutiny, validation, and a rigorous admissibility hearing. In addition to the familiar complaint that methods of calculating the probability of a random match were still controversial, and that no statistical method had been accepted by the relevant scientific communities, despite several appellate court decisions to the contrary, the defense team also planned to argue that the PCR techniques used in the case were novel in the forensic context and therefore not automatically admissible in court.

As they argued in their "Motion to Exclude DNA Evidence,"[54] the use of PCR technology in the forensic context necessitated general acceptance of the methods of collection, preservation, handling, and processing of crime scene samples among molecular biologists who understood the chemical basis of the technique, not just the forensic scientists who used the technique. According to the defense, they were the only people who fully understood the sensitivity of the technique. Because there had not yet been an admissibility hearing in California on this issue, the defense wrote that one must take place in the Simpson case before the DNA evidence could be considered admissible. Further, in the absence of a statistical technique for estimating the chance of such an error, DNA evidence produced using PCR should not be admissible in court.[55] Until well into the trial the defense planned to call Kary Mullis, the inventor of PCR, to testify that the LAPD SID was so poorly run that contamination of their work was almost guaranteed. In the end, Mullis, who was as famous for his drug use and womanizing as he was for his scientific prowess, did not testify because the defense feared that his credibility would be easily challenged by the prosecution. The defense also argued, against the claims of Lander and Budowle, that the ceiling principle proposed in the first NRC report had not yet proven to be an acceptable compromise between sides.[56] Although the motion did not explicitly say so, it seems that the defense's position was that DNA evidence should not be admitted until the NRC issued its forthcoming second report.

In a relatively new strategy for the defense community, Simpson's lawyers argued that a match between two DNA profiles could not be admitted until statistical methods were developed to accurately measure the probability of a random match due to error. The issue of error had always been a part of disputes over DNA evidence, but not from a statistical perspective. In early challenges, defense attorneys and their witnesses disagreed with prosecution claims that DNA profiling gave the right result or no result at all—in other words, they argued that false positives were an ever-present danger because humans were an integral part of the DNA testing process. While this strategy was commonly used, it was generally unsuccessful. After the introduction of TWGDAM standards and protocols in the early 1990s, it was essentially a nonstarter in court. As the court in

State v. Streich (658 A.2d 38, 1995) stated, although DNA profiling could certainly not be considered an error-free technology, "adherence to accepted procedures and controls minimizes error. . . . We cannot find any recent decision under any standard of admissibility which refuses to admit the DNA match result based on the invalidity or risk of error of the underlying technology."[57]

As the debate over DNA evidence turned to statistical issues in the wake of *Yee*, however, critics began to frame error as something that needed to be measured and incorporated into the interpretation of the evidence in order for that evidence to be admissible. The Supreme Court's *Daubert* decision stimulated this turn, since a measurable error rate was one of the main criterion that judges were supposed to use to evaluate the reliability of DNA evidence. The essential argument made by Simpson's legal team was that jurors needed an accurate picture of how often mistakes were made in DNA proofing in order to appropriately assign meaning to the value of a very small random match probability.[58] Although there were debates about how this error rate information should be conveyed to the jury—that is, whether it should be integrated into the statistical calculation with the probability of a random match between two profiles or presented as a separate number—almost all defense experts agreed that it was crucial to the interpretation of evidence. In his testimony in *Yee*, Hartl had stated that any statistical calculation that fails to take error into account was "simply meaningless."[59] Before his truce with Bruce Budowle, Eric Lander testified as a court witness in *United States v. Porter* that "it is simply crazy and scientifically unacceptable to agonize over the exact population frequencies, which might be one in a million, or one in a hundred thousand, or one in ten thousand for the frequency of a genotype in a population, and yet not have actual data for the accuracy, the proficiency of a laboratory's handling of samples. . . . [T]he scientific acceptability of DNA evidence depends on the proficiency of a laboratory being tested such that one can know what the error rate is likely to be, or at least have an upper bound on that error rate."[60] Based partly on this testimony, the *Porter* court was one of the few in the country to hold that a laboratory's performance record must accompany the probability of a random match.[61]

The major problem with this demand was that no agreed-upon method to estimate the frequency of errors in DNA testing existed at the time (or today, for that matter). Although the first NRC report advocated the measurement of error rates, the best the commission could do was to suggest that any valid proficiency test should simulate actual forensic casework as closely as possible. It offered no concrete suggestions for ensuring the realism of the tests, at least in part because no studies had been done on the best way to measure error rates in DNA profiling. In the 1996 NRC report, the new committee abandoned the first committee's commitment to measuring error rates. They concluded that the measurement of error rates was essentially impossible and that "the occurrence of errors can be minimized by scrupulous care in evidence collecting, sample-handling, laboratory procedures, and case review."[62]

The first serious study to evaluate the possibility of developing a nationwide external blind proficiency program was carried out by Joseph L. Peterson and R. E. Gaensslen from the University of Illinois at Chicago (UIC).[63] UIC was contracted to conduct this study because the FBI had received a mandate from Congress in the DNA Identification Act of 1994 to investigate the feasibility of such a program. After constituting a bipartisan National Forensic DNA Review Panel (NFDRP) that included members of all interested parties, including senior FBI officials, academics, and defense lawyers, UIC held several meetings of this panel, surveyed crime laboratories and defense attorneys across the country, and developed a blind proficiency test that was successfully administered to several laboratories that had volunteered to take part in the feasibility study.[64]

The discussions of the NFDRP accurately captured the diversity of views on proficiency testing. Many members of the defense community argued that the best method for detecting error was external blind proficiency testing, in which laboratories received what appeared to be a real case (or an actual case with a known outcome) sent by an outside agency with no indication that it was a proficiency test. Ideally, the laboratory would carry out tests on the evidence samples and then return results to the outside agency. Only after the investigation was complete would the lab find out that it had been tested. By conducting numerous external proficiency tests, proponents of this protocol believed that they could put together a picture of how often errors occur, initially on an overall national scale and ultimately on a laboratory-by-laboratory basis.

In opposition to the proposal for external blind proficiency testing, most forensic scientists and their allies argued that it was impossible to create a truly blind external test because laboratory technicians work so closely with detectives in testing evidence in a case.[65] For example, with limited time and budget, forensic technicians often ask crime scene investigators which pieces of evidence are most likely to contain the genetic materials of the perpetrator of the crime rather than testing every single crime scene sample. They also seek information on whose profiles are likely to turn up, as well as information on which people are considered prime suspects in the case. In other words, they do not approach the analysis of the crime blindly. To create a realistic external blind proficiency test, the organization conducting the test would not only have to fabricate an accurate and complete set of crime scene samples, but would also have to concoct a story and find a law enforcement agency willing to submit the case to a lab and answer any questions that the laboratory might have about the evidence.

Opponents also argued that general validation data had already been published in peer-reviewed journals, proving the reliability of the technique.[66] The biannual internal proficiency tests mandated by TWGDAM were more than adequate to ensure that errors were not taking place. If errors did happen to occur, their cause was quickly determined and any problems were resolved. Finally, opponents were also against providing an overall error rate for all laboratories in the country because it might wrongly make good laboratories look bad and vice versa.[67] The upshot of the claims made by opponents of blind proficiency testing was that errors were so rare in DNA profiling that an expensive and time-consuming proficiency testing scheme was superfluous and a waste of limited resources. Of course, the defense community countered such claims were ridiculous because no studies of potential error rates had ever been carried out. In their view, the prosecution was arguing the lack of evidence of errors was reason enough not to collect data on how often errors may occur.

After a great deal of deliberation among panel members and analysis of the results of the surveys and proficiency tests, the UIC team issued its final recommendations.[68] The report concluded that while blind proficiency testing was indeed possible, it was too fraught with problems to move ahead with large-scale implementation. It was not clear to the majority of panel members that such an expensive program would produce enough improvement in DNA profiling to make the investment worthwhile. In the end, William Thompson was the only panel member who voted in favor of external blind proficiency testing.[69] The panel generally agreed that not enough time had passed for the accreditation system and quality assurance guidelines of the FBI's DNA advisory board to take hold. It concluded that external audits of contested or completed cases might be a better way to ensure the reliability of DNA profiling. Although there was some disagreement about the criteria by which cases would be flagged for audit, all participants agreed that auditing was a good way to examine the entire DNA profiling process, from initial crime processing to statistical interpretation of a match. Agreeing with the conclusions of the 1996 NRC report, the UIC report stated that the purpose of blind proficiency testing was not to estimate error rates (such a task was impossible from a pragmatic standpoint), but rather to point out potential problem areas in the field so that laboratory directors and workers could fix them. Finally, the panel agreed that whenever possible a portion of all evidence samples should be put aside for reanalysis or examination by the defense at a later date.

As the lawyers in the *Simpson* case were gearing up for court in mid-December 1994, a disagreement arose between the prosecution and defense over whether the admissibility of DNA evidence should be discussed in a pretrial hearing or directly in front of the jury. The defense had hoped to fold their challenge to the admissibility of DNA evidence into the main part of the trial in order to speed the trial along and ensure that their strongest criticisms of the prosecution's evidence would be heard by the jury from the start of the trial.[70] Judge Ito, however, sided with the prosecution and rejected this request. He called for a pretrial admissibility hearing to start on 5 January 1995. This decision caused the defense team to shift its strategy considerably. They decided that with the publicity the trial was receiving, it was in their best interest to forgo the kind of admissibility hearing seen in *Castro* and *Yee*. Their decision reflected a pragmatic effort to minimize both the amount of prejudicial news that the jury would receive while waiting for the trial to begin and, according to news reports, the amount of money being spent by Simpson on his defense.[71]

According to several accounts, the defense had also come to believe that there was little hope of Judge Ito's ultimately ruling that the DNA evidence in the case was inadmissible. By the start of the Simpson trial there were very few jurisdictions in which DNA evidence was ruled inadmissible. Indeed, as of June 1995, the appeals from most of the cases described thus far had been rejected and DNA evidence was ruled admissible.[72]

With no admissibility hearing on the horizon, Simpson's DNA lawyers all but abandoned the strategy that had worked so well for them in the past. Throughout the nine-month trial, there was significantly less discussion of the kinds of systemic problems with DNA profiling that they had emphasized in *Castro* and *Yee*. Population substructure and the ceiling principle took a backseat to the practices and contingencies that could lead to misleading DNA profiles, even if one assumed that DNA profiling was a perfectly robust and mature technology. They sought to interrogate the human and institutional dimensions of DNA profiling, leaving the technical core of the system relatively unscathed. Their ultimate argument was that DNA profiling was indeed capable of producing valid and reliable results, but the potential for human error existed every step of the way. Ironically, while Scheck, Neufeld, and Thompson had previously argued that the potential for human error was an inherent part of the technological system, now they sought to turn DNA profiling into a black box that only produced errors when criminalists and laboratory technicians made mistakes. Borrowing a metaphor from computer science, Scheck would repeat countless times in court the phrase "garbage in, garbage out"—in other words, don't blame the technology when your inputs are fatally flawed.

Day after day, the defense team called witnesses, most notably Henry C. Lee, who was chief criminalist for the State of Connecticut at the time of the trial, and John C. Gerdes, the clinical director at a private medical diagnostics company called Immunological Associates in Denver, to attack the process by which evidence was collected, processed, stored, and analyzed. Both witnesses took criminalist Dennis Fung and his trainee Andrea Mazolla to task for their sloppy crime scene practice, including not allowing crime scene samples to properly dry before storing them. They also testified that the LAPD's DNA testing facility was one of the most rundown laboratories in the country and one that failed to follow generally accepted guidelines for operation. Lee and Gerdes sought to convince the jury that there were numerous opportunities for contamination of crime scene samples with Simpson's reference blood sample. On cross-examination, Yamauchi admitted that he had spilled a small amount of Simpson's blood in his workspace and then processed the bloody glove and the swatches containing the blood drops from the crime scene without cleaning his instruments or changing his gloves.[73] While the prosecution argued that the use of controls ruled out any potential contamination from this mishap, the defense argued that these controls were not properly executed.[74] Defense witnesses even presented testimony suggesting that the reference blood samples from Nicole Brown Simpson and

Ronald Goldman were contaminated with O.J.'s blood.[75] Overall, the defense argued that criminalists from the LAPD SID were poorly trained in proper evidence handling and did not understand the dangers of even minute amounts of DNA from one sample contaminating another sample when PCR was involved. When not disparaging Fung or Yamauchi, the defense was busy constructing plausible explanations for why the LAPD detectives might want to frame Simpson. In addition to Fuhrman's racism and his professed disdain for interracial relationships, they also pointed out that many detectives knew about Simpson's previous domestic disputes with Nicole.

THE POWER OF DNA

In the end, the defense's case rested less on imposing scientific doubt about DNA profiling in the minds of the jury than convincing them that they should not trust the crooked cops who processed the crime scene and the bungling technicians at the LAPD laboratory who prepared the samples for DNA testing. This strategy was most famously summed up when Scheck repeatedly asked criminalist Fung ("where is it, Mr. Fung?") to point out the blood on the back gate of the Bundy Avenue crime scene and the bloodstain on Simpson's sock in pictures and video taken the morning after the double murders took place. As defense counsel William C. Thompson pointed out in an article analyzing the DNA evidence in the *Simpson* case, "there were a number of scientific issues in the case, but they had little to do with the fundamental science underlying DNA testing or the details of laboratory procedures for typing DNA, and everything to do with the potential for cross-contamination of samples before they reached the DNA laboratories."[76] The only systemic problem that the defense team pursued was that statistical calculations of the rarity of a match between Simpson's profile and evidence found at the crime scene should not be admitted because they did not incorporate a correction factor for the likelihood of false results due to error or tampering.[77]

Even the *New York Times*, which had published some of the most controversial accounts of the debates over DNA profiling in the late 1980s and early 1990s, was surprised about the lack of a fundamental technical challenge to DNA profiling. In an editorial entitled "The Power of DNA," the *Times* wrote the following:

> While the O. J. Simpson trial stumbles ahead with a dwindling pool of jurors and new revelations bearing on the veracity of a key witness, the most important development in recent weeks may be the respect accorded to the science of DNA blood testing. Even in this bitterly contested murder trial, the principles and potential of DNA testing have not yet been seriously questioned. Indeed, the defense has not generally challenged the validity of the science but has instead charged that blood evidence was mishandled or manipulated by conspiratorial police before it was subjected to DNA analysis.[78]

As Thompson noted, however, a crucial reason why the defense adopted this strategy was because the prosecution had sent out samples to two additional labs for duplicate testing in order to verify results. Because the odds of both facilities making the same mistake numerous times were vanishingly low, the defense had to focus on evidentiary problems at earlier stages of processing.[79] Had all of the DNA testing been carried out by Yamauchi at the LAPD's facilities, one can imagine that the defense case would have taken on a substantially different character, including, perhaps, the admissibility hearing that they had decided not to pursue.

In the end, Orenthal James Simpson was acquitted of all charges against him, setting off an uproarious celebration in black communities around the country and leaving many other people scratching their heads and wondering how anybody could possibly think that Simpson did not commit the murders. As the final verdict suggests, the defense was highly successful in casting doubt over the DNA evidence in the case, so much so that the nongenetic evidence, including the bloody footprints that matched a pair of rare shoes owned by Simpson, was neutralized. The jury ultimately deliberated for less than four hours before proclaiming Simpson not guilty of the murders on 2 October 1995. Although critics accused the defense team of obfuscating and confusing the jury, Simpson's defense team argued that it was the failure of the prosecution to rebut their claims that ultimately swayed the jurors in the case. There were too many chances for cross-contamination, too many chances for evidence to be planted, and too few guidelines in place to prevent such problems from occurring.[80]

The Simpson trial was in many ways the last hurrah in the challenges to DNA profiling in the United States. Despite the vigorous critique of the evidence in the case, Scheck and Neufeld made a strong distinction between the technology itself and its use in this particular case. As Scheck stated in a 2003 interview:

> In terms of the forensic science, there was a silver lining [to the Simpson trial], because we never attacked the validity of DNA technology, because it is a valid technology, it's a revolutionary technology. The problem in the O.J. Simpson case is the way that the evidence was collected. They were using nineteenth-century evidence collection techniques for twenty-first century technology, which had more than the potential, but the reality of producing contaminated results. So in the end, the collection and handling of the evidence was the way that we attacked what was produced in the Simpson case.[81]

Thus, while a few others continued to doubt DNA evidence, by the end of the Simpson trial, Scheck and Neufeld seemed to have been satisfied that all of the important issues surrounding the validity and reliability of DNA profiling had been exposed. In fact, in interviews they point out with pride that the Simpson trial raised the consciousness of crime laboratories and fundamentally altered the way that biological evidence from the crime scene is collected and processed.[82]

THE INNOCENCE PROJECT

In the wake of the Simpson trial, Scheck and Neufeld made public peace with the prosecutors whom they had faced down in court, in the halls of Congress, and at meetings over the past five years. In Scheck's words: "The opponents that we had, who were the DNA specialists—Woody Clark, who is the prosecutor in San Diego; Rock Harmon, who is the prosecutor here in Alameda County—in the end, after the trial was over, we all became friends—we actually knew each other and were friendly before that—but allies after the trial."[83] Neufeld told me that he and Scheck were now also friendly with John Hicks, who was the FBI's laboratory director during the bureau's implementation of its DNA Analysis Unit and now heads New York State's forensic science operations.

Such a shift made sense for Scheck and Neufeld considering that their mission was becoming much bigger than protecting their clients from unreliable evidence. In 1992 they had founded the Innocence Project (IP), a nonprofit legal clinic at the Cardozo School of Law in New York City, where Scheck remains a professor. The nonprofit legal clinic was set up in order to free some of the thousands of wrongfully convicted people languishing in the American prison system. In order to achieve success in this daunting task, however, they needed a form of proof that was so credible and so convincing that prosecutors and law enforcement agents would be unable to disagree with them. They found this truth-teller in DNA.

As the O.J. Simpson murder trial was fading into the collective memory of Americans, Scheck and Neufeld were busy instilling a kind of mythic power to DNA evidence. As Scheck and Neufeld stated in their 2000 book, *Actual Innocence*, which they wrote about the Innocence Project in collaboration with journalist Jim Dwyer:

> Now the fabric of false guilt is laid bare, and the same vivid threads bind a wealthy Oklahoma businessman and a Maryland fisherman: Sometimes eyewitnesses make mistakes, Snitches tell lies. Confessions are coerced or fabricated. Racism trumps the truth. Lab tests are rigged. Defense lawyers sleep. Prosecutors lie. DNA testing is to justice what the telescope is for the stars: not a lesson in biochemistry, not a display of wonders of magnifying optical glass, but a way to see things as they really are. It is a revelation machine. And the evidence says that most likely, thousands of innocent people are in prison.[84]

Put aside were their concerns about the nature of forensic samples; gone, at least for a while, were their fears of laboratory error; and gone was their overall skepticism toward DNA evidence. *Actual Innocence* does not even mention Scheck and Neufeld's earlier experiences with DNA evidence in *Castro* and *Yee*—it is a sanitized version of history with DNA as the triumphant hero. When I asked Neufeld why he and Scheck did not discuss earlier cited problems with the technique, he said, "It is a different book, that's all. No one's hiding anything—obviously our writings are out there—it's just a different book. And the book is not a book about DNA. The book is about what's wrong with the criminal justice system and how we can fix it."[85]

Scheck and Neufeld had long argued that while DNA evidence was still problematic when used for purposes of inclusion, it could be reliably used for exclusionary purposes, since no population genetics or statistics was needed. In other words, while the probability of a random match between two identical profiles needed to be calculated in order to definitively say that those samples came from the same person, a nonmatch required no interpretation. This view, however, ignored the very real problems of contamination, degradation of DNA in samples generally stored in less than pristine conditions, as well as chain of custody issues (i.e., who handles the sample at each stage of the process). Because they carefully reviewed the circumstances of each case before accepting it, however, the Innocence Project rarely had to publicly discuss potential problems with DNA evidence. Indeed, the Innocence Project makes no secret of the fact that it only accepts cases in which DNA evidence can yield conclusive proof of actual innocence.[86]

Over the past fifteen years, the Innocence Project has grown in size and prominence. To date, nearly two hundred people have been exonerated because of its work.[87] The project is now at the center of a national Innocence Network, a group of law schools, journalism schools, and public defender offices that work to free innocent prisoners from jail. The project has also expanded its mission to the legislative arena and has lobbied politicians at the state and local level, asking them to pass legislation that protects the rights of convicted felons to seek post-conviction relief and requires law enforcement officials to comply with requests in these cases. The entire edifice rests on the credibility of DNA profiling as the gold standard of forensic evidence. As such, it is clear why Scheck and Neufeld might be less willing to overtly challenge the technique as they did in the late 1980s and early 1990s.

STR Analysis: The Nail in the Coffin

Just as debates about RFLP-based DNA profiling were coming to a close, the next generation of DNA profiling technology was being developed: short tandem repeat (STR) analysis. Although STRs were first developed in the United States in the early 1990s by Thomas Caskey, their implementation was much faster in Europe. Today, however, almost all American DNA profiling laboratories conduct STR testing. Like RFLPs, STRs are also based on length polymorphisms. However, as the name suggests, the repetitive elements in STRs are much shorter than the sequences identified by RFLP analysis (2–6 base pairs compared to 10–100 base pairs). Another crucial difference between RFLP testing and STR testing is that PCR, rather than using restriction enzymes to fragment samples into discrete bands, identifies and isolates only the locus of interest.

There are numerous major advantages to the use of PCR-based STR testing. First, as mentioned earlier, the process creates millions of copies of the locus of interest, so only minute quantities of a biological material need to be recovered from a crime scene for the analysis to be effective. The second advantage of a PCR-based system is that because no digestion of DNA occurs during analysis, multiple STR regions from the same small sample of biological evidence can be examined simultaneously. This characteristic allows for the third advantage of STR testing: the probes used to identify particular regions of interest can be combined into a single analysis kit. This process, often called "multiplexing" or "multiplex PCR," requires a significant amount of optimization for different probe systems and requires a great deal of work to perfect.[88] The kits greatly simplify the work of the forensic technician, because he or she can simply add DNA to a tube filled with all of the chemicals, enzymes, and molecules needed to do PCR. Fourth, because of various technical developments like capillary electrophoresis (in which DNA can be run through a thin tube for analysis rather than a slab gel) and fluorescent dyes, coupled with a dramatic increase in computing technology, STR analysis can be highly automated. A result of all of these characteristics is that STR analysis can yield results in as little as one day, compared to several days for the fastest RFLP-based processing.

STR analysis, like previous forms of DNA profiling, begins by separating DNA from the crime scene sample. Once isolated, the DNA is subjected to PCR, which is an enzymatic process that replicates a specific region of the genome over and over and over. The process entails heating and cooling DNA at precise temperatures through about thirty cycles in the presence of various chemicals, enzymes, and molecules. The most important components of the PCR mix are the two "primers," or short nucleotide sequences, that are complementary to DNA sequences at either boundary of the region of interest. These primers bind to the region of the genome being targeted for amplification and serve to attract the thermostable enzyme that actually replicates DNA. These primers are labeled with a colored dye that is used to measure the length of each locus when the results of the PCR reaction are analyzed. Called *Taq* polymerase, this enzyme uses the nucleotides (A, C, T, and G) that are also present in the reaction mixture to copy the DNA between the two primers. In addition to these components, various salts and water are also added to the mixture to stabilize the reaction.

The first step of the PCR process involves denaturing the DNA, or separating it into single-stranded molecules. This is done by heating the DNA to high temperatures, approximately 94 degrees Celsius. Once separated, the reaction mixture is cooled to approximately 60 degrees Celsius, when the primers bind to the DNA sequences to which they are complementary. As mentioned above, several separate regions can be analyzed in the same reaction. Next, the mixture is heated to approximately 72 degrees Celsius, when *Taq* polymerase locates the primers and copies the region of DNA between them, forming a double-stranded copy of the original DNA molecule. Once this process is complete, the temperature is once again raised to 94 degrees Celsius, and the cycle, which takes about five minutes to complete, is repeated twenty-nine more times.

At the end of the reiterative PCR cycle, several hundred million copies of the regions of interest are produced from only a few original DNA molecules. The next step in the process is to analyze the results. Today, almost all major crime laboratories use automated machines sold by private companies to do this. The most common version of this apparatus is called the "Genetic Analyzer" and is made by a company called Applied Biosystems.[89] As with RFLP analysis, PCR products are separated with electrophoresis. In this case, however, the PCR product is run through a small capillary tube subjected to an electrical current containing a viscous material through which the DNA fragments pass.[90] Although the analogy is not perfect, one can imagine that the capillary tube is like a one-lane version of the gel used for RFLP testing.[91] The smaller fragments travel faster than the longer fragments and therefore pass through the capillary tube more quickly. As the fragments get close to the end of the tube, they pass through a laser beam, which causes the dye bound to them to fluoresce and emit a unique color. The machine's camera records the time, color, and intensity of light when each DNA fragment passes through this detection window. Based on this information, which is recorded as a series of peaks on an electropherogram, which looks a bit like a Technicolor heart-rate monitor, a specialized computer program interprets this data. Applied Biosystems' unit employs a software program called GeneScan. In the final step of the interpretation process, data from GeneScan-processed electropherograms is imported into another program called Genotyper, which ultimately calls individual alleles. It is important to remember that the computer program must distinguish between the numerous alleles of more than a dozen STR loci from data that is not always perfect. Numerous technical problems can lead to interpretation mishaps and occasional miscalls, most of which involve smaller peaks that occur just before, just after, or in between true peaks.[92] While these errant peaks are relatively easy to resolve in single DNA contributor samples, they are much more troublesome in mixtures, which frequently occur in rape cases. In addition to these problems, there are a range of other issues that require human intervention in the interpretation process.[93]

Because of the complexity of the operation, and despite significant automation, human intervention and judgment is sometimes necessary. This is especially true in cases of mixture—when there is more than one contributor to a particular crime scene sample—and degraded DNA. At present, however, there are no generally accepted or nationally recognized standards for declaring a match between two profiles. As in other realms of DNA profiling and forensic science, there are only guidelines and recommendations.

Thus, DNA analysts at two different labs might make different decisions about whether a peak is a valid allele or a technical artifact, or whether an allelic dropout is caused by degraded DNA or represents nonmatching DNA profiles. Further, there is no strict standard for how many alleles can be missing from one profile for it to match another—the legal system must rely on the DNA analyst to make his or her best judgment based on the facts of the case. According to many defense advocates, most notably William Thompson, these decision are not made objectively but rather are based on the results that the prosecutor or law enforcement is looking for.[94] While this statement is not entirely accurate, since there has been a tremendous amount of research carried out by the forensic community to deal with these problems, and numerous solutions advanced, it is true that most methods have not been subjected to the kind of widespread legal scrutiny like that which occurred in early RFLP cases like *Castro* and *Schwartz*. Perhaps unsurprisingly, in an effort to reduce the potential for human bias by removing overt human intervention in DNA profiling as much as possible, computer programmers are hard at work developing expert systems software to make these calls automatically.[95]

Currently, there are dozens of STR markers available for use, but the FBI has chosen thirteen of them for its DNA database, eight of which are in common with the standard European systems. As such, almost all DNA analyses carried out in the United States contain, at minimum, these thirteen loci. In many cases, however, additional loci are used for increased individualization. These thirteen loci have been incredibly well characterized and all population data is available or referenced on the National Institute of Standards and Technology's Short Tandem Repeat DNA Internet Database.[96] As a result of defense challenges to RFLP-based DNA profiling, the FBI and its allies went to *extraordinary* lengths to ensure that the STR loci chosen were not subject to unusually high levels of substructure in particular populations and were inherited relatively independently from one another. In cases where there was some evidence of substructure or dependence, a correction factor to be plugged into the product rule was developed based on the recommendations of the second NRC report. Although there are generally not as many alleles in a given STR locus when compared to an RFLP locus, their sheer number of loci means that population genetics issues are not nearly as great of a concern as they were with RFLP-based profiling. According to one calculation, when all thirteen loci are used in conjunction with the product rule, the average unadjusted probability of a random match is less than one in a trillion among unrelated individuals. Because of careful attention paid to population genetics issues during the development of STR technology, combined with the statistical corrections for potential substructure prescribed in the second NRC report, there are no longer major concerns about population genetics in DNA profiling.

When first introduced into the legal system, most of the STR systems used by private companies such as Applied Biosystems and Perkin Elmer to develop kits were considered proprietary and protected by trade secret claims. (It should be noted that Lifecodes and Cellmark are now both part of Orchid Biosciences and conduct DNA testing using these systems but were not integral in their creation.) In several early cases, including *People v. Bokin* (California, 1999, unreported), *State v. Pfenning* (Vermont, 2000) and *People v. Shreck* (Colorado, 2000), defense attorneys argued that these kits had not been adequately validated by scientists outside the companies or the forensic community, and that such validation was currently impossible because the companies kept the primer sequences and other information private. Although, the judges in *Pfenning* and *Shreck* initially accepted this argument, the Colorado State Supreme Court rejected the latter decision a year later.[97] Since that decision, courts across the country have rejected the claim that primer sequences need to be publicly available for proper validation. In these cases, judges have argued that the crime laboratories using the technique validate the technology each time they use it, over and above the initial work done by the private companies to ready their STR kits for market. As a Minnesota judge stated in a highly anachronistic analogy, STR technology is "like a Model A Ford. Thousands of owners can tell us it works even if Henry Ford can't or won't explain it."[98]

More recently, the machines used by crime laboratories to process and analyze DNA profiles have become impenetrable black boxes that are protected from defense scrutiny by closely guarded computer algorithms and automation. It remains nearly impossible for people outside of forensics labs to view the data underlying a DNA profile made using the methods described above because the companies that create the proprietary software would only sell the requisite computer programs as a bundle with their DNA analysis machines. The package is incredibly expensive and far exceeds the budget of most defense attorneys and consultants (the program needed to open the data files received during discovery proceedings costs several thousand dollars). Although a few consultants, such as Simon Ford and Wright State biologist Dan Krane, have been able to purchase these programs, most defense attorneys still cannot afford to routinely send questionable DNA cases to them because analysis is time consuming (five to ten hours on average) and expensive. In an effort to redress this problem, Krane founded a company called Forensic Bioinformatics with Thompson, Ford, and two bioinformatics specialists, Travis Doom and Michael Raymer, to provide this expertise to defense attorneys at low cost using custom-built computer software that performs a basic review of DNA evidence meant to flag potential problems.[99]

A MATTER OF TRUST

As David Lazer of Harvard's Kennedy School of Government remarked in his introduction to the 2004 edited collection *DNA and the Criminal Justice System: The Technology of Justice*:

> An earlier volume on the use of DNA in the criminal justice system was titled *DNA on Trial*. . . . DNA technology is no longer on trial; in fact, it has now been rather neatly integrated into the courtroom. This, of course, does not mean that there are no controversial issues around the use of DNA technology in criminal prosecution. What is striking is that the dominant controversies

about DNA technology now revolve around arguments about the competence of the system rather than the reliability of the technology itself. The very precision of DNA technology is, in fact, exhibit 1 in the current trial of the criminal justice system.[100]

By "competence of the system," Lazer is referring to an apparent paradox: Scheck and Neufeld were perfectly willing to challenge the reliability of DNA evidence in the Simpson trial, but also use DNA evidence in postconviction exonerations as if it were entirely unproblematic. Lazer resolves this conundrum by arguing that Scheck and Neufeld have complete faith in the technology of DNA profiling but little faith in the criminal justice system that uses it to convict people. In their view, the system is fueled by racial biases, underfunded defense attorneys, and questionable law enforcement practices.[101]

Lazer then goes on to argue that the *Simpson* case was a "prime example of relative roles of trust in the technology and trust in the system.... Underlying the OJ Simpson case was the issue of trust: not trust in DNA technology, but trust in the system. Did the Los Angeles Police Department handle the samples from Simpson properly? Might they have engaged in a conspiracy to frame Simpson? In the end, the technology was not a cure for distrust in the system."[102] Implicit in this framing is the notion that Scheck and Neufeld had gained trust in the technology of DNA profiling but not in the cops who used it. Yet, this view can only be accepted by reading the meaning of the *Simpson* case in hindsight. As the "Motion to Exclude DNA Evidence" clearly indicates, Simpson's defense team fully intended to argue that the evidence in the case should not be admissible because of unresolved technical controversies. The decision to withdraw the admissibility challenge occurred not because of a fundamental faith in the reliability of the technique, but because of the contingencies of a trial that was beamed live to homes around the country on a daily basis. As such, their strategy of critiquing the system while leaving the technical core of DNA technology unexamined should not be taken as an a priori indication that the technology could be fully trusted. Rather, Scheck and Neufeld's trust in DNA evidence can only be understood in the context of their work with the Innocence Project.

Their newfound trust in DNA evidence can be meaningfully contrasted with the views of William Thompson, the lawyer they brought in as co-counsel to help them draft the motion and cross-examine witnesses on the interpretation of the results of DNA testing. In Thompson's view, the Simpson trial illuminated *yet another problem* that affected the validity and reliability of DNA evidence—that is, how the sample was handled before testing. After the trial, Scheck and Neufeld began proclaiming that handling of crime scene evidence was the *last remaining issue* that needed to be addressed. For Thompson and a vocal but increasingly small group of critics of DNA evidence, three crucial issues remained to be resolved. The first was the need for an agency other than the FBI to regulate and set standards for DNA profiling laboratories. The second, in stark contrast to the findings of all of the major advisory boards, panels, and committees, was the persistent problem of laboratory, human, and interpretation error compromising the reliability of test results. And third, on a related note, the rate of error could not be estimated unless a meaningful blind proficiency testing scheme was developed for DNA laboratories and technicians. More recently, the reprivatization of DNA profiling through the production of kits used to perform STR analysis has raised new concerns that still need to be addressed.

The ability to get their message out has been hampered both by the lack of media interest in the faults of DNA after the overload caused by the Simpson spectacle, as well as a lack of big-name experts who helped them get their message out a few years earlier. Without a doubt, Scheck and Neufeld's decision to champion DNA evidence as the gold standard of forensic science certainly set back the defense's cause when it comes to challenging DNA evidence. Only time will tell if the few remaining skeptics of the validity and reliability of DNA evidence are right that systemic problems remain as strong today as when DNA profiling was introduced into the legal system nearly two decades ago.

CHAPTER 9

The Legacy of History

On 30 October 1998, a sixteen-year-old high-school student named Josiah Sutton and a friend were arrested by the Houston Police Department (HPD) for their involvement in the brutal kidnapping and rape of a local woman named Priscilla Stewart.[1] The woman had been driving in her car a few days after the incident and recognized the two young men as they were walking down the street. When she called the police to report their location, she stated that they were wearing the same hats as the rapists. Although Sutton and his friend bore almost no physical resemblance to Ms. Stewart's earlier description of her attackers (other than the color of their skin), they were required to submit blood and saliva samples so that they could be compared to biological evidence left at the crime scene. While his friend was cleared of any wrongdoing, the Houston Police Department's crime laboratory found that Sutton's DNA was an exact match with that of the rapist. Based almost entirely on this DNA evidence, Sutton was convicted and sentenced to twenty-five years in prison.

After languishing for four and a half years in a Texas penitentiary, where he witnessed acts of brutality and callousness that were previously inconceivable to him, Sutton was released. It turned out that the DNA evidence used to convict him was bogus—the laboratory deliberately took test results that should have been interpreted as exculpatory and twisted them until it appeared that Sutton was one of the two rapists. Although Josiah Sutton did indeed possess a few of the same alleles as one of the rapists, he also possessed alleles that were not present in either of the rapists' profiles. According to William Thompson, the University of California-Irvine professor and defense attorney, the combination of alleles that both Sutton and rapist shared could be found in about one out of fifteen randomly chosen African American men. At Sutton's trial, however, a HPD crime lab employee testified that Sutton's profile matched one of the two rapists exactly, and that the probability of a random match was 1 in 694,000.

The case led to an ongoing investigation into serious deficiencies in the Houston Crime Laboratory's procedures, protocols, and quality control mechanisms. In addition to Sutton, another man named Ronald Cantrell was released from prison when it became clear that he was also convicted as a result of bad DNA tests. Since then, retesting of biological evidence in previously closed cases has yielded numerous examples of DNA work that matched the poor quality in the Sutton investigation. In some of these cases, lawyers are currently seeking post-conviction exoneration. In the immediate aftermath of the scandal, the HPD's DNA laboratory was shuttered and, as of March 2007, remains closed pending the completion of an ongoing investigation. To date, the independent investigation into laboratory problems has identified, among many other issues, forty-three DNA cases analyzed by the HPD "in which there are significant doubts about reliability of the work performed, the validity of the Crime Lab's analytical results, or the correctness of the analysts' reported conclusions."[2]

The discovery of faulty DNA evidence in the *Sutton* case was not made by the internal audits suggested by TWGDAM, investigations taking place as part of ASCLD accreditation, or through any of the recommendations of the DNA Advisory Board put into place by an act of the U.S. Congress. The system that was so highly praised for preventing errors and ensuring the validity and reliability of DNA evidence simply did not work in the case of the Houston Police Department. How was the error revealed? Acting on tips that the HPD's crime lab was turning out shoddy evidence, a local TV station, KHOU, decided to investigate. After gaining access to court transcripts and crime laboratory records, the reporters turned to William Thompson for help in interpreting the documents they had located. In an extensively written report and an interview conducted for KHOU, Thompson blasted the HPD's crime laboratory for not running adequate controls, failing to document their work, blatantly ignoring exculpatory evidence, and reporting DNA test results in a way that did not accurately communicate their meaning to the jury. In the end, Thompson called it the worst laboratory work he had ever seen.[3]

In a highly publicized Virginia case, investigations associated with the postconviction exoneration of death-row inmate Earl Washington Jr. led to the discovery that the State Division of Forensic Sciences Central Laboratory in Richmond had made numerous mistakes in the course of their 2000 review of Washington's 1983 conviction for rape and murder. According to an ASCLD/LAB audit, the Virginia state laboratory had not followed accepted procedures when conducting PCR-based DNA profiling (running thirty-three PCR cycles instead of thirty and modifying the reaction mixture for the test); misinterpreted the results of the test, leading to erroneous exclusion of the victim from the vaginal smear taken from her dead body; and made numerous conclusions that were unsupported by available data.[4] The report also concluded that the laboratory's own mechanisms for catching errors failed in this case since the technical reviewer did not observe the errors in the processes and the final results.[5]

While the isolated (one hopes) mistakes discovered at the Virginia state laboratory were undoubtedly minor when compared with the systematic malfeasance in Houston, they still illustrate some of the inherent problems of forensic science in the legal context. The ASCLD/LAB report noted that the deviations from protocol and the misinterpretation of results were at least partially caused by political pressure from above. According to the scientist who carried out the analysis, "inconclusive results were not an option." The governor wanted to know whether Washington's DNA was found at the crime scene, and he felt the obligation to provide it.[6]

In a rare statement explicitly acknowledging errors in DNA evidence, Peter Neufeld said:

> This laboratory that touts itself as the best DNA laboratory in the country generated erroneous test results in a capital case, twice, using two different DNA methods. The audit reveals not only that the laboratory's most senior DNA analyst, responsible for DNA testing in many of the state's capital cases, made serious errors, but that the laboratory's system to catch these errors completely failed. This audit provides compelling evidence that crime labs cannot police themselves, and that only with the statutory requirement that they be subject to independent, expert oversight can we have faith that appropriate controls are in place.[7]

In his response to the ASCLD audit, Paul Ferrara, the director of the Division of Forensic Sciences, accepted the conclusions of the report but minimized their overall importance. He noted that the mistakes were made on a single subsample of a case carried out over five years before the investigation and that numerous policies had been put into place in the interim to prevent the kinds of problems that were discovered from happening again. Additionally, he noted that even given the mistakes made, the analyst conducting the DNA tests had accurately excluded Earl Washington as a donor of the DNA found at the crime scene.[8]

Unfortunately, the Houston and Virginia incidents are not isolated examples of the potential for error in DNA profiling. Related problems have been discovered in Seattle's crime laboratory, where DNA contamination and other errors were found in at least twenty-three cases.[9] In North Carolina, the state Bureau of Investigations DNA unit has recently come under fire in an exposé by the *Winston-Salem Journal*.[10] Although an audit by ASCLD/LAB cleared the laboratory of wrongdoing in the particular case in question, the organization urged the state to improve the way it collected, handled, and stored biological evidence.[11] According to William Thompson, who tracks crime laboratory errors involving DNA evidence, cross-contamination and sample mislabeling problems have been documented in Pennsylvania, Nevada, and California.[12] Further, the Illinois State Police recently terminated its contract with Bode Technology Group, a private company it had hired to help work through the state's DNA testing backlog, citing shoddy laboratory work and numerous errors.[13] In November 2004, Cellmark, one of the largest private DNA testing labs in the world, fired an analyst for falsifying control data in at least twenty cases.[14] In May 2006, the U.S. Army reported that it was investigating the work of a civilian DNA analyst in more than five hundred military criminal investigations around the country.[15]

The FBI's Office of the Inspector General (OIG) recently issued a scathing two-hundred-page report reviewing the vulnerabilities of the FBI's own protocols and laboratory practices.[16] The impetus for the review was the discovery that a bureau scientist named Jacqueline Blake had been omitting negative control tests from her DNA analyses and falsifying her laboratory notes for more than two years, from March 2000 to April 2002, without being caught. Blake's malfeasance, which resulted in thousands of wasted hours in investigations and rechecking her work, also led to the temporary removal of twenty-nine profiles from the national DNA database.[17] What was so shocking about this story was that Blake was not caught through any of the official mechanisms that were put into place to ensure the validity and reliability of DNA evidence. Rather, her deception was discovered by chance by one of her colleagues who noticed late one evening that the testing results displayed on Blake's computer were inconsistent with the proper processing of control samples.[18]

In addition to trying to figure out how Blake could have gotten away with such blatant scientific misconduct for so long, the OIG also sought to "assess whether the [FBI's DNA analysis] protocols were vulnerable to other abuse and instances of noncompliance."[19] Although the OIG investigation found that the FBI's protocols and procedures were generally sound and appeared to be followed by all analysts other than Blake, it concluded that 31 of 172 sections of FBI's protocols were "significantly vulnerable to inadvertent or willful noncompliance."[20] The report also noted that the there is a great deal of variation in individual practice because the FBI's guidelines are silent on so many issues.[21] While some of these variations "served to mitigate, at least to some degree," the effects of the vulnerabilities that the report identified, the lack of adequate guidelines leaves the FBI's DNA analysis unit "subject to an increased risk of employee error and inadvertent protocol noncompliance."[22]

While these problems are certainly troubling to anyone who cares about justice or science, the real issue is ascribing some sort of meaning to these mistakes. Are they nothing more than the kinds of isolated errors that one must expect from any endeavor involving people, or are they the result of fundamental faults in DNA profiling as a complex technological system at the intersection of science, politics, and law? Further, why is it that all of these errors have been caught by journalists and crusading defense advocates, rather than forensic scientists and their rigorous quality control and quality assurance programs? Are most errors caught well before evidence is introduced in the courtroom, with the above problems being the very small minority that slips through the cracks? Or, are the above problems a very small minority of the numerous errors that actually make their way into the courtroom? Do these recent errors represent an inevitable deterioration of the quality of DNA evidence now that it is not in the public spotlight anymore, or, to borrow the words of Thompson, is it "the emergence and recognition of problems that existed all along but heretofore were successfully hidden"?[23]

Without solid empirical evidence provided by regular case audits or proficiency testing, it is impossible to answer these questions with any certainty. What can be done, however, is to examine the historical record for potential explanations of why errors continue to arise despite the fact that the controversies over the validity and reliability of DNA evidence have been resolved for nearly a decade now. Based on my analysis, despite dramatic improvements in the quality of DNA evidence over the past two decades, several historical factors have prevented it from being even better, beginning with the reduction of disputes surrounding DNA evidence to statistics and population genetics in the early 1990s. This unfortunate situation obscured or minimized other crucial issues, most notably those of error and interpretation of test results. As a result, judges around the country were acutely aware of the population genetics problems associated with DNA profiling and made resolution of these issues the primary criteria for admissibility. Beginning in 1992, all other issues came to be seen as influencing only the weight that jurors should attach to DNA evidence and not whether it should be introduced at all. Thus, the FBI and their allies, who were strongly driven by concerns about the legal admissibility, had to deal with population genetics issues in order to use DNA evidence.

In addition, the FBI's ability to control the expert commissions (i.e., TWG-DAM and the DNA Advisory Board) that set standards for DNA profiling, both in terms of membership and scope, has profoundly affected the forensic community's willingness to engage actively with issues like error. The second NRC committee's deference to these two expert bodies meant that they simply wrote off the first NRC committee's view that error rates must somehow be estimated and integrated into jurors' evaluation of random match probabilities. To be fair, though, there is an inescapable problem with estimating error—it is much harder to investigate than something like allele frequencies in ethnic subpopulations (assuming, of course, that there is agreement on what the relevant subpopulations are). In the population genetics controversy, the defense experts laid down a straightforward challenge to the forensic DNA profiling community: go out into the world and measure allele frequencies for the genes you use to construct DNA profiles in a variety of subpopulations. If you find that alleles are distributed randomly within these populations (i.e., that certain alleles don't tend to associate with one another more often than you would expect if each one was independent of the next), then it is okay to use the product rule. If, however, you find unexpectedly high or low frequencies of certain alleles, or various combinations of alleles appearing together more often than you would expect, then some sort of statistical correction needs to be put into place. Unfortunately, an effective and practical means of measuring error has not yet been developed, and it seems doubtful that one can do so, because the chances for error are constantly changing depending on a host of contingent factors—at the institutional, individual examiner, and evidence levels. There is no such thing as a stable error rate in the way that there is such a thing as an allele frequency that can be assumed to be stable at least for a period of time.

Third, following the traditions of forensic science, the guidelines put out by these commissions and panels are predominantly voluntary. While a certain degree of flexibility in rules for forensic labs is necessary because of the contingencies associated with investigating crime, a voluntary system encourages a situation in which the best laboratories, of which there are many, will follow the guidelines to the letter, while the worst labs will wantonly ignore them. One need only think about the Los Angeles Police Department's crime laboratory circa 1995 and, more recently, Houston, in order to understand the danger of voluntary standards and guidelines. While the ASCLD/LAB accreditation scheme and the quality control required by the FBI to load samples to CODIS probably go a long way to assuring that most laboratories follow the rules, there is no excuse for crime labs in some of the largest cities in the country to be so woefully out of step with accepted practice in forensic science.

Fourth, the majority of the problems associated with DNA profiling, whether real or perceived, have tended to be brought to light by defense attorneys. Even the most ardent proponent of DNA evidence acknowledges that strong defense review is one of the ways that system ensures valid and reliable evidence. However, the defense community has often been prevented from fully doing their job for a variety of reasons, including unwillingness of judges to issue discovery orders, difficulty in finding credible scientists to testify during admissibility hearings, secrecy on the part of crime laboratories, private laboratories shielding their products by claiming trade secrets, and a chronic lack of funding for indigent defense. Perhaps most important, though, defense lawyers who wish to challenge scientific evidence must both become scientifically knowledgeable in their own right and cultivate a stable of highly credible scientists willing to testify on their behalf. This requires sustained commitment on both sides. As the members of the defense and scientific community who initially challenged DNA in the early 1990s have moved on to other issues, they have not been replaced by individuals of the same degree of status and influence in the eyes of the courts and the public. Although people like William Thompson and Dan Krane continue to point out the shortcomings of DNA profiling, they simply do not command the same kind of attention that Scheck, Neufeld, Lander, Hartl, and Lewontin did. While Scheck and Neufeld were deeply involved in the investigation of laboratory errors in the Earl Washington Jr. case in Virginia, they no longer play an active role in critiquing DNA evidence.

While it is inevitable that people move on to other issues over time, the current problems with DNA evidence around the country may not have been as severe had the defense community been able to apply as much pressure to the FBI on the error issue as they had in the population genetics debates. However divisive and inefficient the controversies over DNA evidence were from 1989 to 1994, they encouraged many positive technical and social changes in the DNA profiling regime. From the very beginning, defense challenges forced the private companies to be more accountable for their work and encouraged public agencies to take an active role in developing DNA profiling for the criminal justice system. They also paved the way for increased involvement of the forensic community's leadership in regulating the production and use of DNA profiles. The early problems that the defense illuminated in court, combined with the decision to develop a DNA database, encouraged the FBI to set rigorous standards for producing DNA profiles so that they could be meaningfully compared across jurisdictions. As a result, protocols, probes, reagents, quality control procedures, and quality assurance programs were standardized and disseminated around the country.

The shift from private labs to public labs, which was at least partially stimulated by early defense challenges, made the practice of DNA profiling somewhat more open to defense scrutiny (after the FBI abandoned its policy of shielding its DNA Analysis Unit from discovery to enable self-critical analysis). It also prevented the use of the concept of trade secrecy to shield DNA profiling regimes from defense scrutiny for a brief period of time. Since the advent of STR analysis, most of the profiling process has become highly automated once DNA is extracted from the crime scene samples. Forensic laboratories now use kits, equipment, and analysis tools made by private companies like Applied Biosystems, Perkin Elmer, and Promega based on publicly available marker systems. Although the conditions under which the PCR reaction is run can be manipulated, as we saw in the Earl Washington case, it is conceivable that the entire profiling process can be carried out with little *overt* human interaction. DNA is simply loaded into premixed reaction vials for PCR amplification and is then placed into a machine that uses capillary electrophoresis to identify and measure the lengths of alleles of particular markers. This information is then fed into a computer program that interprets the spectroscopic information and calls the allele lengths that make up the profile. While the probes and population data are now no longer kept secret (in fact, most are freely available on NIST's Web site), the exact design of the PCR kits, and the algorithms used to interpret the output of the computer programs that type DNA profiles, are highly guarded as trade secrets by the companies that sell them and by their customers in forensic laboratories around the country.[24] As such, defense attorneys wishing to challenge the interpretation of the underlying data produced by the profiling machine must find someone with access to the software used to read the raw data. This step is crucial, since there are numerous potential problems that can arise with creating an STR profile that require analysts to make judgment calls.

The creation of a robust DNA profiling system is undoubtedly a great achievement. The value of the technique in the criminal justice system is indisputable. Thousands of criminals have been found and convicted as a result of the technique, and innumerable innocent people have been cleared of any wrongdoing without ever having to step into the courtroom. The growing DNA database has aided in the investigation of numerous cases and potentially even deterred some people from committing crimes in the first place. At the same time, DNA profiling has been used by defense lawyers to secure the release of nearly two hundred wrongfully convicted individuals from prisons around the country.

Yet, it would be a mistake to conclude from this process that DNA profiling has reached a stage of development that we no longer need to be concerned about its validity and reliability. The system that we have today is not the best possible one, but rather the one that the legal system has determined to be valid and reliable enough for use in the criminal justice system. In other words, we have not reached the end of history in the context of DNA profiling. The persistence of significant errors, the lack of a robust proficiency testing scheme, and still active disagreements over how to interpret mixed profiles all suggest that much work remains to be done.

In writing this book, my goal has not been to propose ready-to-implement solutions to the problems I have identified. History amply demonstrates that the successful modification of a complex technological system like DNA profiling occurs only after an extended period of negotiation and incremental change. Instead, my intentions have been much more modest. To begin with, I have highlighted the importance of including a wide variety of stakeholders in the creation of the rules, codes of conduct, and standards for forensic science that will be needed to solve these problems. Because of entrenched interests, and a culture that is wary of outside intervention, the forensic community should not be left to make these choices on its own. I have also demonstrated the need to have more direct regulations and oversight for individual laboratories. Under the current state of affairs, the leaders of the forensic community merely issue guidelines that individual labs need not fully comply with. Although the requirements of admission to the Combined DNA Index System (CODIS) community and various accreditation schemes mitigate this problem to a certain extent, there is still a great deal of room for improvisation, especially in the interpretation of ambiguous test results.

Additionally, I have made it clear that the forensic community must become more willing to admit that error is a potentially serious problem that needs to be addressed systematically. One way to institutionalize this mind-set is the use of error logs in all laboratories, as recommended (but not required) in the DNA Advisory Board's Quality Assurance Standard #14. Another would be the creation of a national commission with rotating membership that could be called on to investigate the causes of DNA (and other forensic) errors when they occur.

Perhaps most important at this stage in the development of the technique, the best way to ensure the validity of results in particular cases is to provide the resources and legal mechanisms that will enable defense attorneys to evaluate and challenge DNA tests in particular cases. Especially important is access, through the mechanism of discovery, to all aspects of the DNA profiling, including raw data, as well as money to pay the few people who own the software programs used to interpret this information. In the case of a newly introduced method of DNA analysis, the set of discovery materials should also include data on validation procedures and the protocols and laboratory notebooks used by technicians. History has shown that a competent defense review of forensic evidence can reveal problems that are either consciously ignored or overlooked by forensic scientists and prosecutors.

It is important to recognize that making the claim that DNA profiling is imperfect is not the same as calling it unreliable, invalid, or inadmissible. As such, our collective focus should not be on arguing over the infallibility of DNA profiling, but rather on developing methods to ensure that when problems do occur, whether they be related to error, fraud, laboratory protocol, or interpretation of results, there is a high likelihood that they will be caught. Currently such mechanisms are ad hoc, voluntary, and not always effective.

As they demonstrated in the resolution of the controversies over population genetics and molecular biological issues, the DNA profiling community is certainly up to the challenge, provided that the concerns and criticisms of the legal and academic communities are taken seriously. If they fail to heed the warnings of those defense attorneys and academics who focus on problems with DNA testing, it is possible that the scandal in Houston, the errors made by the Virginia DNA Laboratory, and the Jacqueline Blake fiasco will be the tip of an ever-growing iceberg.

Finally, I have made clear that the development of the technique was not guided by a scientific method or by any sort of inherent logic—there was no linear, rational pathway from the laboratory to the courtroom. Rather, the technique itself, standards of good science, and the expertise needed to make DNA evidence credible in the legal system, evolved together over the course of a decade. If readers remember only one thing from this book, I hope it is that developing DNA profiling for use in the criminal justice system was just as much about social engineering as it was about getting the science right.

Notes

1. INTRODUCTION

1. *New York v. Wesley [and Bailey]*, 140 Misc.2d 306 (1988); Ricki Lewis, "DNA Fingerprints: Witness for the Prosecution," *Discover*, June 1988, 44–52; Jean L. Marx, "DNA Fingerprinting Takes the Witness Stand," *Science* 240 (1988): 1616–1618; Debra Cassens Moss, "DNA—The New Fingerprints," *ABA Journal* 74 (May 1988): 66–70.

2. U.S. Federal Bureau of Investigation, *Combined DNA Index System*, "Measuring Success," http://www.fbi.gov/hq/lab/codis/success.htm, 15 April 2007.

3. Barry Scheck, Peter J. Neufeld, and Jim Dwyer, *Actual Innocence: Five Days to Execution and Other Dispatches from the Wrongly Convicted* (New York: Doubleday, 2000), 122.

4. Attorney General John Ashcroft, "Attorney General Ashcroft Announces DNA Initiatives," U.S. Department of Justice, 4 March 2002, http://www.usdoj.gov/archive/ag/speeches/2002/030402newsconferncednainitiative.htm.

5. Diane Cardwell, "New York State Draws Nearer to Collecting DNA in All Crimes, Big and Small," *New York Times*, 4 May 2006.

6. Ibid.

7. The Innocence Project, "About the Innocence Project," Benjamin N. Cardozo School of Law, Yeshiva University, http://www.innocenceproject.org/about/index.php.

8. For more information on DNAPrint Genomics, see DNA Print Genomics, "Forensics," http://www.dnaprint.com/welcome/productsandservices/forensics/.

9. For information on this case, see: Arlington National Cemetery, "The Vietnam Unknown Controversy," http://www.arlingtoncemetery.com/vietnam.htm.

10. Michael Lynch, "God's Signature: DNA Profiling, the New Gold Standard in Forensic Science," *Endeavour* 42, no. 2 (2003): 93–97.

11. Simon Cole, "Grandfathering Evidence: Fingerprint Admissibility Rulings from Jennings to Llera Plaza and Back Again," *American Criminal Law Review* 41 (2004): 1189–1276.

12. Michael Saks and Jonathan Koehler, "The Coming Paradigm Shift in Forensic Identification Science," *Science* 309, no. 5736 (2005): 892–895.

13. For the most developed versions of the "culture clash" thesis, see Marcia Angell, *Science on Trial: The Clash of Medical Evidence and the Law in the Breast Implant Case* (New York: W. W. Norton, 1996); Steven Goldberg, *Culture Clash: Law and Science in America*

(New York: New York University Press, 1994); and Lee Loevinger, "Law and Science as Rival Systems," *University of Florida Law Review* 19 (1967): 530–551.

14. Simon Cole points out that the same phenomenon can be seen in the context of fingerprinting. See Simon Cole, "More Than Zero: Accounting for Error in Latent Fingerprint Identification," *Journal of Criminal Law and Criminology* 95, no. 3 (2005): 985–1078.

15. Office of the Independent Investigator for the Houston Police Department Crime Laboratory and Property Room, "Independent Investigator Issues Fifth Report on Houston Police Department Crime Lab," 11 May 2006, http://www.hpdlabinvestigation.org/pressrelease/060511pressrelease.pdf.

2. SCIENCE FOR HIRE

1. "Quantum Chemical Corp Reports Earnings for Qtr to Dec 31," *New York Times*, 30 January 1988, http://query.nytimes.com/gst/fullpage.html?res=9E05EFD71726E630BC4953 DFBE6E958A; and Cellmark Diagnostics, *DNA Fingerprinting* (Germantown, MD, 1988), reprinted in U.S. Senate Committee on the Judiciary, Subcommittee on Constitution, *DNA Identification*, 101st Cong., 1st sess., 1989, 92–114.

2. This case received a great deal of attention in the British media. The details of the case were reported on in various newspapers, including Brian Silcock, "Genes Tell Tales," *Sunday Times*, 3 November 1985, 13, and Andrew Veitch, "Son Rejoins Mother as Genetic Test Ends Immigration Dispute/Ghanian Boy Allowed to Join Family In Britain," *Guardian*, 31 October 1985.

3. Veitch, "Son Rejoins Mother."

4. Ibid.

5. Sheona York, e-mail to author, 19 November 2001.

6. It should be noted that just because we do not yet know the function of so-called junk DNA does not mean that it serves no purpose.

7. Restriction enzymes are thought to play a crucial role in the immune response of bacteria to foreign organisms.

8. For an excellent overview of this process, see Lorne T. Kirby, *DNA Fingerprinting: An Introduction* (New York: Stockton Press, 1990). The process is named after its creator, E. M. Southern. He originally described the entire procedure in E. M. Southern, "Detection of Specific Sequences among DNA Fragments Separated by Gel Electrophoresis," *Journal of Molecular Biology* 98, no. 3 (1975): 503–517.

9. A[lec] J. Jeffreys, V[ictoria] Wilson, and S[wee] L[ay] Thein, "Hypervariable 'Minisatellite' Regions in Human DNA," *Nature* 314, no. 6006 (1985): 67–73.

10. Alec J. Jeffreys, Victoria Wilson, Swee Lay Thein, "Individual-Specific 'Fingerprints' of Human DNA," *Nature* 316, no. 6023 (1985): 76–79.

11. Michael Lynch, "God's Signature: DNA Profiling, The New Gold Standard in Forensic Science," *Endeavour* 27 (2003): 93–97.

12. "DNA Fingerprinting: DNA Probes Control Immigration," *Nature* 319 (1986): 171.

13. Robert Walgate, "Futures: You and Nobody Else: Focus on the Technique of Genetic Fingerprinting," *Guardian*, 8 November 1985.

14. Alec Jeffreys, interview with Michael Lynch, 6 August 1996, lines 1001–1007, O.J. Simpson Murder Trial Papers and DNA Typing Archive, #53/12/3037, Division of Rare and Manuscripts Collections, Cornell University Library (hereafter cited as Simpson MSS), Box 4, Folder 20.

15. Unfortunately, legal records only need to be kept for six years in the United Kingdom, so I was unable to track down any information about this hearing.

16. Z. Wong et al., "Characterization of a Panel of Highly Variable Minisatellites Cloned from Human DNA," *Annals of Human Genetics* 51 (1987): 269–288.

17. Craig Seton, "Life for Sex Killer Who Sent Decoy to Take Genetic Test," *Times*, 23 January 1988, 3.

18. The introduction to Cellmark's first DNA typing manual begins: "Every year, in hundreds of thousands of court cases in the United States, a just decision rests on the ability of judges and juries to establish a person's identity. It happens in paternity suits and criminal cases. Who is the child's father? Is the drop of blood found at the scene of the crime from the defendant now on trial?" The corporate background section highlighted Cellmark's financial connections to its parent company. The manual states that Cellmark "takes advantage" of the more than $2 million a day allotted to research by ICI, with a significant percentage of that money going to "molecular biology and gene probe development," Cellmark Diagnostics, *DNA Fingerprinting*, 13.

19. California Association of Crime Laboratory Directors, *DNA Committee Report #2*, 19 November 1987, 7. Cellmark originally charged $200 per sample for paternity tests (making the average cost for a test involving mother, suspected father, and child $600), plus $500 per day for expert testimony in court. The standard fee for forensic casework was $285 per prepared sample, plus $80 (plus $75 per hour for labor) for sample preparation, plus $500 per day for expert testimony. Telephone consultation was free in both cases. Depositions were also offered either via phone or at Cellmark's offices. The first half hour was free of charge, then $75 per hour for a non-Ph.D., and $150 per hour for a Ph.D., from Cellmark Diagnostics, *General Questions and Answers*, June 1988, personal collection of Arthur Daemmrich.

20. Daniel Garner, interview with author, 24 January 2002.

21. Anna D. Uchman of Cellmark Diagnostics to Attorney, 20 July 1988, case file of *Minnesota v. Schwartz*, 1989 (SIP No. 89903565/C.A. No. 88–3195), Hennepin County District Attorney's Office.

22. Cellmark advertisement, "DNA Fingerprinting Links the Criminal to the Crime," 1988, publication information unknown, personal collection of Arthur Daemmrich.

23. "Quantum Chemical Corp Reports Earnings," *New York Times*, 30 January 1988.

24. Lifecodes, "DNA-Print™ Identification Test" (publicity brochure), 1988, 20–21, personal collection of Arthur Daemmrich.

25. Ivan Balazs, interview with author, 13 February 2002.

26. Michael Baird, interview with author, 19 February 2002.

27. In such situations, geneticists usually denote one allele with a capital letter and the other with a lowercase letter. The Hardy-Weinberg Principle can be extended to systems with multiple alleles, although the calculation of frequencies becomes more complicated. When multiple alleles are present, geneticists often use a combination of numbers and letters to represent each one—for example, C_1, C_2, C_3, etc.

28. Michael Baird, interview with author, 19 February 2002.

29. Ivan Balazs, interview with author, 13 February 2002.

30. Baird interview. Nearly all of Lifecodes' validation experiments were published in two papers appearing back to back in the April 1986 issue of the *Journal of Forensic Sciences*. Some of this work was also presented at the 11th Congress of the Society for Forensic Haemogenetics in Copenhagen in August 1985. See Alan M. Giusti et al., "Application of Deoxyribonucleic Acid (DNA) Polymorphisms to the Analysis of DNA Recovered from Sperm," *Journal of Forensic Sciences* 31, no. 2 (1986): 409–417; and Evan Kanter et al., "Analysis of Restriction Fragment Length Polymorphisms in Deoxyribonucleic Acid

(DNA) Recovered from Dried Bloodstains," *Journal of Forensic Sciences* 31, no. 2 (1986): 403–408. The ultimate conclusion from these two papers was that Lifecodes' technology was successful in identifying individuals from dried bloodstains and biological fluids recovered after sexual activity. This claim was based on numerous experiments carried out in their own laboratories using specimens from simulated rapes and murders.

31. I. Balazs et al., "The Use of Restriction Fragment Length Polymorphisms for the Determination of Paternity" (paper presented at the American Society of Human Genetics, Toronto, Ontario, 1984) *American Journal of Human Genetics Supplement* 36:476.

32. Lifecodes, "DNA-Print™ Identification Test," 6.

33. Lifecodes, publicity materials, "DNA-Print Test Makes Proof-of-Paternity Child's Play," 1988, personal collection of Arthur Daemmrich.

34. Baird interview.

35. Ibid.

36. Ibid.

37. As science studies scholar Simon Cole has pointed out, transcribing identity and making it mobile has long been a goal of law enforcement and the state, Simon Cole, *Suspect Identities: A History of Fingerprinting and Criminal Identification* (Cambridge: Harvard University Press, 2001).

38. *Oklahoma v. BJ Hunt*, 15 September 1987, unreported. According to various press reports, Lifecodes determined that blood found at the crime site was indeed the defendant's. The jury, however, concluded that his presence in the house where the crime occurred did not necessarily prove his guilt. He was acquitted of the charge. This case still represents the first time that a Lifecodes representative testified in a courtroom. See Janny Scott, "Chemists Told of Advances in 'Genetic Fingerprinting,'" *Los Angeles Times*, 8 November 1987.

3. DNA on Trial

1. Ricki Lewis, "DNA Fingerprints: Witness for the Prosecution," *Discover*, June 1988, 45.

2. *Tommie Lee Andrews v. Florida*, 533 So.2d 831, 834 (5th App 1988).

3. Jeffrey L. Ashton, testimony, U.S. Senate Committee on the Judiciary, Subcommittee on Constitution, *DNA Identification*, 101st Cong., 1st sess., 1989, 76.

4. Ibid., 81.

5. Ibid., 76.

6. *Frye v. United States*, 293 F, 1013 (D.C. Cir 1923).

7. Ibid., 1014.

8. For a review of this topic, Paul Giannelli, "Frye v. United States: Background Paper Prepared for the National Conference of Lawyers and Scientists," *Federal Rules Decisions* 99 (1983): 188–201; Irene M. Flannery, "Frye or Frye Not: Should the Reliability of DNA Evidence Be a Question of Weight or Admissibility?" *American Criminal Law Review* 30 (1992): 161–186; and Paul Giannelli, "The Admissibility of Novel Scientific Evidence: Frye v. United States, a Half-Century Later," *Columbia Law Review* 80 (1980): 1197–1250.

9. Giannelli, "The Admissibility of Novel Scientific Evidence," 192–193.

10. *Andrews v. Florida*, 843.

11. Charles T. McCormick, *Handbook of the Law of Evidence* (St. Paul, MN: West Publishing, 1954), 363–364.

12. Florida Evidence Code, sec. 90.702.

13. *Andrews v. Florida*, "Answer Brief of Appellee," 19 May 1988, 1–2.

14. Lewis, "DNA Fingerprints," 50; *Andrews v. Florida*, "Answer Brief of Appellee," 19 May 1988, 5.

15. Hal Uhrig, personal communication with author, 12 September 2002.

16. *Andrews v. Florida*, "Answer Brief of Appellee," 22.

17. *Andrews v. Florida*, "Amicus Curiae" (written by Andre Moennsens), 16 May 1988, 15.

18. *Andrews v. Florida*, 843.

19. Excerpt of trial proceedings in *State of Florida v. Tommie Lee Andrews*, 20 October 1987, 66, as quoted in Laurel Beeler and William R. Weibe, "DNA Identification Tests and the Courts," *Washington Law Review* 63 (1988): 942.

20. *Andrews v. Florida*, "Amicus Curiae," 15. This notion also was reiterated in the Appeals Court decision that DNA evidence was properly admitted by the trial court (see below).

21. *Andrews v. Florida*, "Answer Brief of Appellee," 13.

22. Identical claims were made in numerous other case transcripts that I have examined. See, for example, the testimony of Cellmark's laboratory director, Daniel Garner, in *State of Florida v. Randall S. Jones*, 1308, O.J. Simpson Murder Trial Papers, Division of Rare and Manuscripts Collections, Cornell University Library (hereafter cited as Simpson Archive), Box 2.

23. Lewis, "DNA Fingerprints," 52.

24. *Andrews v. Florida*, 849–850.

25. Ibid., 849.

26. Ibid.

27. Ibid., 850.

28. Ibid., 851.

29. Defense attorneys Barry Scheck and Peter Neufeld, as well as many other commentators, have made this claim in numerous published and unpublished sources. Neufeld told me they did a survey of cases involving DNA evidence for the *Castro Frye* hearing. *People of New York v. Joseph Castro*, 545 N.Y.S.2d 985 (1989).

30. *Martinez v. State of Florida*, 549 So.2d 694 (1989).

31. Ted Gest, "Convicted by Their Own Genes: DNA Fingerprinting Is Facing a Major Legal Challenge from Defense Attorneys and Civil Libertarians," *U.S. News and World Report*, 31 October 1988, 74.

32. *New York v. Wesley [and Bailey]*, 140 Misc.2d 306 (1988).

33. He told investigators that he did not murder the woman but accidentally tripped her, causing her to fall to the floor. He then stated that after noticing that she was bleeding, he attempted to determine whether she was alive by feeling her vaginal area for a pulse. When he felt none, he said he then attempted to administer CPR. Unsuccessful in reviving her, he said he placed her face down and left the apartment. He also told investigators that he did not have sexual intercourse with the woman, but turned his head when somebody else did it. Ibid., 421.

34. Transcript of *New York v. Wesley [and Bailey]* DNA Admissibility Hearing, Day 1, 22 February 1988, 4, personal collection of Jan Witkowski.

35. "Independent Expert Witnesses," Cellmark, 1988, personal collection of Arthur Daemmrich.

36. Richard Roberts, interview with author, 25 April 2002.

37. Michael Baird, testimony in *New York v. Wesley [and Bailey]*, 32, personal collection of Jan Witkowski.

38. Ibid., 35.

39. Judge Judy Kaye made a similar argument in her concurring opinion to the appellate decision in *New York v. Wesley [and Bailey]*, 462. She wrote: "We do not agree that the eight steps of forensic DNA analysis, then in its infancy, were shown to have been accepted

as reliable within the scientific community. Rather, the standard for general acceptance of the new techniques was seen as commensurate as the standards set by Lifecodes." I only became aware of her statement, however, after I had completed my analysis of the trial.

40. Much testimony in *New York v. Castro* case would demonstrate that Lifecodes had not actually been following its own protocol to the letter.

41. Baird testimony, 51.

42. It is interesting to note that the four steps laid out by Baird are similar in many respects to the Supreme Court's standards for legal admissibility set forth five years later in the 1993 *Daubert* decision.

43. Baird testimony, 53–54.

44. Ibid., 55–56. As we shall see, Lifecodes' claim that their racial databases were in HWE was challenged in *New York v. Castro*. Ultimately, the efficacy of the HWE test for substructure was also challenged and shown to be lacking in scientific rigor.

45. Baird testimony, 58.

46. Ibid., 57.

47. Ibid., 58.

48. Ibid., 59.

49. Ibid., 60.

50. *Frye* hearing, *New York v. Wesley[and Bailey]*, 61–63.

51. Ibid., 75.

52. Ibid., 96.

53. Ibid., 74–75.

54. *New York v. Wesley [and Bailey]*, 329; and Richard Borowsky, e-mail to author, 19 September 2002. This same claim about physical linkage and linkage equilibrium also was made in *Andrews v. Florida*. Linkage equilibrium represents the extent to which the alleles of different loci tend to associate during the process of gamete (sperm or egg) formation. If allele A1 and allele B2 are associated with one another no more than one would expect them to be associated at random, they are said to be in linkage equilibrium. If they are associated more than one would expect them to be at random, they are said to be in linkage disequilibrium. In order for the product rule to be valid, alleles being multiplied together must be independent of one another. If alleles A1 and B2 are associated with one another in a nonrandom pattern, this makes it more likely that an individual with A1 will also have B2. Thus, a DNA profile of A1, B2 would be more common than one would expect in a population in which they are in linkage disequilibrium.

55. Neville Colman, testimony in *New York v. Wesley [and Bailey]*, 404.

56. M. Baird et al., "Allele Frequency Distribution of Two Highly Polymorphic DNA Sequences in Three Ethnic Groups and Its Application to the Determination of Paternity," *American Journal of Human Genetics* 39 (1986): 489–501.

57. Ibid., 406.

58. D2S44, D12S1, D17S79, and DXYS14.

59. *Frye* hearing, 52.

60. Baird testimony, 204.

61. *Frye* hearing, 205–207.

62. Ibid., 219–220. In invoking the notion of "innocent until proven guilty" to explain the peer review process to Judge Harris, Roberts assimilates basic legal concepts into his courtroom explanations of science.

63. Richard Roberts, testimony in *New York v. Wesley [and Bailey]*, 222.

64. Ibid., 224.

65. *New York v. Wesley [and Bailey]*, 321.

66. *New York v. Wesley*, 633 N. .E.2d 451, 457 (New York Court of Appeals, 1994).

67. *New York v. Wesley [and Bailey]*, 322.

68. Borowsky e-mail.

69. *New York v. Wesley [and Bailey]*, 318.

70. Ibid.

71. Ibid., 329.

72. Ibid.

73. Ibid., 317.

74. Ibid., 308.

75. Michael Lynch, "The Discursive Production of Uncertainty: The OJ Simpson 'Dream Team' and the Sociology of Knowledge Machine," *Social Studies of Science* 28, no. 5–6 (1998): 829–868; and Sheila Jasanoff, *Science at the Bar: Law, Science, and Technology in America* (Cambridge: Harvard University Press, 1995).

76. Brian Wynne, "Establishing the Rules of Laws: Constructing Expert Authority," in *Expert Evidence: Interpreting Science in the Law*, ed. Brian Wynne and Roger Smith (New York: Routledge, 1989), 34. This point also has been made in work on science fraud and misconduct. See, for example, William Broad and Nicholas Wade, *Betrayers of the Truth* (New York: Simon and Schuster, 1982).

77. At the same time, many of these scholars also argue that the closure mechanisms need to be examined from time to time to make sure the decisions made are not arbitrary or overtly biased.

78. Notice that in this case, the "lag" that took place between the time that the prosecution had a well-developed stable of experts and the defense bar established the same was within law, not between law and science.

4. Challenging DNA

1. Sheindlin highlights this phenomenon in the *Castro* judgment. See *New York v. Joseph Castro*, 545 NYS.2d 985, 996 (1989).

2. Anthony Pearsall, "DNA Printing: The Unexamined 'Witness' in Criminal Trials," *California Law Review* 77 (1989): 665–703.

3. Roger Parloff, "How Barry Scheck and Peter Neufeld Tripped up the DNA Experts," *American Lawyer*, December 1989, 53.

4. All information about the circumstances surrounding the trial is taken from Parloff, "How Barry Scheck and Peter Neufeld Tripped up the DNA Experts," and *New York v. Joseph Castro*.

5. See Scheck's testimony in U.S. House Committee on Judiciary, Subcommittee on Civil and Constitutional Rights, *FBI Oversight and Authorization Request for Fiscal Year 1990 (DNA Identification)*, 101st Cong., 1st sess., 1989; and Peter J. Neufeld and Neville Coleman, "When Science Takes the Witness Stand," *Scientific American* 262, no. 5 (1990): 46–53.

6. Both Neufeld and Scheck made this point to me in conversations and personal communications. Neufeld also made the point in Neufeld and Coleman, "When Science Takes the Witness Stand."

7. Bureau of National Affairs, "Geneticist, Defense Lawyers Debate Merits of DNA Typing," *BNA Criminal Practice Manual* 3 (1989): 261.

8. Peter Neufeld, interview with author, 27 February 2002.

9. Peter J. Neufeld and Barry Scheck, "Factors Affecting the Fallibility of DNA Profiling: Is There Less Than Meets the Eye?" *Expert Evidence Reporter* 1, no. 4 (1989): 93–97.

10. *Tommie Lee Andrews v. State of Florida*, 533 So.2d 831, 849–850 (1988).

11. Neufeld and Scheck, "Factors Affecting the Fallibility of DNA Profiling," 94–95.

12. Barry Scheck, testimony, U.S. House Committee on Judiciary, Subcommittee on Civil and Constitutional Rights, *FBI Oversight and Authorization Request for Fiscal Year 1990 (DNA Identification)*, 101st Cong., 1st sess., 1989, 412.

13. See chapter 3.

14. The three organizers were Jack Ballantyne, director of the Suffolk County (N.Y.) Office of the Medical Examiner; George Sensabaugh, a professor of forensic science at University of California, Berkeley; and Jan Witkowski, director of the Banbury Center.

15. George Sensabaugh, interview with author, 6 November 2002.

16. Parloff, "How Barry Scheck and Peter Neufeld Tripped up the DNA Experts," 53.

17. Ibid. Lander also discusses this interaction in his testimony in *New York v. Castro*.

18. Discovery refers to the period before a trial when the parties in a legal dispute ask for, and exchange, documents and information pertinent to the issue under dispute. As we shall see, private corporations were reluctant to comply with the demands placed upon them by defendants and often sought protection from these requests on the grounds that the information they were being asked to hand over was proprietary (and therefore a trade secret).

19. Roger Lewin, "DNA Typing on the Witness Stand," *Science* 244, no. 4908 (1989), 1034; and Parloff, "How Barry Scheck and Peter Neufeld Tripped up the DNA Experts," 53.

20. Lander refused to be compensated for his time because he believed it would damage his reputation as unbiased. Parloff, "How Barry Scheck and Peter Neufeld Tripped up the DNA Experts," 52.

21. Ibid., 53.

22. Eric Lander, testimony, *New York v. Castro*, vol. 13, 3061–3062, Simpson MSS, Box 3.

23. Michael Baird, interview with Saul Halfon and Arthur Daemmrich, 14 July 1994, Simpson Archive, Box 2, Folder 27.

24. Lander biography at Whitehead Institute, "Lander Biography," http://www.wi.mit.edu/news/genome/lander.html.

25. Karen Hopkin, "Eric S. Lander, Ph.D.," Howard Hughes Medical Institute, http://www.hhmi.org/lectures/2002/lander.html.

26. Eric Lander testimony, 3012.

27. Ibid., 3019.

28. Peter Neufeld, direct examination of Eric Lander in *New York v. Castro*, vol. 13, 3024, Simpson Archive, Box 3.

29. Ibid., 3030.

30. Eric Lander testimony, 3036–3037.

31. Ibid., 3041.

32. Peter Neufeld, direct examination, 3040.

33. *New York v. Castro*, 986. Judge Sheindlin did not declare any witness to have explicit expertise in forensic DNA typing.

34. Eric S. Lander, "DNA Fingerprinting on Trial," *Nature* 339, no. 6225 (1989): 502. This information can also be found at various places in the *New York v. Castro* testimony.

35. Michael Baird, testimony, *New York v. Castro*, vol. 2, 502, 603–611, Simpson Archive, Box 2.

36. Howard Cooke, interview with author, 20 February 2002.

37. Lander, "DNA Fingerprinting on Trial," 502.

38. This standard was published in M[ichael] Baird et al., "Allele Frequency Distribution of Two Highly Polymorphic DNA Sequences in Three Ethnic Groups and Its Application

to the Determination of Paternity," *American Journal of Human Genetics* 39 (1986): 489–501, as well as several other articles published by Lifecodes employees between 1986 and 1989.

39. Baird testimony, vol. 1, 397–398.

40. This issue emerged from the fact that it is impossible to size a particular band with 100 percent accuracy, primarily due to measurement error. To account for this error, Lifecodes did a test in which they ran the exact same DNA fragment seventy times in duplicate and had two technicians independently measure each band. After averaging the measurements, they determined that the error in size measurements, or standard deviation, was 0.6% of the fragment size. Using 3 s.d. as their standard, the size of two bands had to be within 1.8% of the mean size of the band.

41. Lander, testimony, vol. 13, 3707–3708.

42. See testimony in *New York v. Castro*, vol. 8, 2346, Simpson Archive, Box 3. Scheck made this hand motion despite the fact that Baird had already stated that Lifecodes used visual observation several times during his earlier testimony.

43. Alan Giusti, testimony in *New York v. Castro*, vol. 4, 1410–1415, Simpson Archive, Box 3.

44. Lander, testimony, vol. 14, 3212.

45. Barry Scheck, cross-examination of Michael Baird, *New York v. Castro*, vol. 8, 2216–2217, Simpson Archive, Box 3.

46. *New York v. Castro*, 996.

47. Baird testimony, vol. 8, 2220.

48. These phrases are taken from an exchange between Neufeld and Lander in *New York v. Castro*, vol. 14, 3198, Simpson Archive, Box 3.

49. Ibid.

50. Ivan Balazs et al., "Human Population Genetic Studies of Five Hypervariable DNA Loci," *American Journal of Human Genetics* 44 (1989): 182–190.

51. Lander, "DNA Fingerprinting on Trial." In the wake of Lander's claim, Bernie Devlin, Neil Risch, and Kathryn Roeder published an article refuting the notion that there was an excess of homozygosity in any of Lifecodes' population databases and that Lander's result was more apparent than real. They argued that the occurrence of homozygosity was caused primarily by the fact that small fragments run off the gel during electrophoresis and are therefore not detected during the Southern blot process. See B[ernie] Devlin, Neil Risch, and Kathryn Roeder, "No Excess of Homozygosity at Loci Used for DNA Fingerprinting," *Science* 249, no. 4975 (1990): 1416–1420.

52. As quoted in Parloff, "How Barry Scheck and Peter Neufeld Tripped up the DNA Experts," 55.

53. Baird, interview with Saul Halfon and Arthur Daemmrich.

54. Risa Sugarman, testimony, *New York v. Castro*, vol. 14, 3401, 3416–3419, Simpson Archive, Box 3.

55. Pablo Rubinstein, testimony, *New York v. Castro*, vol. 6, 1979–1980, Simpson Archive, Box 3.

56. Ibid., 1970.

57. Lander, testimony, vol. 16, 3905–3907.

58. Ibid., 3914–3917.

59. Ibid., 3916.

60. Ibid., 3917.

61. Richard Roberts, interview with author, 25 April 2002.

62. Ibid.

63. Ibid.

64. Ibid.

65. Lewin, "DNA Typing on the Witness Stand," 1034–1035.

66. For two press accounts of the extrajudicial meeting of experts, see Harold M. Schmeck, "DNA Findings Are Disputed by Scientists," *New York Times*, 25 May 1989; and Lewin, "DNA Typing on the Witness Stand."

67. Carl Dobkin et al., "Statement of the Independent Expert Scientists Having Testified in the Frye Hearing in *People v. Castro*," 11 May 1989, 1, personal collection of Richard C. Lewontin.

68. Ibid., 2.

69. Ibid.

70. Lewin, "DNA Typing on the Witness Stand," 1035. Roberts made essentially the same comment to me when I interviewed him on 25 April 2002.

71. *New York v. Castro*, 997 n. 12.

72. These procedures stated that "the proponent, whether defense or prosecution, must give discovery to the adversary, which must include" copies of autorads, laboratory notebooks, copies of quality control tests run on samples in question, copies of reports issued to proponents, a written report from the laboratory laying out the actual methodology used to evaluate samples in question and to calculate statistical probabilities, a copy of the population frequency database used in this calculation, "a certification by the testing lab that the same rule used to declare a match was used to determine the allele frequency in the population," a statement on any potential or actual contamination, a statement about any potential or actual degradation of samples, a statement about any "other observed defects or laboratory errors, the reasons therefore and results thereof," as well as chain of custody documents. The procedure also stated that the proponent must serve an intent to offer DNA evidence as soon as possible, and that the proponent had the burden of showing that the tests were conducted in a scientifically reliable way.

73. *New York v. Castro*, 987.

74. Ibid., 988.

75. Ibid., 987.

76. Ibid., 988.

77. Ibid., 995.

78. *New York v. Castro* transcript, vol. 7, 2159–2190, Simpson Archive, Box 3.

79. *New York v. Castro*, 990.

80. In his testimony, Lander stated that he did not believe that Sheindlin's prong two test had been met since there was still disagreement about the best methods for doing DNA typing. However, he said that by the end of 1989 he expected the FBI and the scientific community to agree upon a generally accepted method for doing DNA typing. Lander also believed that the population genetics issues would be resolved around the same time. Although he qualified his statements by saying that he did not want to be held to a December 31 date, he did claim to "see all of the usual indicia for scientific resolution coming in motion right now, and all of those are designated target dates by the end of the year." See Lander, testimony, vol. 14, 3377–3379; quotation, 3383.

81. *New York v. Castro*, 990.

82. Ibid., 985.

83. Ibid.

84. According to Parloff, Sugarman stated that both she and her co-counsel believed that "it was a better service to the criminal justice system that this happened." See Parloff,

"How Barry Scheck and Peter Neufeld Tripped up the DNA Experts," 55. The latter quotation was from Edward McCarthy, a spokesman for the Bronx district attorney, in Robert D. McFadden, "Reliability of DNA Testing Challenged by Judge's Ruling," *New York Times,* 15 August 1989.

85. K. C. McElfresh, "DNA Fingerprinting (Letter to the Editor)," *Science* 246, no. 4927 (1989): 192.

86. McFadden, "Reliability of DNA Testing."

87. Michael Baird, interview with author, 19 February 2002.

88. Baird, interview with Saul Halfon and Arthur Daemmrich. Specifically, they stopped using the weighted Gaussian method to calculate the frequency of a particular fragment within the population database. Under this system, alleles were given increasingly less weight as they approached the outer limits of the floating $\pm 2/3$ s.d. around the fragment size. Under the straight binning system, all alleles that fit within the floating bin were counted equally. The defense objected to the weighted Gaussian method because it made fragments seem rarer in the population database than the nonweighted system.

89. Parloff, "How Barry Scheck and Peter Neufeld Tripped up the DNA Experts," 56.

90. Ibid.

91. For an excellent discussion of major DNA evidence cases, see Thomas M. Fleming, "Admissibility of DNA Identification Evidence," *A.L.R.4th* 84 (1991): 313.

92. Colin Norman, "Caution Urged on DNA Fingerprinting," *Science* 245, no. 4919 (1989): 699.

93. Alun Anderson, "Judge Backs Technique," *Nature* 340, no. 6235 (1989): 582.

94. McFadden, "Reliability of DNA Testing Challenged by Judge's Ruling."

95. Parloff, "How Barry Scheck and Peter Neufeld Tripped up the DNA Experts," 56.

96. McElfresh, "DNA Fingerprinting (Letter to the Editor)."

97. Jim Dawson, "Attacker of Woman at Ramp Left 'DNA Fingerprints' at Scene," *Star Tribune,* 21 June 1988; and Sgt. Bernard Bottema (Minneapolis Police Department) to Cellmark Diagnostics, 24 June 1988, case files, *State of Minnesota v. Thomas Schwartz,* 1989 (SIP No. 89903565/C.A. No. 88–3195), access provided by Hennepin County District Attorney's Office (hereafter cited as *Schwartz* Case Files).

98. Patrick Sullivan, interview with author, 8 August 2001.

99. Ibid.

100. *State of Indiana v. Frank E. Hopkins,* (1988), CCR-86–428. Hopkins was accused of murdering a woman named Sharon Lapp in her home in May 1985. Although Lapp was also raped during the commission of this crime, Hopkins was not being tried for that offense. At the time of the trial, Hopkins was serving a 100-year sentence for rape and burglary in Oregon and was extradited to Indiana to stand trial.

101. Jerry Shackelford, "DNA Test Error Admitted in Lapp Case," *Fort Worth Journal-Gazette,* 23 November 1988; Jerry Shackelford, "Procedure Varied in Lapp DNA Test," *Fort Worth Journal-Gazette,* 24 November 1988.

102. Sullivan interview. See also William C. Thompson and Simon Ford, "DNA Typing: Promising Forensic Technique Needs Additional Validation," *Trial,* September 1988, 55–64.

103. Ibid., 62.

104. William C. Thompson and Simon Ford, "DNA Typing: Acceptance and Weight of the New Genetic Identification Tests," *Virginia Law Review* 75 (1989): 58.

105. Thompson and Ford, "DNA Typing: Promising Forensic Technique," 63.

106. Ibid.

107. Ibid., 64.

108. Ibid.

109. Ibid.

110. See "State's Memorandum on the Admissibility of DNA Scientific Tests," *Minnesota v. Schwartz*, 8 February 1989, 1–8, *Schwartz* Case Files.

111. Ibid., 6–7.

112. In addition to witnesses mentioned in the text, there were Robin Cotton (of Cellmark), Dale Dykes (of the Minneapolis Memorial Blood Bank), and Harry Orr (mentioned earlier).

113. "State's Memorandum," 14. Note that all of these witnesses come from medical research.

114. "Defendant's Memorandum in Opposition to State's Motion to Relitigate Frye Hearing," *Minnesota v. Schwartz*, 4 January 1990, 9, *Schwartz* Case Files.

115. "State's Petition for Rehearing," *Minnesota v. Schwartz*, 13 November 1989, 2–3, *Schwartz* Case Files.

116. See chapter 7 for a discussion of the role of the FBI in the development and standardization of DNA typing. Bruce Budowle et al., "An Introduction to the Methods of DNA Analysis Under Investigation in the FBI Laboratory," *Crime Laboratory Digest* 15, no. 1 (1988): 19.

117. California Association of Crime Laboratory Directors, *Position on DNA Typing of Forensic Samples*, 20 November 1987, 2, personal collection of William C. Thompson.

118. Michael J. Davis, "Findings of Fact, Conclusions of Law, Order for Judgment," *Minnesota v. Schwartz*, 17 February 1989, 31, *Schwartz* Case Files.

119. Minnesota Rules of Criminal Procedure, sec. 28.03.

120. *State of Minnesota v. Thomas Robert Schwartz*, 447 N.W.2d 422, 424 (1989).

121. Patrick Sullivan, interview with author, 8 August 2001.

122. Steve Redding, interview with author, 14 December 2001.

123. Sullivan interview.

124. *Minnesota v. Schwartz*, 426.

125. Ibid.

126. Ibid., 427.

127. Ibid., 428.

128. Ibid.

129. Margaret Zack, "Hennepin County Drops DNA Test of Murder Suspect," *Star Tribune*, 11 January 1990.

130. This initially loose network would grow to become the "DNA Task Force," founded and organized by Barry Scheck and Peter Neufeld under the auspices of the National Association of Criminal Defense Attorneys. Barry Scheck and Peter J. Neufeld, "DNA Task Force Report," *The Champion*, June 1991, 13–21.

131. In other words, they served a similar function to the defense as Lifecodes and Cellmark employees performed for prosecutors. As one prominent scientist at Cellmark (who wishes not be quoted by name) told me in a January 2002 interview, the company regularly had to coach prosecutors on how to conduct a *Frye* hearing. Cellmark scientist, interview with author, 23 January 2002.

132. Lawrence Mueller, interview with author, 19 February 2002. *State of Washington v. Cauthron*, Snohomish County No. 88–1–01253–3 [1989]; appeal decided 1991, 846 P.2d 502.

133. See "Order Compelling Discovery," *Washington v. Cauthron*, 10 March 1989, personal collection of William C. Thompson. A fifth scientist, geneticist Randy Libby from University of Washington, also agreed to testify on behalf of the defense.

134. See "Amended Order Compelling Discovery," *Washington v. Cauthron*, 25 March 1989, 2, personal collection of William C. Thompson.

135. See, for example, Seymour N. Geisser, "Statistics, Litigation, and Conduct Unbecoming," in *Statistical Science in the Courtroom*, ed. Joseph L. Gastwirth (New York: Springer-Verlag, 2000), 79; and William C. Thompson, "Evaluating the Admissibility of New Genetic Identification Tests: Lessons from the "DNA War," *Journal of Criminal Law and Criminology* 84, no. 1 (1993): 72.

136. Ibid., 27.

137. Ibid.

138. *Washington v. Cauthron*, 502.

139. After deliberations, the Supreme Court ruled that forensic DNA analysis met the requirements of *Frye*, but that the prosecution had not provided sufficient expert testimony to demonstrate that Cellmark's probability statistics were admissible (*Washington v. Cauthron*, 516). All of the defense's other criticisms of the quality of DNA evidence in the case were deemed to go to the weight of the evidence, rather than its admissibility (512). On retrial, Cauthron was reconvicted of the five counts of rape.

140. *Washington v. Cautheron [sic]*, no. 41191-8-I (Washington Court of Appeals of the State of Washington, 1999).

141. *People of California v. Lynda Axell*, 1 Cal.Rptr.2d 411 (Cal. Court of Appeal, 1991); *State of Delaware v. Pennell*, 584 A.2d 513 (Del., 1989).

142. *Delaware v. Pennell*, 520.

143. Ibid.

144. *California v. Axell*, 421.

145. *California v. Axell* and *Delaware v. Pennell*.

146. Other than a minor setback in a December 1989 case, *State of Maine v. McLeod* (Cumberland City, Maine Superior Court, No. CR-89–62), in which the prosecutor withdrew evidence because of a series of misunderstandings over the procedure used by Lifecodes to correct for band-shifting, defense lawyers were unable to persuade judges in other cases to rule that Lifecodes' DNA evidence was inadmissible. For more on McLeod, see Bureau of National Affairs, "DA Faults Lifecodes' DNA Test, Withdraws Results, Drops Case," *BNA Criminal Practice Manual* 4 (1990): 3–6; Bureau of National Affairs, "Rugged Cross-Examination Exposes Flawed DNA Tests," *BNA Criminal Practice Manual* 4 (1990): 31–38; and Alun Anderson, "DNA Fingerprinting on Trial," *Nature* 342, no. 6252 (1989): 844.

147. Robert Manor, "DNA 'Fingerprinting' Questioned; Geneticist Says Test May Be Less Reliable Than First Believed," *St. Louis Post-Dispatch*, 15 October 1989; and Larry Thompson, "A Smudge on DNA Fingerprinting?; N.Y. Case Raises Questions about Quality Standards, Due Process," *Washington Post*, 26 June 1989.

148. *CBS This Morning*, transcript, 5 February 1990, Lexis-Nexis Academic Universe.

5. Public Science

1. On obligatory passage points, see John Law, "Technology, Closure and Heterogeneous Engineering: The Case of Portuguese Expansion," in *The Social Construction of Technological Systems*, ed. Wiebe Bijker et al. (Cambridge: MIT Press, 1987); and Bruno Latour, *Science in Action* (Cambridge: Harvard University Press, 1987).

2. Mary Ann Giordano, "DNA Test Pose New Dilemma for Courts," *Manhattan Lawyer*, 3 January 1989, 1.

3. Howard Harris to Lawrence T. Kurlander, 10 November 1987, Cold Spring Harbor Laboratory Archive, Box Containing Jan Witkowksi's Materials Related to DNA

Fingerprinting and the Banbury Conference on "DNA Technology and Forensic Science," [uncataloged], (hereafter cited as Cold Spring Harbor Laboratory Archive).

4. Ibid.

5. Anonymous, interview with author, 25 April 2002.

6. New York State Forensic DNA Analysis Panel, "Report of the New York State Forensic DNA Analysis Panel" (Albany, N.Y., 6 September 1989), iv. A copy of this document can be found in U.S. Senate Committee on the Judiciary and U.S. House Committee on the Judiciary, House Subcommittee on the Civil and Constitution Rights, *Forensic DNA Analysis*, 13 June 1991, House Serial 30/Senate Serial J-102–47 (Washington, DC: GPO, 1992), 203–275.

7. NY CLS Exec §995 (1994, ch. 737). For more on New York State legislation, see George H. Barber and Mira Gur-Arie, *New York's DNA Data Bank and Commission on Forensic Science* (New York: Matthew Bender, 1994); Bureau of National Affairs, "New York Law Would Regulate Forensic DNA Testing Labs," *BNA Criminal Practice Manual* 4 (1990): 315–318; and Bureau of National Affairs, "Landmark DNA Law Stalled," *BNA Criminal Practice Manual* 4 (1990): 491–492.

8. NY CLS Exec §995 (1994, ch. 737), §995-e.

9. John Hicks, testimony, U.S. Senate Committee on the Judiciary, Subcommittee on Constitution, *DNA Identification,* 101st Cong., 1st sess., House Serial 30/Senate Serial J-101–47 (Washington, D.C.: GPO, 1992), 65.

10. James J. Kearney, testimony, *State of Minnesota v. Jobe*, 23 August 1990, 9, case files of *State of Minnesota v. Larry Lee Jobe*, 1990 [SIP No. 88903565/C.A. No. 88–3301]; Hennepin County District Court), access provided to author by Hennepin County District Attorney's Office.

11. Ibid.

12. Ibid.

13. Bruce Budowle et al., "An Introduction to the Methods of DNA Analysis under Investigation in the FBI Laboratory," *Crime Laboratory Digest* 15, no. 1 (1988): 16. This group consisted of Budowle, F. Samuel Baechtel, Harold Deadman, Randall S. Murch, and four technicians.

14. Ibid., 19.

15. Ibid., 20.

16. Eric Lander, testimony, U.S. House Committee on Judiciary, Subcommittee on Civil and Constitutional Rights, *FBI Oversight and Authorization Request for Fiscal Year 1990 (DNA Identification)*, 101st Cong., 1st sess. (Washington, DC: GPO, 1990), 371.

17. Barry Scheck, open discussion in, Jack Ballantyne, et al., eds., *DNA Technology and Forensic Science*, vol. 32, *Branbury Report* (Cold Spring Harbor, N.Y.: Cold Spring Harbor Laboratory Press, 1989), 97.

18. Budowle et al., "An Introduction to the Methods of DNA Analysis under Investigation in the FBI Laboratory," 8.

19. Ibid., 9.

20. Hicks made this statement at least twice. See John W. Hicks, "FBI Program for the Forensic Application of DNA Technology," in *DNA Technology and Forensic Science*, 209; and Hicks, testimony.

21. John Hicks, interview with author, 21 March 2003.

22. Ibid.

23. See, e.g., ibid. and Harold Deadman, interview with author, 27 March 2003.

24. HaeIII was a four-base cutter (meaning that it cleaved DNA at the specific four nucleotide GGCC motif) rather than a five-base cutter like Hinf I (GANTC, with N being

any nucleotide), or a six-base cutter like Pst I (CTGCAG). Bruce Budowle *et al.*, "Hae III—A Suitable Restriction Endonuclease for Restriction Fragment Length Polymorphism Analysis of Biological Evidence Samples," *Journal of Forensic Sciences* 35, no. 3 (1990): 532.

25. The authors also pointed out that Hinf I was unable to cleave DNA when methylation of the first base in the restriction site occurs (the attachment of a methyl group to the nucleotide occurs most commonly in the dinucleotide CG). Such susceptibility was dangerous because methylation occurs in an unpredictable fashion, and analysis of different tissues of the same individual could potentially produce incompatible DNA profiles. Budowle et al. also reported that Hinf I could produce partially digested DNA under certain conditions, Budowle et al., "Hae III—A Suitable Restriction Endonuclease," 533.

26. Daniel Garner, interview with author, 24 January 2002.

27. Initial trainees included Special Agent Dwight Adams and three technicians. Lawrence Presley, Hal Deadman, and several additional technicians joined them a few months later.

28. Dwight E. Adams, "Validation of the Procedure for DNA Analysis: A Summary," *Crime Laboratory Digest* 15, no. 4 (1988): 85–87.

29. John W. Hicks, "DNA Profiling: A Tool for Law Enforcement," *FBI Law Enforcement Bulletin*, August 1988, 3–4.

30. Jan Bashinski, interview with author, 7 February 2002.

31. Hicks interview.

32. Ibid.

33. Roger T. Castonguay, "Message from the Assistant Director in Charge of the FBI Laboratory," *Crime Laboratory Digest* 15, supplement (1988): 1.

34. Randall Murch, "Summary of the [FBI] DNA Technology Seminar," *Crime Laboratory Digest* 15, no. 3 (1988): 79–85.

35. Castonguay, "Message from the Assistant Director," 2.

36. Hicks interview.

37. TWGDAM publicity material, undated, in FBI Academy, *A Resource Manual Compiled from the Legal Aspects of Forensic DNA Analysis Seminar*, 26–28 February 1990, 94, Simpson Archive, Box 2, Folder 43.

38. William Sessions to Representative Don Edwards, 9 August 1989, 2, in U.S. House, *FBI Oversight and Authorization Request for Fiscal Year 1990 (DNA Identification)*, 800–804, quotation, 801.

39. Dwight Adams, interview with author, 9 July 2002, and Hicks interview.

40. TWGDAM Participants List, undated, in FBI Academy, *A Resource Manual compiled from the Legal Aspects of Forensic DNA Analysis Seminar*, 112.

41. Sessions to Edwards, 8 August 1989, 3; Sessions to Edwards, 9 August 1989, 2, in U.S. House, *FBI Oversight and Authorization Request for Fiscal Year 1990 (DNA Identification)*, 800–804, quotation, 802.

42. U.S. Senate Committee on the Judiciary, Subcommittee on Constitution, *DNA Identification*, 101st Cong., 1st sess. (1989); and U.S. House, *FBI Oversight and Authorization Request for Fiscal Year 1990 (DNA Identification)*.

43. U.S. Congress and Office of Technology Assessment (OTA), *Genetic Witness: Forensic Uses of DNA Tests* (Washington, DC: GPO, 1990), iii.

44. For more information on the OTA, see U.S. Congress, Office of Technology Assessment, "The OTA Legacy: 1972–1995," http://www.wws.princeton.edu/~ota/.

45. OTA, *Genetic Witness: Forensic Uses of DNA Tests*, iv.

46. Ibid., 10.

47. Ibid., 12.

48. Ibid., 12–13.

49. Ibid., 13.

50. Ibid., 14.

51. Phillip J. Bereano, testimony, U.S. House, *FBI Oversight and Authorization Request for Fiscal Year 1990* (DNA Identification), 387.

52. Ibid., 415.

53. Jeff Brown to Don Edwards, 9 August 1989, 5, in U.S. House, *FBI Oversight and Authorization Request for Fiscal Year 1990 (DNA Identification)*, 807–808, quotation, 428.

54. Ibid., 5–7, in U.S. House, FBI *Oversight and Authorization Request for Fiscal Year 1990 (DNA Identification)*, 428–430.

55. William S. Sessions, "Invited Editorial," *Journal of Forensic Science* 34, no. 5 (1989): 1051.

56. Brown to Edwards, 29 November 1989, 1, in *U.S. House, FBI Oversight and Authorization Request for Fiscal Year 1990 (DNA Identification)*, 807–808, quotation, 807.

57. John W. Hicks to Evan A. Davis (Counsel to Governor Cuomo), 17 July 1990, 6; quotation, 2, Cold Spring Harbor Laboratory Archive. Letter cc-ed to John Poklemba, Dawn Herkenham, Howard Harris, and Robert Horn (NYS Police Crime Laboratory).

58. Ibid., 2.

59. Ibid., 2–3.

60. Ibid., 3.

61. Ibid., 4.

62. Lander, testimony, 415.

63. Ibid.

64. Dwight E. Adams et al., "Deoxyribonucleic Acid (DNA) Analysis by Restriction Length Fragment Polymorphisms of Blood and Other Bodily Fluid Stains Subjected to Contamination and Environmental Insults," *Journal of Forensic Sciences* 36, no. 5 (1991): 1284–1298.

65. Bruce Budowle et al., "Fixed-Bin Analysis for Statistical Evaluation of Continuous Distributions of Allelic Data from VNTR Loci for Use in Forensic Comparisons," *American Journal of Human Genetics* 48 (1991): 841–855; and Bruce Budowle et al., "A Preliminary Report on Binned General Population Data on Six VNTR Loci in Caucasians, Blacks, and Hispanics from the United States," *Crime Laboratory Digest* 18, no. 1 (1991): 10–26.

66. John Hicks, open discussion, in Ballantyne et al., *DNA Technology and Forensic Science*, 101.

67. Leslie Roberts, "Hired Guns or True Believers?" *Science* 257 (1992): 735.

68. Stephen Daiger et al., "Analysis of DNA Typing Data for Forensic Applications: A Research Proposal for the National Institute of Justice Program in Forensic Sciences and Criminal Justice Technology, 1990–1992 (NIJ-90-IJ-CX-0038; total award: $291,317), 1, personal collection of Richard C. Lewontin; made public in *United States v. Yee, et al.*). Ranajit Chakraborty's CV at http://www.identigene.com/SWIMX/docs/Chakraborty-CV.PDF. Daiger, Chakraborty and Boerwinkle received another $105,000 to sustain this work throughout 1992–1994 (NIJ-92-IJ-CX-K024).

69. Daiger et al., "Analysis of DNA Typing Data," 1.

70. Ibid.

71. It should be noted that by 1991 Bruce Weir would point out that the actual issue is whether the population database represents the population of alternative likely suspects. This is a subtle but important distinction from its matching the suspect's subpopulation.

72. Daiger et al., "Analysis of DNA Typing Data," 9.

73. Barry Scheck and Peter Neufeld, "Defendant's Reply Memorandum and Exhibits," *United States of America v. Yee, et al.*, 30 March 1992, 18, personal collection of Richard C. Lewontin.

74. Sheldon Krimsky, affidavit, *United States v. Yee, et al.*, 26 March 1992, paragraphs 22–25, personal collection of Richard C. Lewontin.

75. Ibid., paragraph 26.

76. Lawrence Presley, testimony, *State of Ohio v. Amos Lee*, 5 December 1990, 446–447, Simpson Archive, Box 1, Folder 21. It should be noted that protocols, no matter how rigorous, retain a certain amount of "interpretive flexibility" at the local level (i.e., in order to obtain the same product all the time, numerous minor changes to practice must be made to counteract problems specific to a particular forensic sample or laboratory situation). See Kathleen Jordan and Michael Lynch, "The Dissemination, Standardization, and Routinization of Molecular Biological Technique," *Social Studies of Science* 28, no. 5–6 (1998): 773–800.

77. See Linda Derksen, "Towards a Sociology of Measurement: The Meaning of Measurement Error in the Case of DNA Profiling," *Social Studies of Science* 30, no. 6 (2000): 803–845.

78. See Budowle et al., "Fixed-Bin Analysis," 844.

79. Ibid.

80. See K. L. Monson and Bruce Budowle, "A System for Semi-Automated Analysis of DNA Autoradiograms," in *Proceedings of an International Symposium on Forensic Aspects of DNA Analysis* (Washington, DC: GPO, 1989), Simpson Archive, Box 2, Folder 43.

81. Thus, a match was found at 699 bp, and the two closest bins were 650–700 bp, which appeared in 1 in 100 individuals, and 701–750 bp, which appeared in 1 in 10 individuals, the FBI would use the 1 in 10 figure for calculating the probability of a random match for a particular profile.

82. This point was made to me by numerous FBI researchers during interviews and is also present in numerous court cases. See, e.g., "Government's Response to Defense's Discovery Motion," in *United States v. Yee*, 2 January 1990, paragraph 4, personal collection of Richard C. Lewontin.

83. See Bruce Budowle's testimony in *United States v. Yee*, 86, as well as several other early cases involving DNA evidence, personal collection of Richard C. Lewontin.

84. In November 1989, a member of the DNA Analysis Unit sent a memo to the FBI's Legal Council Division seeking guidance on what it should do with the results of proficiency tests. The memo asked if it would be possible to use the "principle of self-critical analysis" as a justification for denying discovery requests for this material and for destroying it once it was no longer of use to the bureau. (John W. Hicks [actual author's initials "J.L.M. Hicks," presumably written by James L. Mudd] to Mr. Davis [FBI Legal Counsel], 21 November 1989, personal collection of William C. Thompson.) FBI Legal Counsel advised against destroying the material and advised that the materials be permanently maintained in the DNA administrative control file, and further argued that this information should be subject to disclosure. (Legal Counsel to Assistant Director, Laboratory Division, 20 April 1990, personal collection of William C. Thompson.) Ultimately, FBI Legal Counsel concluded that "we must emphasize that the FBI is seeking to establish a system that will be acceptable in all 50 states and the federal system. Therefore, we must proceed cautiously since each FBI decision will be scrutinized by many different judicial systems under many different standards" (8). To wantonly destroy proficiency testing data, or to

take any action that appeared to go against the guidelines set forth by TWGDAM (which said that raw materials from proficiency testing should be kept for a minimum of one year), would leave the FBI open to the kind of decision that occurred in the *Schwartz* case.

85. James J. Kearney, testimony, *State of Iowa v. Smith*, 18 December 1989, vol. 2, 57, 65, personal collection of William C. Thompson.

86. Ibid., 64.

87. Ibid., 67–68.

88. Science studies scholars have noted similar rhetoric in contexts as diverse as seventeenth-century English natural philosophy and late-twentieth-century regulatory science. See Steven Shapin and Simon Schaffer, *Leviathan and the Air-Pump* (Princeton, N.J.: Princeton University Press, 1985); and Sheila Jasanoff, "Contested Boundaries in Policy-Relevant Science," *Social Studies of Science* 17 (1987): 195–230.

89. *State of New Mexico v. Anderson*, 848 P.2d 531 (1989); and 853 P.2d 135 (N.M. Court of Appeals, 1993).

90. See William C. Thompson, memo to NACLD DNA Conference Participants RE: Proficiency Testing Errors by the FBI," 22 February 1990, 1–2, personal collection of William C. Thompson.

91. John W. Hicks to K. W. Nimmick, 22 December 1989, personal collection of William C. Thompson.

92. *United States v. Two Bulls*, 918 F.2d 56, 56 (8d App., 1990).

93. Ibid.

94. Judge Henry H. Kennedy Jr. of the D.C. Superior Court made this observation in his decision to admit the FBI's DNA matching procedure, but rejected the admissibility of the FBI's method for calculating the probability of a random match, in *United States v. Porter* (*United States v. Porter*, 618 A.2d 629 [D.C. Sup., 1992]). See also Judge Kennedy, "Order and Memorandum," in *Porter*, 20 September 1991, 59–60, personal collection of Richard C. Lewontin.

6. THE DNA WARS

1. See Thomas F. Gieryn, *Cultural Boundaries of Science: Credibility on the Line* (Chicago: University of Chicago Press, 1999).

2. They did so by increasing the range of fragment lengths that fit within each size bin, thus decreasing the rarity of any particular fragment within the bin.

3. Leslie Roberts, "Science in Court: A Culture Clash," *Science* 257, no. 5071 (1992): 732. See also Stephen Labaton, "DNA Fingerprinting Showdown Expected in Ohio," *New York Times*, 22 June 1990.

4. Barry Scheck and Peter Neufeld, affidavit, *United States v. Yee, et al.* (Case No. 3:89 CR 0720, U.S. District Court, Northern District of Ohio), 8 February 1991, 1, personal collection of Richard C. Lewontin.

5. James R. Wooley, "Government's Response to Motion for New Trial," in *United States v. Yee, et al.*, 20 February 1992, personal collection of Richard C. Lewontin.

6. Scheck and Neufeld, affidavit.

7. James Carr, "Magistrate's Report," *United States v. Yee, et al.*, 26 October 1990, 134 FRD. 161, 174.

8. Peter Neufeld and Barry Scheck, "Defendants Post-Hearing Memorandum on the Inadmissibility of Forensic DNA Evidence," *United States v. Yee*, 3, personal collection of Richard C. Lewontin. A similar sentiment was expressed to me by William Thompson, who told me that he has come to see admissibility challenges as the only lever to influence

the way that DNA typing is carried out. The only way to put pressure on DNA labs to change their methodology was to come to court and say that their methods, techniques, and protocols were not generally accepted in the scientific community. He called such challenges "awkward" and "crude," but said there was no other way for nonforensic scientists who are not in collaboration with the FBI to have an impact. William C. Thompson, interviews with author, 6 February 2002, 1 March 2002.

9. Neufeld and Scheck, "Defendants Post-Hearing Memorandum," 5.

10. Ibid., 8.

11. Ibid., 9–10.

12. Ibid., 13.

13. Ibid., 95.

14. For more on matching criteria, see Linda Derksen, "Towards a Sociology of Measurement: The Meaning of Measurement Error in the Case of DNA Profiling," *Social Studies of Science* 30, no. 6 (2000): 803–845.

15. Paul Hagerman, "Loading Variability and the Use of Ethidium Bromide: Implications for the Reliability of the FBI's Methodology for Forensic DNA Typing Criteria ("Expert's Report," in *United States v. Yee, et al.*)," 2, personal collection of Richard C. Lewontin.

16. Neufeld and Scheck, "Defendants Post-Hearing Memorandum," 63.

17. There were no fixed bin sizes in the FBI's system. Instead, the bins instead were constructed on an *ad hoc* basis by adding the standard deviation (2.5 percent) to each side of measured fragment. So, if a band in the crime scene DNA sample measured 1,000 base pairs, the FBI's bin used for probability statistics would be 9,975–1,025. The FBI would then use their database to calculate the percentage of fragments in this range compared to all other sizes. The private companies, on the other hand, calculated the match probabilities based on predetermined bins with predetermined frequencies.

18. Conrad Gilliam, testimony, *United States v. Yee, et al.*, vol. 18, 42, personal collection of Richard C. Lewontin.

19. Daniel Hartl, "Expert Report," *United States v. Yee et al.*, 2, personal collection of Richard C. Lewontin.

20. Ibid., 3.

21. Ibid., 6.

22. Ibid.

23. Carr, "Magistrate's Report," 208.

24. Bruce Budowle and John Stafford (legal counsel), "Response to Expert Report by D. L. Hartl (Submitted in the Case of *United States v. Yee*)," *Crime Laboratory Digest* 18, no. 3 (1991): 101.

25. Ibid., 104.

26. Carr, "Magistrate's Report," 209.

27. Neufeld and Scheck, "Defendants Post-Hearing Memorandum," addendum 1, 1.

28. See Carr's discussion in Carr, "Magistrate's Report," 209–210. It should be noted that Carr was less impressed by the testimony of FBI scientists. In his decision, Magistrate Carr noted that "Dr. Budowle did not respond persuasively to Dr. D'Eustachio's criticisms, and he refused to acknowledge the potential significance or merit of a competent scientist's critique and to consider the desirability for further experimentation."

29. Carr, "Magistrate's Report," 188–203.

30. Ibid., 195.

31. Ibid.

32. Ibid., 202.
33. Ibid.
34. Ibid.
35. Ibid., 203.
36. Ibid., 818.
37. Peter Neufeld, interview with author, 27 February 2002.
38. See chapter 1 for a review of population genetics and statistical issues.
39. Bruce Budowle et al., "Fixed-Bin Analysis for Statistical Evaluation of Continuous Distributions of Allelic Data from VNTR Loci for Use in Forensic Comparisons," *American Journal of Human Genetics* 48 (1991): 841–855; and Ranajit Charkraborty and Stephen P. Daiger, "Polymorphisms at VNTR Loci Suggest Homogeneity of the White Population of Utah," *Human Biology* 63, no. 5 (1991): 571–587.
40. Richard C. Lewontin, testimony, *United States v. Yee, et al.*, vol. 16, 51–53, 117–130, personal collection of Richard C. Lewontin.
41. Richard C. Lewontin, "Expert's Report," *United States v. Yee et al.*, 3–6, personal collection of Richard C. Lewontin; Neufeld and Scheck, "Defendant's Post-Hearing Memorandum," 70; and R[ichard] C. Lewontin and Daniel L. Hartl, "Population Genetics in Forensic DNA Typing," *Science* 254, no. 5039 (1991): 1745–1750. For a recent review of the literature on mate choice, see Matthijs Kalmijn, "Intermarriage and Homogamy: Causes, Patterns, Trends," *Annual Review of Sociology* 24 (1998): 395–421.
42. For a list of the literature cited by Lewontin, see Lewontin and Hartl, "Population Genetics in Forensic DNA Typing," 1750, references 19–23.
43. Ibid., 1747.
44. Ibid.
45. Daniel Hartl, *Basic Genetics* (Boston: Jones and Bartlett Publishers, 1991).
46. Daniel Hartl, interview with author, 4 April 2003.
47. Daniel Hartl, "Expert's Report," 9.
48. Budowle and Stafford (legal counsel), "Response to Expert Report by D. L. Hartl (Submitted in the Case of United States v. Yee)."
49. See Conneally's testimony (vol. 4a, 163 and vol. 5, 105–109); Kidd's testimony (vol. 13, 426–427; vol. 14, 120); as well as Daiger's testimony (vol. 7, 214), personal collection of Richard C. Lewontin.
50. Carr, magistrates report, 186. Carr reported that, in an effort to show that there was no substantial substructure in the Caucasian databases, both Kidd and Caskey made visual inspections of various Caucasian databases and concluded that they were "remarkably similar."
51. Under normal use of the product rule, when a homozygous pattern is found one simply assumes that there are two copies of that allele and multiplies the probability of that particular allele by itself to determine the probability that those two alleles would appear together. In an effort to be conservative, many DNA typing labs chose to double the frequency just in case the homozygote was apparent and not real. For instance, using the product rule, the frequency of a homozygous pattern for an allele that accounted for 10 percent of the total alleles in the relevant population would be 1/100, while the figure for the conservative 2p would be 1/20.
52. Carr, "Magistrate's Report," 182.
53. Ken Kidd, testimony, *United States v. Yee et al.*, vol. 13, 376, personal collection of Richard C. Lewontin.
54. Carr, "Magistrate's Report," 182.

55. See Derksen, "Towards a Sociology of Measurement," 843.

56. See Lewontin and Hartl, "Population Genetics in Forensic DNA Typing," for a well-developed explanation of this argument.

57. Carr, "Magistrate's Report," 204.

58. Ibid., 205.

59. Ibid. I could be wrong in this interpretation, but it was the closest standard that I could construct to provide a framework for Carr's rather bizarre discussion of which experts persuaded him the most.

60. Ibid.

61. Ibid.

62. Ibid.

63. Ibid., 205–206.

64. Ibid., 206.

65. *United States v. Yee et al.*; *United States v. Bond, et al.*, 12 F3d 540 (6th Circ. App., 1994).

66. In one of the few cases to contradict the decision in *Yee*, the Superior Court of the District of Columbia determined in the 1991 case of *United States v. Porter* (618 A.2d 629 [DC App., 1992]) that there was indeed a significant dispute in the scientific community over the validity of the FBI's population genetics assumptions. The court ruled that DNA evidence in the case was inadmissible because the FBI's methodology for calculating probability statistics was not generally accepted. Thus, the *Porter* court believed, contra Magistrate Carr, that probability statistics were an integral part of DNA evidence and therefore went to its admissibility, not just its weight.

67. Eric S. Lander, "Invited Editorial: Research on DNA Typing Catching up with Courtroom Application [Plus Responses from Numerous Individuals and Lander's Reply]," *American Journal of Human Genetics* 48 (1991): 819–823.

68. Charles J. Epstein, "Editorial: The Forensic Applications of Molecular Genetics— the *Journal's* Responsibilities," *American Journal of Human Genetics* 49 (1991): 697–698.

69. Eric S. Lander, "Lander Reply," *American Journal of Human Genetics* 49 (1991): 899–903.

70. James R. Wooley, "A Response to Lander: The Courtroom Perspective," *American Journal of Human Genetics* 49 (1991): 892–893.

71. C. Thomas Caskey, "Comments on DNA-Based Forensic Analysis," *American Journal of Human Genetics* 49 (1991): 893–895.

72. Bruce Budowle et al., "A Preliminary Report on Binned General Population Data on Six VNTR Loci in Caucasians, Blacks, and Hispanics from the United States," *Crime Laboratory Digest* 18, no. 1 (1991): 10–26.

73. Lander, "Lander Reply," 901.

74. In Lander's view, "to determine whether VNTR allele frequencies vary significantly among ethnic groups, it should be sufficient to collect perhaps a dozen or so well-separated ethnic population samples. If there are no significant differences, there is probably little cause for concern. If there are significant differences, statistical calculations could reflect the observed degree of variability. In any case, it seems important to know the answer." Lander, "Lander Reply," 902.

75. For more on these allegations, see Gina Kolata, "Critic of 'Genetic Fingerprinting' Tests Tells of Pressure to Withdraw Paper," *New York Times*, 20 December 1991; Leslie Roberts, "Fight Erupts over DNA Fingerprinting," *Science* 254, no. 5039 (1991): 1721–1723; Leslie Roberts, "Was Science Fair to Its Authors?" *Science* 254, no. 5039 (1991): 1722.

76. Ranajit Charkraborty and Kenneth K. Kidd, "The Utility of DNA Typing in Forensic Work," *Science* 254, no. 5039 (1991): 1735–1739.

77. Kenneth Kidd, interview with author, 30 May 2002, and e-mail to author, 18 May 2003.

78. Kidd interview.

79. Charkraborty and Kidd, "The Utility of DNA Typing in Forensic Work," 1735.

80. Ibid.

81. Ibid., 1736.

82. Ibid., 1737.

83. Ibid.

84. Lewontin and Hartl, "Population Genetics in Forensic DNA Typing," 1749.

85. Ibid., 1739.

86. Roberts, "Fight Erupts over DNA Fingerprinting."

87. Roberts, "Was Science Fair to Its Authors?"

88. Shannon Brownlee, "Courtroom Genetics," *U.S. News and World Report*, 27 January 1992, 60–61.

89. Editorial, "Quashing the DNA Debate," *Sacramento Bee*, 28 December 1991.

90. Kolata, "Critic of 'Genetic Fingerprinting' Tests."

91. See, e.g., Roberts, "Fight Erupts over DNA Fingerprinting"; Roberts, "Science in Court: A Culture Clash"; and William C. Thompson, "Evaluating the Admissibility of New Genetic Identification Tests: Lessons from the 'DNA War,'" *Journal of Criminal Law and Criminology* 84, no. 1 (1993): 22–104.

92. Edward Humes, "DNA War," *L.A. Times Magazine*, 29 November 1992, 29.

93. As we shall see, many academic critics of the FBI came to believe that Lander's research program may have at least partially affected his decision in late 1994 to coauthor an article with Budowle arguing that the controversies surrounding forensic DNA evidence had been solved thanks to the diligent and conservative work of the FBI.

94. Leslie Roberts, "Prosecutor v. Scientist: A Cat and Mouse Relationship," *Science* 257 (1992): 733.

95. Ibid.

96. Ibid.

97. Rockne Harmon to Editor of *Science*, 27 February 1991, personal collection of William C. Thompson.

98. William C. Thompson to James X. Dempsey, 24 January 1992, personal collection of William C. Thompson.

99. Laurence Mueller, interview with author, 19 February 2002.

100. These letters were mentioned in Roberts, "Prosecutor v. Scientist: A Cat and Mouse Relationship."

101. See, e.g., Mueller interview.

102. Leslie Roberts, "Hired Guns or True Believers?" 735.

103. Ibid.

104. Ray White, interview with author, 1 April 2003.

105. Thompson, interview, 1 March 2002.

106. Richard C. Lewontin, interview with author, 5 March 2003.

107. Hartl, interview.

108. Ibid.

7. THE DEBATE IN WASHINGTON

1. U.S. House Senate Committee on the Judiciary, U.S. House Committee on the Judiciary, Subcommittee on Civil and Constitutional Rights, *Forensic DNA Analysis*, 102nd Cong.,

1st sess., (1991), House Serial 30/Senate Serial J-102–47 (Washington, DC: GPO, 1992) (hereafter cited as *Forensic DNA Analysis*).

2. H.R. 339, 3 January 1991, in *Forensic DNA Analysis*, 131–135.

3. S.1355, in *Forensic DNA Analysis*, 289–296.

4. Jay V. Miller, "Analysis of DNA Identification Act of 1991 (S.1355)," in *Forensic DNA Analysis*, 286–288, 287. Hicks passed along nineteen letters in total to the congressional joint committee. Although there is no need to comment on each one, some general patterns emerged: a slim majority of respondents argued that the FBI should not serve in a regulatory capacity, but should continue to provide research and training support to crime labs. Of these individuals, most suggested ASCLD as a good alternative. A large minority suggested that the FBI was the best organization to regulate and set standards for the forensic community. There was almost unanimous opinion that forensic science should be regulated from within the forensic community, though. There was also unanimous support for federal financial assistance to local crime labs. Many respondents said that the FBI forensic DNA lab should have to abide by the same standards as the local crime labs.

5. Paul Ferrara to John Hicks, 15 August 1991, in *Forensic DNA Analysis*, 308–310; quotation, 308.

6. Rod Caswell to John Hicks, 15 August 1991, in *Forensic DNA Analysis*, 306–307; quotation, 306.

7. Ibid.

8. Richard L. Tanton, testimony in *Forensic DNA Analysis*, 59–60; quotation, 60.

9. See chapter 7, note 62, for more information on New York State legislation.

10. Richard L. Tanton, statement in "Forensic DNA Analysis Joint Hearing," 68. It is interesting to note that when the admissibility of DNA evidence was at stake, the forensic community was quick to argue that forensic DNA analysis was no different than medical DNA testing. Yet, when it came to regulation, the story was precisely the opposite.

11. Richard L. Tanton to Don Edwards, 15 July 1991, in *Forensic DNA Analysis*, 152–154; quotation, 153.

12. Tanton, statement in *Forensic DNA Analysis*, 68.

13. Senator Paul Simon, statement in *Forensic DNA Analysis*, 68.

14. Barry Scheck, statement in *Forensic DNA Analysis*, 122.

15. John Hicks to Don Edwards, 10 July 1991, in *Forensic DNA Analysis*, 156–159; quotation, 158–159.

16. 42 USC 136, §14131–14135. The DNA Identification Act was a small part of the reauthorization of the Omnibus Crime Bill, which covered numerous aspects of law enforcement.

17. William C. Thompson, "Accepting Lower Standards: The National Research Council's Second Report on Forensic DNA Evidence," *Jurimetrics Journal* 37 (1997): 405.

18. National Research Council, *DNA Technology in Forensic Science* (Washington, DC: National Academy Press, 1992) (hereafter cited as *NRC 1992*). For more on the NAS, see its Web site at http://www.nas.edu.

19. Eric S. Lander, "DNA Fingerprinting on Trial," *Nature* 339, no. 6225 (1989): 595.

20. James K. Stewart and Roger T. Castonguay to John Burris, 31 May 1989, National Academy of Sciences, National Research Council Archives, Committee on Life Sciences Board on Biology Collection (hereafter cited as NAS-NRC Archives), Box 1, File: "CLS:BB:Comm. On DNA Typing—Proposals, 1988–1989."

21. *NRC 1992*, viii.

22. George Sensabaugh, interview with author, 9 November 2002.

23. Leslie Roberts, "DNA Fingerprinting: Academy Reports," *Science* 256, no. 5055 (1992): 300.

24. *NRC 1992*, "Biographical Information on Committee Members," 173–175. Representatives of the legal community included a judge (Jack B. Weinstein of the U.S. District Court in Brooklyn) and a law professor (Richard Lempert of the University of Michigan). The forensic science community was well represented but not dominant, with Paul Ferrara (from Virginia's Division of Forensic Sciences), Henry C. Lee (from the Connecticut State Police), as well as George Sensabaugh (from University of California, Berkeley) participating. The best represented group was medical geneticists. In addition to Victor McCusick (from Johns Hopkins School of Medicine), who was chair of the committee and was widely considered to be a "founder" of the field, also participating were Tom Caskey, Haig Kazazian (also from Johns Hopkins), and Mary-Clair King (from University of California, Berkeley). While not explicitly fitting into the category of medical genetics, Michael Hunkapiller (from Applied Biosystems), Thomas Marr (from Cold Spring Harbor), and Eric Lander also devoted their careers to issues at the intersection of genetics and medicine. Because the panel was also charged with examining the ethical, legal, and moral issues associated with DNA databanks, the committee also contained two bioethicists: Ruth Macklin (from Albert Einstein College of Medicine) and Phillip J. Reilly (an MD/JD who directed the Shriver Center for Mental Retardation). It should be noted that the two members of the committee who had ties to the biotechnology industry, Tom Caskey and Michael Hunkapiller, resigned from the study as a result of "conflict of interest" allegations.

25. See, e.g., Linda Derksen, "Towards a Sociology of Measurement: The Meaning of Measurement Error in the Case of DNA Profiling," *Social Studies of Science* 30, no. 6 (2000): 822–823.

26. NAS, NRC, Committee on Life Sciences, Board on Biology, Proposal No. 89–046(d), *An Evaluation of the Application of DNA Technology in Forensic Science*, July 1989, NAS-NRC Archives, Box 1, File: "CLS:BB:Comm. On DNA Typing—Proposals, 1988–1989."

27. Sensabaugh interview.

28. *NRC 1992*, x. Interestingly, this conclusion was never explicitly stated in the text of the report. Instead, it appeared in a joint statement by all of the members of the committee in response to an erroneous summary of the report in a *New York Times* article published two days before the report was released to the public.

29. *NRC 1992*, 97.

30. Ibid., 15.

31. Ibid., 15–16.

32. Ibid., 102–103.

33. Ibid., 71.

34. Ibid., 89, 99, 104, 105, 106.

35. Ibid., 16–17.

36. Ibid., 56–63.

37. See chapter 6 for a lengthy discussion of population genetics issues.

38. *NRC 1992*, 82.

39. Ibid., 80.

40. This point was made to me by Lander and other committee members.

41. Eric S. Lander, "Lander Reply," *American Journal of Human Genetics* 49 (1991): 899–903.

42. Thus, if for "Locus 1," the frequencies for allele "B" were 1 percent, 7 percent, and 14 percent in Populations 1, 2 and 3, respectively, 14 percent would be used as the ceiling

frequency in all calculations involving allele B. If the frequencies for allele "b" were 2 percent, 3 percent, and 4 percent in Populations 1, 2, and 3, respectively, 5 percent would be used as the frequency because it is larger than the actual frequencies. Thus, if a Bb heterozygote were found in a particular individual, then the frequency would be calculated as 2(0.14)(0.05).

43. *NRC 1992*, 84.

44. Ibid., 14–15. According to the report, the upper 95 percent confidence limit is found using the following formula: $p + 1.96 \sqrt{p(1-p)/N}$, where p is the observed frequency and N is the number of chromosomes studied.

45. B. Devlin et al., "Statistical Evaluation of DNA Fingerprinting: A Critique of the NRC's Report," *Science* 259, no. 5096 (1993): 749. Many other critics of the NRC report also made this claim.

46. Ibid., 837.

47. N. E. Morton, "DNA in Court," *European Journal of Human Genetics* 1 (1993): 172–178; and Roberts, "DNA Fingerprinting: Academy Reports."

48. FBI, "FBI's Response to the Report by the Committee on DNA Technology in Forensic Science" (Washington, DC: FBI/DOJ, 1992), 13.

49. Ibid., 14.

50. Ibid.

51. Ibid., 11.

52. Ibid., 12.

53. Joel E. Cohen, "The Ceiling Principle Is Not Always Conservative in Assigning Genotype Frequencies for Forensic DNA Testing," *American Journal of Human Genetics* 51 (1992): 1165–1168.

54. Roberts, "DNA Fingerprinting: Academy Reports."

55. Daniel L. Hartl and R. C. Lewontin, "DNA Fingerprinting Report (Letter to the Editor)," *Science* 260, no. 5107 (1993): 473–474.

56. Ibid.

57. It is interesting to note that Krane, together with William C. Thompson and Lawrence Mueller, recently opened up a consulting firm called Forensics Bioinformatics (http://www.bioforensics.com) that provides statistical and population genetics assistance and reanalysis to the defense.

58. D. E. Krane et al., "Genetic Differences at Four DNA Typing Loci in Finnish, Italian, and Mixed Caucasian Populations," *PNAS* 89 (1992): 10583–10587.

59. Ibid.

60. Although it is beyond the scope of this chapter, it should be noted that Krane et al.'s conclusions were vigorously challenged and refuted by the FBI and its allies. Indeed, after reanalyzing their data, Bruce Budowle and colleagues attributed Krane et al.'s conclusions primarily to sampling error. Krane and his colleagues responded by arguing that Budowle's reanalysis was flawed. This debate continued for several years. For a detailed historical review of the debate over the validity of Krane et al.'s analysis, see Bruce Budowle and K. L. Monson, "Clarification of Additional Issues Regarding Statistics and Population Substructure Effects on Forensic DNA Profile Frequency Estimates," paper presented at the Sixth International Symposium on Human Identification, Scottsdale, Ariz., 1995, http://www.promega.com/geneticidproc/ussymp6proc/budow.htm.

61. Richard Lempert, "DNA, Science and the Law: Two Cheers for the Ceiling Principle," *Jurimetrics Journal* 34 (1993): 41–57; and Eric S. Lander, "DNA Fingerprinting: The NRC Report (Letter to the Editor)," *Science* 260, no. 5112 (1993): 1221.

62. Lempert, "DNA, Science and the Law," 47.

63. Ibid.

64. Lander, "DNA Fingerprinting: The NRC Report (Letter to the Editor)."

65. Eric Lander, interview with author, 5 May 2003.

66. Stephen Hilgartner, *Science on Stage: Expert Advice as Public Drama* (Stanford: Stanford University Press, 2000).

67. Ibid., 23–24.

68. Ibid.

69. Federal News Service, "National Research Council Press Conference Transcript," 14 April 1992, Lexis-Nexis Academic Universe.

70. Gina Kolata, "US Panel Seeking Restriction on Use of DNA in Courts; Lab Standards Faulted; Judges Are Asked to Bar Genetic 'Fingerprinting' Until Basis in Science Is Stronger," *New York Times*, 14 April 1992.

71. Ibid.

72. Federal News Service, "National Research Council Press Conference Transcript."

73. Ibid.

74. Ibid.

75. Gina Kolata, "Chief Says Panel Backs Courts' Use of a Genetic Test; *Times* Account in Error; Report Urges Strict Standards, but No Moratorium on DNA Fingerprinting for Now," *New York Times*, 15 April 1992.

76. Ibid.

77. David H. Kaye, "DNA Evidence: Probability, Population Genetics, and the Courts," *Harvard Journal of Law and Technology* 7 (1993): 103.

78. See: *People v. Barney* and *People v. Howard* (10 Cal.Rptr.2d 731; California 1992); *United States v. Porter* (618 A.2d 629; DC 1992).

79. See especially, *People v. Barney* and *People v. Howard* (10 Cal.Rptr.2d 731; California Court of Appeals; August 1992), *People v. Wallace* (17 Cal.Rptr.2d 721; California 1992); *People v. Pizarro* (12 Cal.Rptr.2d 436; California 1992); *Commonwealth v. Langanin* (596 N.E.2d 311; Massachusetts 1992); and *State v. Vandebogart* (616 A.2d 843; New Hampshire 1992).

80. *People v. Barney* and *Howard*, 743.

81. Ibid.

82. Ibid., 744.

83. Ibid., 745. The court held that the trial court's error was harmless in light of other overwhelming evidence of both defendants' guilt.

84. Judge Henry H. Kennedy, order, *United States v. Kevin Eugene Porter*, 20 September 1991, unpublished, 90, personal collection of Richard C. Lewontin.

85. *United States v. Porter*, 618 A.2d 629 (1992), 631.

86. See, e.g., *State v. Bible* (858 P.2d 1152; Arizona 1993); and *State v. Cauthron* (846 P.2d 502; Wash. State 1993).

87. James R. Wooley and Rockne Harmon, "Forensic DNA Brouhaha: Science or Debate," *American Journal of Human Genetics* 51 (1992): 1164–1165.

88. Bruce Weir and more than one hundred attendees of the Second International Symposium on the Forensic Aspects of DNA to Burton Singer, 1 April 1993, NAS-NRC Archives, Box 6.

89. Al Lazen to Frank Press (e-mail, cc: Eric Fischer, Phil Smith, Paul Gilman and Jim Wright), 10 April 1993, NAS-NRC Archives, Box 6, Folder: "Comt. On DNA Typing: An Update."

90. Ibid.

91. William Sessions to Frank Press, 16 April 1993, NAS-NRC Archives, Box 6, Folder: "Comt. On DNA Typing: An Update."

92. National Research Council, "Proposal for 'DNA Forensic Science: An Update,'" 1993, 1, NAS-NRC Archives, Box 6, Folder: "Cmte. on DNA Forensic Science: An Update."

93. Ibid.

94. John W. Hicks to Richard Rau, 27 May 1993 (also sent to Eric Fisher via fax), NAS-NRC Archives, Box 6, Folder: "Comt. On DNA Typing: An Update."

95. NIJ is the agency within the Department of Justice that funded extramural research.

96. Hicks to Rau.

97. Ibid.

98. Ibid.

99. National Research Council, "DNA Forensic Science: An Update (Proposal No. 94-Cls-02)," 1993, NAS-NRC Archives, Box 6, Folder: "Cmte on DNA Forensic Science: An Update."

100. National Research Council, *The Evaluation of Forensic DNA Evidence* (Washington, DC: National Academy Press, 1996 (hereafter cited as *NRC 1996*). For more on the NAS, see its website at http://www.nas.edu.

101. *NRC 1992*, 70–71.

102. *NRC 1996*, 37.

8. The DNA Wars Are Over

1. National Research Council, *The Evaluation of Forensic DNA Evidence* (Washington, DC: National Academy Press, 1996), 122 (hereafter cited as *NRC 1996*).

2. Ibid.

3. Ibid., chap. 4.

4. *State v. Pierce*, 597 N.E.2d 107, 115 (Ohio, 1992).

5. For more on *Daubert*, see Margaret Berger, "Expert Testimony: The Supreme Court's Rules," *Issues in Science and Technology* 16, no. 4 (2000): 57–63.

6. *Daubert v. Merrell Dow*, 509 U.S. 579, 592–593.

7. Ibid., 593.

8. Ibid., 594.

9. Ibid.

10. Ibid.

11. Lloyd Dixon and Brian Gill, *Changes in the Standards for Admitting Expert Evidence in Federal Civil Cases since the Daubert Decision* (Santa Monica, CA: RAND Institute for Civil Justice, 2001).

12. Project on Scientific Knowledge and Public Policy, "Daubert: The Most Influential Supreme Court Ruling You've Never Heard Of," report, Tellus Institute, Boston, June 2003.

13. *United States v. Bonds*, 12 F.3d 540 (Sixth Circuit, 1993), 562.

14. Ibid., 564–565.

15. *United States v. Bonds*, 553.

16. See, e.g., *United States. v. Bonds*; Wesley, California cases, *Comm v. Langanin* (MA 1994); *State v. Copeland* (WA 1999 922 P.2D 1304); and Minnesota cases. See also Barry Scheck, "DNA and Daubert," *Cardozo Law Review* 15 (1994): 1959.

17. For an evaluation of Daubert that suggests how it could lead to heightened judicial review, see Scheck, "DNA and Daubert," 1959–1997.

18. The product rule was ruled inadmissible in these jurisdictions in the following cases: *State v. Bible*, 856 P.2d 1152 (Arizona, 1993); *State v. Vandebogart*, 616 A.2d 485

(New Hampshire, 1992); *State v. Cauthron*, 846 P.2d 502 (Washington State, 1993); *United States v. Porter*, 618 A.2d 629 (Washington, DC, 1992).

19. Johnnie Cochran, "Closing Arguments in *People v. Simpson*," transcript, 28 September 1995, 47793–48036, http://walraven.org/simpson/sep28.html (5 May 2003).

20. *Court TV* Crime Library, "Hit the Road Jack," http://www.crimelibrary.com/notorious_murders/famous/simpson/jack_6.html (8 February 2006).

21. Henry C. Lee and Frank Tirnady, *Blood Evidence: How DNA Is Revolutionizing the Way We Solve Crimes* (Cambridge, Mass.: Perseus Publishing, 2003), 265, quoting *US News and World Report* list.

22. The kits used to detect sequence polymorphisms in the DQ-alpha gene, and the five "Polymarker" genes, were both developed by Perkin Elmer, a biotechnology company, in 1991 and 1993 respectively.

23. A nice introduction to DQ-alpha and Polymarker can be found in Donald E. Riley, "DNA Testing: An Introduction for Non-Scientists, An Illustrated Explanation," http://www.scientific.org/tutorials/articles/riley/riley.html.

24. Eric S. Lander, "DNA Fingerprinting Dispute Laid to Rest," *Nature* 371, no. 6500 (1994): 735–738.

25. See especially Gina Kolata, "Two Chief Rivals in the Battle Over DNA Evidence Now Agree on Its Use," *New York Times*, 27 October 1994; and R[ichard] C. Lewontin, "Forensic DNA Typing Dispute," *Nature* 372, no. 6205 (1994): 398.

26. Lewontin, "Forensic DNA Typing Dispute Laid to Rest."

27. Eric Lander, interview with author, 5 May 2003.

28. Lander, "DNA Fingerprinting Dispute Laid to Rest," 735.

29. Ibid.

30. Ibid.

31. Ibid., 738. In their view, the FBI already employed more than adequate conservative safeguards, such as combining bins that contained less than five alleles for calculating random match probabilities and using 2p instead of p2 when calculating the frequency of a profile with a single allele.

32. Ibid.

33. U.S. Federal Bureau of Investigation, *VNTR Population Data: A Worldwide Study*, 3 vols. (Washington, DC: Federal Bureau of Investigation, 1993). Much of the content of this report was summarized in two published papers: B[ruce] Budowle et al., "The Assessment of Frequency Estimates of Hae III-Generated VNTR Profiles in Various Reference Databases," *Journal of Forensic Science* 39 (1994): 319–352; B[ruce] Budowle et al., "Evaluation of Hinf I-Generated VNTR Profile Frequencies Determined Using Various Ethnic Databases," *Journal of Forensic Science* 39 (1994): 988–1008.

34. FBI, *VNTR Population Data: A Worldwide Study*, vol. 1, 1–2.

35. Ibid.

36. D[aniel] L. Hartl, "Forensic DNA Typing Dispute," *Nature* 372, no. 6505 (1994): 398–399.

37. Ibid., 398.

38. B[ruce] Budowle et al., "A Reassessment of Frequency Estimates of PvuII-Generated VNTR Profiles in a Finnish, an Italian, and a General U.S. Caucasian Database: No Evidence for Ethnic Subgroups Affecting Forensic Estimates," *American Journal of Human Genetics* 55 (1994): 533–539.

39. Hartl, "Forensic DNA Typing Dispute," 398–399.

40. Ibid., 399.

41. Lander, "DNA Fingerprinting Dispute Laid to Rest," 735.

42. *NRC 1992*, 51.

43. *NRC 1992*, 52.

44. Daniel Hartl, interview with author, 4 April 2003.

45. *State v. Carlson*, 267 N.W.2d 170 (1978), *State v. Boyd*, 331 N.W.2d 480 (1983), and *State v. Kim*, 398 N.W.2d 544 (1987).

46. As quoted in *State v. Kim*, 547.

47. Dan Hartl, testimony in *People of Minnesota v. Bloom*, 14 (Minn. Sup., 1994), Simpson MSS, Box 1.

48. Ibid., 9–10.

49. Ibid., 30.

50. Ibid., 38.

51. Ibid., 68–71.

52. Ibid., 76.

53. Ibid., 124.

54. Barry C. Scheck, Peter J. Neufeld, and William C. Thompson, "Motion to Exclude DNA Evidence," in *People of California v. Orenthal James Simpson*, Case No. BA097211, L.A. County Superior Court, 4 October 1994, personal collection of Richard Lewontin. The prosecution wrote three opposition briefs to the defense's Motion: Gil Garceti, Lisa Kahn, and Rockne Harmon, "People's Opposition to the Defendant's Motion to Exclude DNA RFLP Evidence," in *California v. Orenthal James Simpson*, 1 November 1994, personal collection of Richard Lewontin. For an analysis of the defense's motion, see Michael Lynch, "The Discursive Production of Uncertainty: The O.J. Simpson 'Dream Team' and the Sociology of Knowledge Machine," *Social Studies of Science* 28, no. 5–6 (1998): 829–868.

55. Scheck, "Motion to Exclude DNA Evidence," 16.

56. Ibid., 52.

57. *State v. Streich*, 658 A.2d 38 (Vermont, 1995).

58. This argument was advanced in National Research Council, *DNA Technology in Forensic Science* (Washington, DC: National Academy Press, 1992), 88, as well as in Richard Lempert, "Some Caveats Concerning DNA as Criminal Identification Evidence: With Thanks to the Reverend Bayes," *Cardozo Law Review* 13 (1991): 303–341; Jonathan Koehler, "On Conveying the Probative Value of DNA Evidence: Frequencies, Likelihood Ratios, and Error Rates," *University of Colorado Law Review* 67 (1996): 859–886; Jonathan Koehler, "Error and Exaggeration in the Presentation of DNA Evidence," *Jurimetrics Journal* 34 (1993): 21–39; Michael Saks and Jonathan Koehler, "What DNA 'Fingerprinting' Can Teach the Law about the Rest of Forensic Science," *Cardozo Law Review* 13 (1991): 361–372; Judith McKenna, Joe Cecil, and Pamela Coukos, "Reference Guide on Forensic DNA Evidence," in *Reference Manual on Scientific Evidence*, ed. Federal Judicial Center (Washington, DC: Federal Justice Center, 1994), 275–489; William C. Thompson, "Accepting Lower Standards: The National Research Council's Second Report on Forensic DNA Evidence," *Jurimetrics Journal* 37 (1997): 405–424; William C. Thompson, "Subjective Interpretation, Laboratory Error and the Value of Forensic DNA Evidence: Three Case Studies," *Genetica* 96 (1995): 153–168.

59. Daniel Hartl, "Expert's Report," in *United States v. Yee, et al.*, 4, personal collection of Richard C. Lewontin.

60. Eric Lander, testimony as a court's witness in *United States v. Porter*, (D.C. Superior Court, 1991, No. 91-CO-1277), 46; judgment in Porter can be found at 618 A.2d 629 (1992).

61. *United States v. Porter*, 618 A.2d 629 (1992).

62. *NRC 1996*, 86–87; quotation, 87.

63. Joseph L. Peterson and R. E. Gaensslen, *Developing Criteria for Model External DNA Proficiency Testing: Final Report* (Chicago: University of Illinois at Chicago, 2001).

64. Ibid.

65. Some of the problems associated with truly blind tests can be found in Peterson, *Developing Criteria*, 95.

66. See, e.g., John Hick's comments at the third National Forensic DNA Review Panel meeting, which was convened to evaluate various proficiency testing schemes in the wake of the DNA Identification Act of 1994; comments summarized in Peterson, *Developing Criteria*, 149.

67. *NRC 1996*, 86.

68. See executive summary of Peterson, *Developing Criteria*.

69. William C. Thompson, personal communication, 1 February 2006.

70. David Margolick, "Simpson Lawyers Ask to Forgo DNA Hearing," *New York Times*, 14 December 1994.

71. David Margolick, "Simpson Defense Drops DNA Challenge," *New York Times*, 5 January 1995; William C. Thompson, personal communication, 21 March 2006.

72. For a complete summary of admissibility of DNA evidence by jurisdiction, see NRC 1996, tables 6.1 and 6.2, 205–209.

73. Official Transcript, "Examination of Colin Yamauchi," *People v. Simpson*, No. BA 097211, 1995 WL 324772, 11. (L.A. Cty. Sup., 30 May 1995).

74. William C. Thompson, "DNA Evidence in the OJ Simpson Trial," *University of Colorado Law Review* 67, no. 4 (1996): 833.

75. Ibid.

76. Ibid., 841.

77. For more on the importance of error rates in probability calculations, see Koehler, "On Conveying the Probative Value of DNA Evidence," 859–886. For an alternative view, see *NRC 1996*.

78. "The Power of DNA Evidence," editorial, *New York Times*, 28 May 1995, 10.

79. Thompson, "DNA Evidence in the OJ Simpson Trial," 844.

80. Ibid.

81. Barry Scheck, interview with Harry Kriesler as part of "Conversations with History Series at UC-Berkeley," University of California, Berkeley, *DNA and the Criminal Justice System*, 25 July 2003, http://globetrotter.berkeley.edu/people3/Scheck/scheck-con3.html.

82. See, e.g., ibid.

83. Ibid.

84. Barry Scheck, Peter J. Neufeld, and Jim Dwyer, *Actual Innocence: Five Days to Execution and Other Dispatches from the Wrongly Convicted* (New York: Doubleday, 2000), xv.

85. Peter Neufeld, interview with author, 27 February 2002.

86. See the first paragraph of the Innocence Project, http://www.innocenceproject.org/.

87. Ibid.

88. Different probe systems have different hybridization properties and thus bind to their target sequences at different temperatures. Additionally, care must be taken that the various primers being used in a multiplex reaction do not have a significant degree of sequence similarity, and also that loci being labeled with the same color dye have large enough size differences to avoid confusing the computer program that analyses test results.

89. More information on various models of the Genetic Analyzer can be found at Applied Biosystems, http://www.appliedbiosystems.com.

90. Older machines tended only to process a single sample at a time through single capillary, while newer machines speed this work up because they have multiple capillaries. The ABI Prism 3100, for instance, has sixteen capillaries, while the 3700 has ninety-six. See John M. Butler, *Forensic DNA Typing: Biology, Technology, and Genetics of STR Markers*, 2d ed. (New York: Elsevier Academic Press, 2005), 358–361.

91. It should be noted that STR analysis can also be carried out using a polyacrilamide gel similar to that used for RFLP testing. Once the samples are run on the gel, the interpretation process involves the same basic technology as the system described for capillary runs. Hitachi makes the most widely used scanner for sizing STRs from gels, called the FMBIO.

92. For more information on potential sources of error, see Butler, *Forensic DNA Typing*.

93. Butler, *Forensic DNA Typing*. William C. Thompson et al., "Evaluating Forensic DNA Evidence: Essential Elements of a Competent Defense Review," *Champion* 27, no. 3 (2003): 16–25.

94. Thompson, "Evaluating Forensic DNA Evidence."

95. Butler, *Forensic DNA Typing*, 424–425.

96. Available at National Institute of Standards and Technology, Chemical Science and Technology Laboratory, "Short Tandem Repeat DNA Internet Database," http://www.cstl.nist.gov/div831/strbase/.

97. In Re: *People v. Shreck*, 22 P.3d 68 (Col. Sup., 2001). For more on this issue, see Kim Herd and Adrianne Day, American Prosecutors Research Institute, "An Update on STR Admissibility," *Silent Witness* 5, no. 3 (2000), http://www.ndaa.org/publications/newsletters/silent_witness_volume_5_number_3_2000.html.

98. *State v. Dishmon* (Minnesota 2000, unreported), 15.

99. For more on Forensic Bioinformatics, see http://www.bioforensics.com/index.html.

100. David Lazer, "Introduction: DNA and the Criminal Justice System," in *DNA and the Criminal Justice System*, ed. David Lazer (Cambridge: MIT Press, 2004), 3–4.

101. Ibid., 4.

102. Ibid.

9. THE LEGACY OF HISTORY

1. For a brief review of the Josiah Sutton case, see The Innocence Project, "Case Summary: Josiah Sutton," Benjamin N. Cardozo School of Law, Yeshiva University, 2001, http://www.innocenceproject.org/case/display_profile.php?id=145.

2. Office of the Independent Investigator for the Houston Police Department Crime Laboratory and Property Room, "Independent Investigator Issues Fifth Report on Houston Police Department Crime Lab," 11 May 2006, http://www.hpdlabinvestigation.org/pressrelease/060511pressrelease.pdf.

3. University of California, Irvine, "As Texas Convict Is Exonerated, DNA Expert Offers Inside Look at the Case Overhauling Justice System," 7 April 2003, http://today.uci.edu/news/release_detail.asp?key=987.

4. ASCLD-LAB, "ASCLD-LAB Limited Scope Interim Inspection Report: Commonwealth of Virginia Division of Forensic Science Central Laboratory," 9 April 2005, http://www.dfs.virginia.gov/services/forensicBiology/externalReviews.cfm.

5. Ibid., 17.

6. Ibid., 14.

7. Innocence Project, "Historic Audit of Virginia Crime Lab Errors in Earl Washington Jr.'s Capital Case," 6 May 2005, http://www.innocenceproject.org/docs/VACrimeLabPressRelease.pdf.

8. Commonwealth of Virginia, Department of Forensic Science, "DFS Responses to ASCLD-LAB DNA Audit Report," undated, http://www.dfs.virginia.gov/services/foresicBiology/esternalReviewDFSResponse.pdf.

9. *Seattle Post-Intelligencer* Special Reports, "Errors in Evidence," 2004, http://www.hpdlabinvestigation.org/pressrelease/060511pressrelease.pdf.

10. See Phoebe Zerwick, "DNA Mislabeled in Murder Case," *Winston-Salem Journal*, 28 August 2005, and related stores, "Crime and Science," *Winston-Salem Journal*, http://www.journalnow.com/servlet/Satellite?pagename=WSJ%2FMGArticle%2FWSJ_BasicArticle&c=MGArticle&cid=1031784716491.

11. Phoebe Zerwick, "N.C. Told to Refine Handling, Storing of Crime Evidence," *Winston-Salem Journal* 19 November 2005, http://www.journalnow.com/servlet/Satellite?pagename=WSJ/MGArticle/WSJ_BasicArticle&c=MGArticle&cid=1128768250886.

12. William C. Thompson, "Tarnish on the Gold Standard: Recent Problems in Forensic DNA Testing," *Champion* (January/February 2006): 14–20.

13. Gretchen Ruethling, "Illinois State Police Cancels Forensic Lab's Contract, Citing Errors," *New York Times*, 20 August 2005.

14. Laura Cadiz, "Md.-Based DNA Lab Fires Analyst over Falsified Tests," *Baltimore Sun*, 18 November 2004.

15. Tim McGlone, "Nearly 500 Military DNA Cases under Investigation," *Virginian-Pilot,* 17 May 2006.

16. U.S. Dept. of Justice, Office of the Inspector General, *The FBI DNA Laboratory, A Review of Protocol and Practice Vulnerabilities* (Washington, DC: Office of the Inspector General, 2004), ii. At the time the report was issued, twenty had been restored.

17. Ibid.

18. Ibid.

19. Ibid., 3.

20. Ibid.

21. Ibid., vi. According to science studies scholars Michael Lynch and Kathleen Jordan, such deviations or improvisations from established protocols are a normal part of scientific practice.

22. Ibid.

23. Thompson, "Tarnish on the Gold Standard."

24. See Jennifer N. Mellon, "Manufacturing Convictions: Why Defendants Are Entitled to the Data Underlying Forensic DNA Kits," *Duke Law Journal* 51 (2001): 1097–1137.

Bibliography

MAJOR CASES CITED

Commonwealth of Massachusetts v. Robert W. Curnin, 565 N.E.2d 440 (Supreme Judicial Court Massachusetts, 1991).

Commonwealth of Massachusetts v. Thomas J. Lanigan, 641 N.E.2d 1342 (Supreme Judicial Court Massachusetts, 1994).

Daubert v. Merrell Dow 509 U.S. 579 (U.S. Supreme Court, 1993).

Fernando Martinez v. State of Florida, 549 So.2d 694 (Florida Court Appeals, 1989).

Frank Hopkins v. State of Indiana, 597 N.E.2d 1297 (Supreme Court of Indiana, 1991).

Frye v. United States of America, 293 F. 1013 (D.C. Circuit Court, 1923).

General Electric v. Joiner, 522 U.S. 136 (U.S. Supreme Court, 1997).

In re Silicone Gel Breast Implant Products Liability Litigation, 174 F. Supp. 2d 1242 (Northern District of Alabama, 2001).

In the Matter of Baby Girl S, an infant, 532 N.Y.S.2d 634 (Surrogates Court of New York, 1988).

James Robert Caldwell v. The State of Georgia, 393 S.E.2d 436 (Supreme Court of Georgia, 1990).

Kenneth S. Cobey v. State of Maryland, 559 A.2d 391 (Court of Appeals, 1989).

Kumho Tire v. Carmichael, 119 S. Ct. 1167 (U.S. Supreme Court, 1999).

People of California v. Brown, 40 Cal.3d 512 (California, 1985).

People of California v. Jeffrey Allen Wallace, 17 Cal.Rptr.2d 721 (California Court of Appeals, 1st District, 1993).

People of California v. Lynda Patricia Axell, 1Cal.Rptr.2d 411 (California Court of Appeals, 2nd District, 1991).

People of California v. Orenthal James Simpson (Los Angeles County Superior Court, 1995).

People of California v. Pizarro, 12 Cal.Rptr.2d 436 (California, 1992).

People of California v. Ralph Edwards Barney and Kevin O'Neal Howard, 10 Cal.Rptr.2d 731 (California Court of Appeals, 1st District, 1992).

People of California v. Reilly, 196 Cal. App. 3d (California Court of Appeals, San Diego, 1987).

People of Illinois v. Flemming and Watson (Cook County Court, 1990, consolidated, unpublished).

People of Illinois v. Robert E. Stremmel II, 630 N.E.2d 1301 (Appeals Court Illinois, 1994).

People of New York v. George Wesley and Cameron Bailey, 533 N.Y.S.2d 643 (Albany County Court, 1988).

People of New York v. Joseph Castro, 545 N.Y.S.2d 985 (Bronx County Supreme Court, 1989).

People of New York v. Roderick Keene, 591 N.Y.S.2d 733 (Queens County Supreme Court, 1992).

State of Delaware v. Steven B. Pennell, 584 A.2d 513 (Delaware Supreme Court, 1989).

State of Kansas v. Washington, 622 P.2d 986 (Kansas, 1979).

State of Michigan v. Young, 308 N.W.2d 194 (Michigan, 1981).

State of Minnesota v. Boyd, 331 N.W.2d 480 (Supreme Court of Minnesota, 1983).

State of Minnesota v. Carlson, 276 N.W.2d 170 (Supreme Court of Minnesota, 1978).

State of Minnesota v. Daryl Duane Alt, 504 N.W.2d 38 (Court of Appeals, 1993).

State of Minnesota v. Joon Kyu Kim, 398 N.W.2d 544 (Supreme Court of Minnesota, 1987).

State of Minnesota v. Larry Lee Jobe, 486 N.W.2d 407 (Supreme Court of Minnesota, 1992).

State of Minnesota v. Thomas Robert Schwartz, 447 N.W.2d 422 (Supreme Court of Minnesota, 1989).

State of Minnesota v. Troy Bradley Bloom, 516 N.W.2d 159 (Supreme Court of Minnesota, 1994).

State of New Hampshire v. Vandebogart, 616 A.2d 843 (New Hampshire, 1992).

State of Washington v. Richard C. Cauthorn, 846 P.2d 502 (Supreme Court of Washington, 1993).

Timothy Spencer v. Commonwealth of Virginia, 384 S.E.2d 253 (1989).

Tommie Lee Andrews v. State of Florida, 533 So.2d 841 (Florida Court Appeals, 1988); appeal of: *State of Florida v. Tommie Lee Andrews*, unpublished (Orange County Circuit Court, 1987).

United States of America v. John Ray Bonds, Mark Verdi, and Steven Wayne Yee, 12 F.2d 540 (U.S. Court of Appeals, 6th Circuit, 1993).

United States of America v. Kevin E. Porter, 618 A.2d 629 (D.C. Court of Appeals, 1992).

United States of America v. Matthew Sylvester Two Bulls, 918 F.2d 56 (U.S. 8th Circuit Court of Appeals, 1990).

United States of America v. Randolph Jakobetz, 747 F.Supp. 250 (U.S. District Court—Vermont, 1990).

United States of America v. Stephen Wayne Yee, et al., 134 F.R.D. 161 (U.S. District Court for Northern District of Ohio, 1991 [adopting Magistrate's Report]).

INTERVIEWS

Adams, Dwight (director of FBI Laboratory), phone interview with author, 9 July 2002.

Amos, Bill (University of Cambridge), interview with author in Cambridge, UK, 22 June 2001.

Baird, Michael (Lifecodes Corporation), phone interview with author, 19 February 2002.

———. Interview with Saul Halfon and Arthur Daemmrich in Stamford, CT, 14 July 1994 (deposited in Cornell DNA Typing Archive).

———. Phone interview with Arthur Daemmrich, 21 March 1995 (deposited in Cornell DNA Typing Archive).

Balazs, Ivan (Lifecodes Corporation), phone interview with author, 13 February 2002.

Balding, David (Imperial College, London), interview with author in London, 26 June 2001.

Bashinski, Jan (retired chief, Bureau of Forensic Services, California Department of Justice), phone interview with author, 7 February 2002.

Blomberg, Bonnie (University of Miami Medical School), phone interview with author, 28 May 2002.

Borowsky, Richard (New York University School of Medicine), e-mail to author, 19 September 2002.

Carlile, Alex (former MP, member of the House of Lords), phone interview with author, 18 November 2001.

Chakraborty, Ranajit (Center for Genome Information, University of Cincinnati Medical Center), phone interview with author, 11 June 2002.

Coleman, Howard (Genelex Corporation), phone interview with author, 23 April 2001.

Cooke, Graham (barrister) interview with author in London, 10 September 2001.

Cooke, Howard (UK Medical Research Council), phone interview with author, 20 February 2002.

Cotton, Robin (Cellmark Corporation), interview with author in Germantown, MD, 23 January 2002.

———. Interview with Arthur Daemmrich in Germantown, MD, 20 June 1994 (deposited in Cornell DNA Typing Archive).

———. Phone interview with Arthur Daemmrich, 17 March 1995 (deposited in Cornell DNA Typing Archive).

Deadman, Harold (former agent in the FBI DNA Analysis Unit), phone interview with author, 27 March 2003.

Donnelly, Peter (Oxford University), interview with author in Oxford, 25 June 2001.

Evett, Ian (Forensic Science Service), phone interview with author, 24 October 2001.

Fedor, Thomas (Serological Research Institute), phone interview with author, 21 May 2001.

Ferrara, Paul (Virginia State Department of Criminal Justice Services), interview with author in Richmond, VA, 10 November 2003.

Flaherty, Lorraine (Wadsworth Center, New York State Department of Health), phone interview with author, 25 April 2002.

Freeman, Mike (lawyer, former Hennepin County Attorney), phone interview with author, 11 February 2002.

Garner, Daniel D. (U.S. Department of Justice, formerly of Cellmark Diagnostics), interview with author in Washington, DC, 24 January 2002.

———. (Cellmark Diagnostics), interview with Arthur Daemmrich, 20 June 1994 (deposited in Cornell DNA Typing Archive).

Geisser, Seymour (Department of Statistics, University of Minnesota), interview with author in Minneapolis, 18 January 2002.

Gill, Peter (Forensic Science Service), interview with author in Birmingham, UK, 25 June 2001.

Hadkiss, Chris (Forensic Science Service), informal interview with author in London, June 2001; phone interview with author, 26 June 2002.

———. Interview with Michael Lynch, Ruth McNally, and Pat Daly, 11 May 1996 (on file with Michael Lynch at Cornell University).

Harmon, Rockne (Alameda County District Attorney's Office), interview with author in Oakland, CA, 7 July 2003.

Hartl, Daniel (Department of Organismal and Evolutionary Biology, Harvard University), interview with author in Cambridge, MA, 4 April 2003.

Hicks, John (former director of FBI Laboratory, currently director of Forensic Services, New York State Division of Criminal Justice Services), phone interview with author, 21 March 2003.

Iverson, Jim (supervisor, Minnesota Bureau of Criminal Apprehension), interview with author in St. Paul, 16 January 2002.

Jeffreys, Sir Alec (University of Leicester), interview with author in Leicester, England, 13 September 2001.

———. Interview with Michael Lynch in Leicester, 8 June 1996 (deposited in the Cornell DNA Typing Archive).

Jonakait, Randy (New York Law School), phone interview with author, 28 November 2002.

Kahn, Phyllis (Minnesota House of Representatives), phone interview with author, 6 February 2002.

Keith, Sandy (former Minnesota Supreme Court Justice), phone interview with author, 20 February 2002.

Kidd, Ken (Yale University), phone interview with author, 30 May 2002.

Kuo, Margaret (formerly with the Orange County Sheriff-Coroner), phone interview with author, 20 October 2002.

Lander, Eric (Whitehead Institute), phone interview with author, 5 May 2003.

Lewontin, Richard C. (Department of Organismal and Evolutionary Biology, Harvard University), interview with author in Cambridge, MA, 5 March 2003.

Lincoln, Patrick (British Academy of Forensic Science), interview with author in London, 10 September 2001.

Mueller, Laurence (University of California–Irvine), phone interview with author, 19 February 2002.

Neufeld, Peter (defense lawyer, cofounder of the Innocence Project), interview with author in Brooklyn, NY, 27 February 2002.

Nichols, Richard (Queen Mary and William College, University of London), interview with author in London, 4 September 2001.

Patel, Dinesh (Forensic Science Service), interview with author in London, 7 September 2001.

Petrillo, Michael (Lifecodes Corporation), interview with Saul Halfon and Arthur Daemmrich in Stamford, CT, 14 July 1994 (deposited in Cornell DNA Typing Archive).

Poklemba, John J. (former commissioner of the New York State Division of Criminal Justice Services), phone interview with author, 22 April 2002.

Pownall, Orlando (barrister and queen's counsel), phone interview with author, 24 October 2001.

Presley, Laurence (former chief of the FBI DNA Analysis Unit, currently with National Medical Services Laboratory), phone interview with author, 31 March 2003.

Redding, Steve (assistant Hennepin County attorney), phone interview with author, 14 December 2001.

Reeder, Dennis J. (National Institute of Standards and Technology), interview with Arthur Daemmrich in Gaithersburg, MD, 4 January 1995 (deposited in the Cornell DNA Typing Archive).

Roberts, Richard (New England Biolabs), phone interview with author, 25 April 2002.

Rutnik, Douglas (lawyer), e-mail to author, 24 September 2002.

Scheck, Barry (Cardozo School of Law), e-mail to author, 5 December 2001.

Sensabaugh, George (University of California, Berkeley), phone interviews with author in Coventry, U.K., 30 October 2002, 6 November 2002, and 9 November 2002.

Shaler, Robert (New York City Office of the Chief Medical Examiner), phone interview with author, 24 October 2002.

Steventon, Beverly (Coventry University School of Law), interview with author in Coventry, U.K., 6 September 2001.

Stolorow, Mark (Cellmark Diagnostics), phone interview with author, 6 December 2002.

Sullivan, Patrick (Hennepin County Defender's Office) phone interview with author, 8 August 2001.

Thompson, William C. (University of California–Irvine), interviews with author in Irvine, 28 February and 1 March 2002; brief phone interview, 23 April 2001.

Uhrig, Hal (lawyer), phone interview with author, June 2002.

Van de Kamp, John (lawyer, former California state attorney general), phone interview with author, 2 June 2002.

Weaver, Charlie (commissioner of the Minnesota Department of Public Safety), phone interview with author, 11 February 2002.

Weir, Bruce (Departments of Genetics and Statistics, North Carolina State University), phone interview with author, 1 June 2002.

White, Ray (University of Utah), phone interview with author, 2 April 2003.

Wraxall, Brian (Serological Research Institute), phone interview with author, 29 October 2002.

York, Sheona (barrister), e-mail to author, 19 November 2001.

Published Material

Adams, Dwight E. "Validation of the Procedure for DNA Analysis: A Summary." *Crime Laboratory Digest* 15, no. 4 (1988): 85–87.

Adams, Dwight E., Lawrence A. Presley, Bruce Budowle, Alan M. Giusti, F. Samuel Baechtel, and others. "Deoxyribonucleic Acid (DNA) Analysis by Restriction Length Fragment Polymorphisms of Blood and Other Bodily Fluid Stains Subjected to Contamination and Environmental Insults." *Journal of Forensic Sciences* 36, no. 5 (1991): 1284–1298.

Anderson, Alun. "DNA Fingerprinting on Trial." *Nature* 342, no. 6252 (1989): 844.

———. "Judge Backs Technique." *Nature* 340, no. 6235 (1989): 582.

Applied Biosystems. http://www.appliedbiosystems.com, 17 August 2006.

Arlington National Cemetery. "The Vietnam Unknown Controversy." http://www.arlingtoncemetery.com/vietnam.htm, 16 August 2006.

ASCLD/LAB. "ASCLD/LAB Limited Scope Interim Inspection Report: Commonwealth of Virginia Division of Forensic Science Central Laboratory." 9 April 2005. http://www.dfs.virginia.gov/services/forensicBiology/externalReviews.cfm. Accessed 8 March 2006.

Ashcroft, John. "Attorney General Ashcroft Announces DNA Initiatives." U.S. Department of Justice. 4 March 2002. http://www.usdoj.gov/archive/ag/speeches/2002/030402newsconferncednainitiative.htm. Accessed 14 August 2006.

Baird, Michael, Ivan Balazs, Alan Giusti, L. Miyazaki, L. Nicholas, K. Wexler, E. Kanter, J. Glassberg, F. Allen, P. Rubinstein, and others. "Allele Frequency Distribution of Two Highly Polymorphic DNA Sequences in Three Ethnic Groups and Its Application to the Determination of Paternity." *American Journal of Human Genetics* 39 (1986): 489–501.

Balazs, Ivan, Michael Baird, Mindy Clyne, and Ellie Meade. "Human Population Genetic Studies of Five Hypervariable DNA Loci." *American Journal of Human Genetics* 44 (1989): 182–190.

Balazs, Ivan, K. Wexler, Alan M. Giusti, and Michael Baird. "The Use of Restriction Fragment Length Polymorphisms for the Determination of Paternity." Paper presented at the American Society of Human Genetics. Toronto, 1984.

Ballantyne, Jack, George Sensabaugh, and Jan Witkowski, eds. *DNA Technology and Forensic Science*. Vol. 32, *Branbury Report*. Cold Spring Harbor, N.Y.: Cold Spring Harbor Laboratory Press, 1989.

Barber, George H., and Mira Gur-Arie. *New York's DNA Data Bank and Commission on Forensic Science*. New York: Matthew Bender, 1994.

Beeler, Laurel, and William R. Wiebe. "DNA Identification Tests and the Courts." *Washington Law Review* 63 (1988): 881–955.

Berger, Margaret A. "Expert Testimony: The Supreme Court's Rules." *Issues in Science and Technology* 16, no. 4 (2000): 57–63.

Broad, William, and Nicholas Wade. *Betrayers of the Truth*. New York: Simon and Schuster, 1982.

Brownlee, Shannon. "Courtroom Genetics." *U.S. News and World Report*, 27 January 1992, 60–61.

Budowle, Bruce, and John Stafford (Legal Counsel). "Response to 'Population Genetic Problems in the Forensic Use of DNA Profiles' (Submitted in the Case of United States v. Yee)." *Crime Laboratory Digest* 18, no. 3 (1991): 109–112.

Budowle, Bruce, and K. L. Monson. "Clarification of Additional Issues Regarding Statistics and Population Substructure Effects on Forensic DNA Profile Frequency Estimates." Paper presented at the Sixth International Symposium on Human Identification, Scottsdale, Arizona, 1995. http://www.promega.com/geneticidproc/ussymp6proc/budow.htm. Accessed 1 September 2006.

Budowle, Bruce, Keith L. Monson, Alan M. Giusti, and B. L. Brown. "The Assessment of Frequency Estimates of Hae III-Generated VNTR Profiles in Various Reference Databases." *Journal of Forensic Science* 39 (1994): 319–352.

Budowle, Bruce, Keith L. Monson, Alan M. Giusti, and B. L. Brown. "Evaluation of Hinf I-Generated VNTR Profile Frequencies Determined Using Various Ethnic Databases." *Journal of Forensic Science* 39 (1994): 988–1008.

Budowle, Bruce, F. Samuel Baechtel, Ron M. Fourney, Dwight E. Adams, Lawrence A. Presley, Harold Deadman, and Keith L. Monson. "Fixed-Bin Analysis for Statistical Evaluation of Continuous Distributions of Allelic Data from VNTR Loci for Use in Forensic Comparisons." *American Journal of Human Genetics* 48 (1991): 841–855.

Budowle, Bruce, John S. Waye, Gary G. Shutler, and F. Samuel Baechtel. "Hae III—A Suitable Restriction Endonuclease for Restriction Fragment Length Polymorphism Analysis of Biological Evidence Samples." *Journal of Forensic Sciences* 35, no. 3 (1990): 530–536.

Budowle, Bruce, Harold Deadman, Randall Murch, and F. Samuel Baechtel. "An Introduction to the Methods of DNA Analysis under Investigation in the FBI Laboratory." *Crime Laboratory Digest* 15, no. 1 (1988): 8–21.

Budowle Bruce, Keith L. Monson, K. Anoe, F. Samuel Baechtel, and D. Bergman. "A Preliminary Report on Binned General Population Data on Six Vntr Loci in Caucasians, Blacks, and Hispanics from the United States." *Crime Laboratory Digest* 18, no. 1 (1991): 10–26.

Budowle, Bruce, Keith L. Monson, and Alan M. Giusti. "A Reassessment of Frequency Estimates of PvuII-Generated VNTR Profiles in a Finnish, an Italian, and a General U.S. Caucasian Database: No Evidence for Ethnic Subgroups Affecting Forensic Estimates." *American Journal of Human Genetics* 55 (1994): 533–539.

Bureau of National Affairs. "DA Faults Lifecodes' DNA Test, Withdraws Results, Drops Case." *BNA Criminal Practices Manual* 4 (1990): 3–6.

———. "Geneticist, Defense Lawyers Debate Merits of DNA Typing." *BNA Criminal Practices Manual* 3 (1989): 259–262.

———. "Landmark DNA Law Stalled." *BNA Criminal Practices Manual* 4 (1990): 491–492.

———. "New York Law Would Regulate Forensic DNA Testing Labs." *BNA Criminal Practices Manual* 4 (1990): 315–318.

———. "Rugged Cross-Examination Exposes Flawed DNA Tests." *BNA Criminal Practices Manual* 4 (1990): 31–38.

Butler, John M. *Forensic DNA Typing: Biology, Technology, and Genetics of STR Markers.* 2nd ed. New York: Elsevier Academic Press, 2005.

Cadiz, Laura. "Md.-based DNA Lab Fires Analyst over Falsified Tests." *Baltimore Sun*, 18 November 2004.

California Association of Crime Laboratory Directors (CACLD). "DNA Committee Report 2." CACLD, 1987.

Cardwell, Diane. "New York State Draws Nearer to Collecting DNA in all Crimes, Big and Small." *New York Times*, 4 May 2006.

Caskey, C. Thomas. "Comments on DNA-Based Forensic Analysis." *American Journal of Human Genetics* 49 (1991): 893–895.

Castonguay, Roger T. "Message from the Assistant Director in Charge of the FBI Laboratory." *Crime Laboratory Digest* 15, Supplement (1988): 1–2.

CBS This Morning. Transcript. 5 February 1990. Lexis-Nexis Academic Universe.

Charkraborty, Ranajit, and Kenneth K. Kidd. "The Utility of DNA Typing in Forensic Work." *Science* 254, no. 5039 (1991): 1735–1739.

Charkraborty, Ranajit, and Stephen P. Daiger. "Polymorphisms at VNTR Loci Suggest Homogeneity of the White Population of Utah." *Human Biology* 63, no. 5 (1991): 571–587.

Cochran, Johnnie. "Closing Arguments in *People v. Simpson*." Transcript. 28 September 1995, 47793–48036. http://walraven.org/simpson/sep28.html. Accessed 5 May 2003.

Cohen, Joel E. "The Ceiling Principle Is Not Always Conservative in Assigning Genotype Frequencies for Forensic DNA Testing." *American Journal of Human Genetics* 51 (1992): 1165–1168.

Cole, Simon. "More Than Zero: Accounting for Error in Latent Fingerprint Identification." *Journal of Criminal Law and Criminology* 95, no. 3 (2005): 985–1078.

———. "Grandfathering Evidence: Fingerprint Admissibility Rulings from Jennings to Llera Plaza and Back Again." *American Criminal Law Review* 41 (2004): 1189–1276.

———. *Suspect Identities: A History of Fingerprinting and Criminal Identification*. Cambridge: Harvard University Press, 2001.

Commonwealth of Virginia. Department of Forensic Science. "DFS Responses to ASCLD-LAB DNA Audit Report." Undated. http://www.dfs.virginia.gov/services/foresicBiology/esternalReviewDFSResponse.pdf. Accessed 8 March 2006.

Court TV. Crime Library. "Hit the Road Jack." http://www.crimelibrary.com/notorious_murders/famous/simpson/jack_6.html. Accessed 8 February 2006.

Dawson, Jim. "Attacker of Woman at Ramp Left 'DNA Fingerprints' at Scene." *(Minneapolis) Star Tribune*, 21 June 1988.

Derksen, Linda. *Agency and Structure in the History of DNA Profiling: The Stabilization and Standardization of a New Technology*. Ph.D. diss., University of California, San Diego, 2003.

———. "Towards a Sociology of Measurement: The Meaning of Measurement Error in the Case of DNA Profiling." *Social Studies of Science* 30, no. 6 (2000): 803–845.

Devlin, Bernie, Neil Risch, and Kathryn Roeder. "No Excess of Homozygosity at Loci Used for DNA Fingerprinting." *Science* 249, no. 4975 (1990): 1416–1420.

———. "Statistical Evaluation of DNA Fingerprinting: A Critique of the NRC's Report." *Science* 259, no. 5096 (1993): 748–749.

Dixon, Lloyd, and Brian Gill. *Changes in the Standards for Admitting Expert Evidence in Federal Civil Cases Since the Daubert Decision*. Santa Monica, CA: RAND Institute for Civil Justice, 2001.

"DNA Fingerprinting: DNA Probes Control Immigration." *Nature* 319 (1986): 171.

DNA Print Genomics. "Forensics." http://www.dnaprint.com/welcome/productsandservices/forensics/. Accessed 15 June 2006.

Epstein, Charles J. "Editorial: The Forensic Applications of Molecular Genetics—The *Journal*'s Responsibilities." *American Journal of Human Genetics* 49 (1991): 697–698.

"Errors in Evidence." Special Report. *Seattle Post-Intelligencer*. 2004. http://www.hpd-labinvestigation.org/pressrelease/060511pressrelease.pdf. Accessed 9 March 2006.

Federal News Service. "National Research Council Press Conference Transcript." 1 April 1992, Lexis-Nexis Academic Universe.

Flannery, Irene M. "Frye or Frye Not: Should the Reliability of DNA Evidence Be a Question of Weight or Admissibility?" *American Criminal Law Review* 30 (1992): 161–186.

Fleming, Thomas M. "Admissibility of DNA Identification Evidence." *A.L.R. 4th* 84 (1991): 313.

Forensic Bioinformatics. http://www.bioforensics.com/index.html. Accessed 17 August 2006.

Geisser, Seymour N. "Statistics, Litigation, and Conduct Unbecoming." In *Statistical Science in the Courtroom*, ed. Joseph L. Gastwirth. New York: Springer-Verlag, 2000.

Gest, Ted. "Convicted By Their Own Genes: DNA Fingerprinting Is Facing a Major Legal Challenge from Defense Attorneys and Civil Libertarians." *U.S. News and World Report*, 31 October 1988, 74.

Giannelli, Paul. "The Admissibility of Novel Scientific Evidence: Frye v. United States, a Half-Century Later." *Columbia Law Review* 80 (1980):1150–1197.

———. "Frye v. United States: Background Paper Prepared for the National Conference of Lawyers and Scientists." *Federal Rules Decisions* 99 (1983): 188–201.

Gieryn, Thomas F. *Cultural Boundaries of Science: Credibility on the Line*. Chicago: University of Chicago Press, 1999.

Giordano, Mary Ann. "DNA Test Pose New Dilemma for Courts." *Manhattan Lawyer*, 3 January 1989, 1.

Giusti, Alan M., Michael Baird, S. Pasquale, Ivan Balazs, and J. Glassberg. "Application of Deoxyribonucleic Acid (DNA) Polymorphisms to the Analysis of DNA Recovered from Sperm." *Journal of Forensic Sciences* 31, no. 2 (1986): 409–417.

Hartl, Daniel L. *Basic Genetics*. Boston: Jones and Bartlett Publishers, 1991.

———. "Forensic DNA Typing Dispute." *Nature* 372, no. 6505 (1994): 398–399.

Hartl, Daniel L., and Richard C. Lewontin. "DNA Fingerprinting Report (Letter to the Editor)." *Science* 260 (1993): 473–474.

Herd, Kim, and Adrianne Day. American Prosecutors Research Institute. "An Update on STR Admissibility." *Silent Witness* 5, no. 3 (2000). http://www.ndaa.org/publications/newsletters/silent_witness_volume_5_number_3_2000.html. Accessed 17 August 2006.

Hicks, John W. "DNA Profiling: A Tool for Law Enforcement." *FBI Law Enforcement Bulletin*, August 1988, 3–4.

———. "FBI Program for the Forensic Application of DNA Technology." In *DNA Technology and Forensic Science*, ed. Jack Ballantyne. Cold Spring Harbor, N.Y.: Cold Spring Harbor Laboratory Press, 1989.

Hilgartner, Stephen. *Science on Stage: Expert Advice as Public Drama*. Stanford: Stanford University Press, 2000.

Hopkin, Karen. "Eric S. Lander, Ph.D." Howard Hughes Medical Institute. http://www.hhmi.org/lectures/2002/lander.html. Accessed February 2003.

Humes, Edward. "DNA War." *L.A. Times Magazine*, 29 November 1992, 29.

Innocence Project. "About the Innocence Project." Benjamin N. Cardozo School of Law. Yeshiva University. http://www.innocenceproject.org/about/index.php. Accessed 16 August 2006.

———. "Case Summary: Josiah Sutton," Benjamin N. Cardozo School of Law. Yeshiva University. 2001. http://www.innocenceproject.org/case/display_profile.php?id=145. Accessed 8 March 2006.

————. "Historic Audit of Virginia Crime Lab Errors in Earl Washington Jr.'s Capital Case." Benjamin N. Cardozo School of Law. Yeshiva University. 6 May 2005. http://www.innocenceproject.org/docs/VACrimeLabPressRelease.pdf. Accessed 14 March 2006.

————. Homepage of Benjamin N. Cardozo School of Law. Yeshiva University. http://www.innocenceproject.org/. Accessed 1 March 2006.

Jasanoff, Sheila. "Contested Boundaries in Policy-Relevant Science." *Social Studies of Science* 17 (1987): 195–230.

————. *Science at the Bar: Law, Science, and Technology in America*. Cambridge: Harvard University Press, 1995.

————. "The Eye of Everyman: Witnessing DNA in the Simpson Trial." *Social Studies of Science* 28 (1998): 713–740.

Jeffreys, Alec J., Victoria Wilson, and Swee Lay Thein. "Hypervariable 'Minisatellite' Regions in Human DNA." *Nature* 314 (1985): 67–73.

————. "Individual-Specific 'Fingerprints' of Human DNA." *Nature* 316 (1985): 76–79.

Jordan, Kathleen, and Michael Lynch. "The Dissemination, Standardization, and Routinization of Molecular Biological Technique." *Social Studies of Science* 28 (1998): 773–800.

Kalmijn, Matthijs. "Intermarriage and Homogamy: Causes, Patterns, Trends." *Annual Review of Sociology* 24 (1998): 395–421.

Evan Kanter, Michael Baird, Robert Shaler, and Ivan Balazs. "Analysis of Restriction Fragment Length Polymorphisms in Deoxyribonucleic Acid (DNA) Recovered from Dried Bloodstains." *Journal of Forensic Sciences* 31, no. 2 (1986): 403–408.

Kaye, David H. "DNA Evidence: Probability, Population Genetics, and the Courts." *Harvard Journal of Law and Technology* 7 (1993): 101–117.

Kirby, Lorne T. *DNA Fingerprinting: An Introduction*. New York: Stockton Press, 1990.

Koehler, Jonathan. "Error and Exaggeration in the Presentation of DNA Evidence." *Jurimetrics Journal* 34 (1993): 21–39.

————. "On Conveying the Probative Value of DNA Evidence: Frequencies, Likelihood Ratios, and Error Rates." *University of Colorado Law Review* 67 (1996): 859–886.

Kolata, Gina. "Critic of 'Genetic Fingerprinting' Tests Tells of Pressure to Withdraw Paper." *New York Times*, 20 December 1991.

————. "Chief Says Panel Backs Courts' Use of a Genetic Test; Times Account in Error; Report Urges Strict Standards, but No Moratorium on DNA Fingerprinting for Now." *New York Times*, 15 April 1992.

————. "Two Chief Rivals in the Battle over DNA Evidence Now Agree on Its Use." *New York Times*, 27 October 1994.

————. "US Panel Seeking Restriction on Use of DNA in Courts; Lab Standards Faulted; Judges Are Asked to Bar Genetic 'Fingerprinting' until Basis in Science Is Stronger." *New York Times*, 14 April 1992.

Krane, D. E., R. W. Allen, S. A. Sawyer, D. A. Petrov, and Daniel L. Hartl. "Genetic Differences at Four DNA Typing Loci in Finnish, Italian, and Mixed Caucasian Populations." *PNAS* 89 (1992): 10583–10587.

Labaton, Stephen. "DNA Fingerprinting Showdown Expected in Ohio." *New York Times*, 22 June 1990.

Lander, Eric S. "DNA Fingerprinting on Trial." *Nature* 339, no. 6225 (1989): 501–505.

————. "DNA Fingerprinting: The NRC Report (Letter to the Editor)." *Science* 260, no. 5112 (1993): 1221.

————. "Invited Editorial: Research on DNA Typing Catching up with Courtroom Application [Plus Responses from Numerous Individuals and Lander's Reply]." *American Journal of Human Genetics* 48 (1991): 819–823.

————. "Lander Reply." *American Journal of Human Genetics* 49 (1991): 899–903.

Lander, Eric S., and Bruce Budowle. "DNA Fingerprinting Dispute Laid to Rest." *Nature* 371, no. 6500 (1994): 735–738.

Latour, Bruno. *Science in Action.* Cambridge: Harvard University Press, 1987.

Law, John. "Technology, Closure and Heterogeneous Engineering: The Case of Portuguese Expansion." In *The Social Construction of Technological Systems,* ed. Wiebe Bijker, Trevor Pinch, and Thomas P. Hughes. Cambridge: MIT Press, 1987.

Lazer, David. "Introduction: DNA and the Criminal Justice System." In *DNA and the Criminal Justice System,* ed. David Lazer. Cambridge: MIT Press, 2004.

Lee, Henry C., and Frank Tirnady. *Blood Evidence: How DNA Is Revolutionizing the Way We Solve Crimes.* Cambridge, Mass.: Perseus Publishing, 2003.

Lempert, Richard. "DNA, Science and the Law: Two Cheers for the Ceiling Principle." *Jurimetrics Journal* 34 (1993): 41–57.

————. "Some Caveats Concerning DNA as Criminal Identification Evidence: With Thanks to the Reverend Bayes." *Cardozo Law Review* 13 (1991): 303–341.

Lewin, Roger. "DNA Typing on the Witness Stand." *Science* 244, no. 4908 (1989): 1033–1035.

Lewis, Ricki. "DNA Fingerprints: Witness for the Prosecution." *Discover,* June 1988, 44–52.

Lewontin, Richard C. "Forensic DNA Typing Dispute." *Nature* 372, no. 6205 (1994): 398.

Lewontin, Richard C., and Daniel L. Hartl. "Population Genetics in Forensic DNA Typing." *Science* 254, no. 5039 (1991): 1745–1750.

Loevinger, Lee. "Law and Science as Rival Systems." *University of Florida Law Review* 19 (1967): 530–551.

Lynch, Michael. "The Discursive Production of Uncertainty: The OJ Simpson 'Dream Team' and the Sociology of Knowledge Machine." *Social Studies of Science* 28, no. 5–6 (1998): 829–868.

————. "God's Signature: DNA Profiling, The New Gold Standard in Forensic Science." *Endeavour* 42, no. 2 (2003): 93–97.

Manor, Robert. "DNA 'Fingerprinting' Questioned; Geneticist Says Test May Be Less Reliable Than First Believed." *St. Louis Post-Dispatch,* 15 October 1989.

Margolick, David. "Simpson Defense Drops DNA Challenge." *New York Times,* 5 January 1995.

————. "Simpson Lawyers Ask to Forgo DNA Hearing." *New York Times,* 14 December 1994.

Marx, Jean L. "DNA Fingerprinting Takes the Witness Stand." *Science* 240, no. 4859 (1988): 1616–1618.

McElfresh, K. C. "DNA Fingerprinting (Letter to the Editor)." *Science* 246, no. 4927 (1989): 192.

McFadden, Robert D. "Reliability of DNA Testing Challenged by Judge's Ruling." *New York Times,* 15 August 1989.

McGlone, Tim. "Nearly 500 Military DNA Cases under Investigation." *Virginian-Pilot,* 17 May 2006.

McKenna, Judith, Joe Cecil, and Pamela Coukos. "Reference Guide on Forensic DNA Evidence." In *Reference Manual on Scientific Evidence,* ed. Federal Judicial Center. Washington, DC: Federal Justice Center, 1994.

Morton, N. E. "DNA in Court." *European Journal of Human Genetics* 1 (1993): 172–178.

Moss, Debra Cassens. "DNA—The New Fingerprints." *ABA Journal* 1988, 66–70.

Murch, Randall. "Summary of the [FBI] DNA Technology Seminar." *Crime Laboratory Digest* 15, no. 3 (1988): 79–85.

National Institute of Standards and Technology. Chemical Science and Technology Laboratory. "Short Tandem Repeat DNA Internet Database." http://www.cstl.nist.gov/div831/strbase/. Accessed 17 August 2006.

National Research Council. *DNA Technology in Forensic Science.* Washington, D.C.: National Academy Press, 1992.

———. *The Evaluation of Forensic DNA Evidence.* Washington, DC: National Academy Press, 1996.

Neufeld, Peter J., and Neville Coleman. "When Science Takes the Witness Stand." *Scientific American* 262, no. 5 (1990): 46–53.

Neufeld, Peter J., and Barry Scheck. "Factors Affecting the Fallibility of DNA Profiling: Is There Less Than Meets the Eye?" *Expert Evidence Reporter* 1, no. 4 (1989): 93–97.

Norman, Colin. "Caution Urged on DNA Fingerprinting." *Science* 245, no. 4919 (1989): 699.

Office of the Independent Investigator for the Houston Police Department Crime Laboratory and Property Room. "Independent Investigator Issues Fifth Report on Houston Police Department Crime Lab." 11 May 2006. http://www.hpdlabinvestigation.org/press-release/060511pressrelease.pdf. Accessed 11 May 2006.

Parloff, Roger. "How Barry Scheck and Peter Neufeld Tripped up the DNA Experts." *American Lawyer*, December 1989, 50–56.

Pearsall, Anthony. "DNA Printing: The Unexamined 'Witness' in Criminal Trials." *California Law Review* 77 (1989): 665–703.

Peterson, Joseph L., and R. E. Gaensslen. *Developing Criteria for Model External DNA Proficiency Testing: Final Report.* Chicago: University of Illinois at Chicago, 2001.

"The Power of DNA Evidence." Editorial. *New York Times*, 28 May 1995, 10.

Project on Scientific Knowledge and Public Policy. "Daubert: The Most Influential Supreme Court Ruling You've Never Heard Of." Report. June 2003. Boston: Tellus Institute.

"Quantum Chemical Corp Reports Earnings for Qtr to Dec 31." *New York Times*, 30 January 1988. http://query.nytimes.com/gst/fullpage.html?res=9E05EFD71726E630BC4953DFBE6E958A. Accessed 14 August 2006.

"Quashing the DNA Debate." Editorial. *Sacramento (CA) Bee*, 28 December 1991.

Riley, Donald E. "DNA Testing: An Introduction for Non-Scientists, An Illustrated Explanation." http://www.scientific.org/tutorials/articles/riley/riley.html. Accessed 8 February 2006.

Roberts, Leslie. "DNA Fingerprinting: Academy Reports." *Science* 256, no. 5055 (1992): 300–301.

———. "Fight Erupts over DNA Fingerprinting." *Science* 254, no. 5039 (1991): 1721–1723.

———. "Hired Guns or True Believers?" *Science* 257 (1992): 735.

———. "Prosecutor v. Scientist: A Cat and Mouse Relationship." *Science* 257 (1992): 733.

———. "Science in Court: A Culture Clash." *Science* 257, no. 5071 (1992): 732–736.

———. "Was Science Fair to Its Authors?" *Science* 254 (1991): 1722.

Ruethling, Gretchen. "Illinois State Police Cancels Forensic Lab's Contract, Citing Errors." *New York Times*, 20 August 2005.

Saks, Michael J., and Jonathan Koehler. "The Coming Paradigm Shift in Forensic Identi-fication Science." *Science* 309, no. 5736 (2005): 892–895.

———. "What DNA 'Fingerprinting' Can Teach the Law about the Rest of Forensic Sci-ence." *Cardozo Law Review* 13 (1991): 361–372.

Scheck, Barry. "DNA and Daubert." *Cardozo Law Review* 15 (1994): 1959–1997.

———. Interview with Harry Kriesler as part of "Conversations with History Series at UC-Berkeley." *DNA and the Criminal Justice System.* 25 July 2003. http://globetrotter. berkeley.edu/people3/Scheck/scheck-con3.html. Accessed 17 August 2006.

Scheck, Barry, Peter J. Neufeld, and Jim Dwyer. *Actual Innocence: Five Days to Execution and Other Dispatches from the Wrongly Convicted.* New York: Doubleday, 2000.

———. "DNA Task Force Report." *Champion*, June 1991, 13–21.

Schmeck, Harold M. "DNA Findings Are Disputed by Scientists." *New York Times*, 25 May 1989.

Scott, Janny. "Chemists Told of Advances in 'Genetic Fingerprinting.'" *Los Angeles Times*, 8 November 1987, 2.

Sessions, William S. "Invited Editorial." *Journal of Forensic Science* 34, no. 5 (1989): 1051.

Seton, Craig. "Life for Sex Killer Who Sent Decoy to Take Genetic Test." *Times* (London), 23 January 1988.

Shackelford, Jerry. "DNA Test Error Admitted in Lapp Case." *Fort Worth Journal-Gazette*, 23 November 1988.

———. "Procedure Varied in Lapp DNA Test." *Fort Worth Journal-Gazette*, 24 November 1988.

Shapin, Steven, and Simon Schaffer. *Leviathan and the Air-Pump: Hobbes, Boyle, and the Experimental Life.* Princeton: Princeton University Press, 1985.

Silcock, Brian. "Genes Tell Tales." *Sunday Times* (London), 3 November 1985.

Southern, E. M. "Detection of Specific Sequences among DNA Fragments Separated By Gel Electrophoresis." *Journal of Molecular Biology* 98, no. 3 (1975): 503–517.

Thompson, Larry. "A Smudge on DNA Fingerprinting?; N.Y. Case Raises Questions about Quality Standards, Due Process." *Washington Post*, 26 June 1989.

Thompson, William C. "Accepting Lower Standards: The National Research Council's Second Report on Forensic DNA Evidence." *Jurimetrics Journal* 37 (1997): 405–424.

———. "DNA Evidence in the OJ Simpson Trial." *University of Colorado Law Review* 67, no. 4 (1996): 827–857.

———. "Evaluating the Admissibility of New Genetic Identification Tests: Lessons from the 'DNA War.'" *Journal of Criminal Law and Criminology* 84, no. 1 (1993): 22–104.

———. "Subjective Interpretation, Laboratory Error and the Value of Forensic DNA Evi-dence: Three Case Studies." *Genetica* 96 (1995): 153–168.

———. "Tarnish on the Gold Standard: Recent Problems in Forensic DNA Testing." *Champion*, January/February (2006): 14–20.

Thompson, William C., and Simon Ford. "DNA Typing: Acceptance and Weight of the New Genetic Identification Tests." *Virginia Law Review* 75 (1989): 45–108.

———. "DNA Typing: Promising Forensic Technique Needs Additional Validation." *Trial*, September 1988, 56–64.

Thompson, William C., Simon Ford, Travis Doom, Michael Raymer, and Dan E. Krane. "Evaluating Forensic DNA Evidence: Essential Elements of a Competent Defense Review." *Champion* 27, no. 3 (2003): 16–25.

University of California–Irvine. "As Texas Convict Is Exonerated, DNA Expert Offers Inside Look at the Case Overhauling Justice System." 7 April 2003. http://today.uci. edu/news/release_detail.asp?key=987. Accessed 8 March 2006.

U.S. Congress. Office of Technology Assessment. *Genetic Witness: Forensic Uses of DNA Tests.* Washington, D.C.: GPO, 1990.

———. "The OTA Legacy: 1972–1995." http://www.wws.princeton.edu/~ota/. Accessed 2 June 2006.

U.S. Department of Justice. Office of the Inspector General. *The FBI DNA Laboratory, A Review of Protocol and Practice Vulnerabilities.* Washington, DC: Office of the Inspector General, 2004.

U.S. Federal Bureau of Investigation. *Combined DNA Index System.* "Measuring Success." http://www.fbi.gov/hq/lab/codis/success.htm. Accessed 15 June 2006.

———. "FBI's Response to the Report by the Committee on DNA Technology in Forensic Science." Quantico, Va.: Federal Bureau of Investigation, 1992.

———. *VNTR Population Data: A Worldwide Study.* 3 vols. Quantico, Va.: Federal Bureau of Investigation, 1993.

U.S. House Committee on Judiciary. Subcommittee on Civil and Constitutional Rights. *FBI Oversight and Authorization Request for Fiscal Year 1990 (DNA Identification).* 101st Cong., 1st sess. Washington, DC: GPO, 1990.

U.S. Senate Committee on the Judiciary. U.S. House Committee on the Judiciary. Subcommittee on Civil and Constitutional Rights. *Forensic DNA Analysis.* 102nd Cong., 1st sess., (1991), House Serial 30/Senate Serial J-102–147. Washington, DC: GPO, 1992.

———. Subcommittee on Constitution. *DNA Identification.* 101st Cong., 1st sess., House Serial 30/Senate Serial J-101–147. Washington, DC: GPO, 1992.

Veitch, Andrew. "Son Rejoins Mother as Genetic Test Ends Immigration Dispute/Ghanian Boy Allowed to Join Family in Britain." *Guardian* (London), 31 October 1985.

Walgate, Robert. "Futures: You and Nobody Else: Focus on the Technique of Genetic Fingerprinting." *Guardian* (London), 8 November 1985.

Whitehead Institute, "Lander Biography." http://www.wi.mit.edu/news/genome/lander.html. Accessed February 2003.

Wong, Zilla, Victoria Wilson, Ila Patel, S. Povey, and Alec J. Jeffreys. "Characterization of a Panel of Highly Variable Minisatellites Cloned from Human DNA." *Annals of Human Genetics* 51 (1987): 269–288.

Wooley, James R. "A Response to Lander: The Courtroom Perspective." *American Journal of Human Genetics* 49 (1991): 892–893.

Wooley, James R., and Rockne Harmon. "Forensic DNA Brouhaha: Science or Debate." *American Journal of Human Genetics* 51 (1992): 1164–1165.

Wynne, Brian. "Establishing the Rules of Laws. Constructing Export Authority." In *Expert Evidence: Interpreting Science in the Law,* ed. Brian Wynne and Roger Smith. New York: Routledge, 1989.

Zack, Margaret. "Hennepin County Drops DNA Test of Murder Suspect." *(Minneapolis) Star Tribune,* 11 January 1990.

Zerwick, Phoebe. "DNA Mislabeled in Murder Case." *Winston-Salem Journal,* 28 August 2005.

———. "N.C. Told to Refine Handling, Storing of Crime Evidence." *Winston-Salem Journal,* 19 November 2005.

Index

About the Author

Jay D. Aronson is an assistant professor of history at Carnegie Mellon University in Pittsburgh, Pennsylvania.